D0830989

HORSE FEEDING
AND NUTRITION
Second Edition

ANIMAL FEEDING AND NUTRITION

A Series of Monographs and Treatises

Tony J. Cunha, Editor

Distinguished Service Professor Emeritus
University of Florida
Gainesville, Florida

and

Dean Emeritus, School of Agriculture
California State Polytechnic University
Pomona, California

HORSE FEEDING
AND NUTRITION
Second Edition

Tony J. Cunha
University of Florida
Gainesville, Florida
and
California State Polytechnic University
Pomona, California

ACADEMIC PRESS, INC.

Harcourt Brace Jovanovich, Publishers
San Diego   New York   Boston
London   Sydney   Tokyo   Toronto

Copyright © 1991, 1980 by Academic Press, Inc.
All Rights Reserved.
No part of this publication may be reproduced or transmitted in any form or
by any means, electronic or mechanical, including photocopy, recording, or
any information storage and retrieval system, without permission in writing
from the publisher.

Academic Press, Inc.
San Diego, California 92101

United Kingdom Edition published by
Academic Press Limited
24–28 Oval Road, London NW1 7DX

Library of Congress Cataloging-in-Publication Data

Cunha, T. J. (Tony Joseph), date
    Horse feeding and nutrition / by Tony J. Cunha. -- 2nd ed.
        p.   cm.
    Includes bibliographical references and index.
    ISBN   0-12-196561-9   (alk. paper)
    1. Horses--Feeding and feeds. 2. Horses--Nutrition--Requirements.
    I. Title.
    SF285.5.C86   1991
    636.1'085--dc20   90-652
                                                                      CIP

Printed in the United States of America
90 91 92 93   9 8 7 6 5 4 3 2 1

This book is dedicated in appreciation to my parents and to my wife Gwen and my daughters Becky, Sharon, and Susan. It is also dedicated to former teachers, Vard Shepard, Gilbert Hutchings, Ralph W. Phillips, Russ Rasmussen, Gus Bohstedt, and Paul Phillips who were very influential in my professional development and to the late Mr. and Mrs. F. T. Roseberry of Los Banos, California who were helpful in a time of need. Special thanks are given to my daughter, Sharon Buddington, Lilburn, Georgia who performed an excellent job in typing the entire manuscript, and to R. F. Johnson, Professor Emeritus, California Polytechnic State University, San Luis Obispo for his assistance in producing some figures for the text.

Contents

Foreword

This is the second edition of *Horse Feeding and Nutrition*. The first edition was the fourth book in a series on animal feeding and nutrition that includes 11 books published to date. As the amount of scientific literature expands, a continuing need exists for revision of the books in the series to keep them up to date.

Horse Feeding and Nutrition is a departure from the other books in the animal feeding and nutrition series because its focus is not on raising horses as a means of providing protein for human consumption. In the United States, horses are used primarily for riding, racing, recreation, pleasure, work, and similar activities. In many countries of the world, especially in Europe, horses are also used as a source of meat for human consumption. In some countries, horse meat sells for more than beef and amounts to 4–6% of total meat consumption. With the focus on stronger, healthier, and better performing animals this book will be of value to those interested in horses throughout the world.

New feed and food crops, improved methods of production and processing, increased productivity of animals and crops, changes in animal products including less fat in meat and milk, longer shelf-life requirements of animal food products, and a myriad of new technological developments have resulted in a need to reevaluate nutrient requirements and supplementation continually.

The cost of feed represents 60–80% of the total cost of producing the various classes of animals. Therefore, it is very important to be up to date on feeding and nutrition and to use well-balanced diets. New developments in feeding and nutrition are quite numerous each year. Improved use of by-product feeds, new developments with feeds and feed additives, vitamins, minerals, amino acids, fatty acids, volatile fatty acids, and other nutrients make it imperative that one evaluates new findings and determines their possible roles in improving animal diets. In addition, the current trend is toward feeding animals in closer confinement. All of this places more pressure on the adequacy of the diet used. The volume of scientific literature is increasing rapidly each year. Moreover, its interpretation is becoming more complex. This increases the need for summarizing and interpreting these new developments in up-to-date books such as those in this series. This necessitates that top scientists and authorities in the field collate into one volume all of the available information on the feeding and nutrition of each species of farm animal.

Tony J. Cunha

Preface

This is the second edition of *Horse Feeding and Nutrition* which was originally published in 1980. The book is entirely revised since a great deal of new research information has become available since the book was first written. Three new chapters have been added. They are Chapter 20 on "Feeding and Health-Related Problems"; Chapter 21 on "The Complexity of Proper Bone Formation"; and Chapter 22 on "Exercise Physiology."

In the first chapter the future of the horse industry is discussed. Chapter 2 reviews the art, science, and myths of feeding horses. Chapter 3 discusses the problems involved in supplying an adequate level of nutrients in horse diets. The digestion of feeds is discussed in Chapter 4. Chapters 5–10 contain concise, up-to-date summaries on vitamins, minerals, protein, amino acids, carbohydrates, fiber, fatty acids, fat, volatile fatty acids, energy, and water. The nutrient requirements of the horse are discussed and compared with National Research Council (NRC) recommendations. Deficiency symptoms of nutrients are discussed. The needs of the horse for various nutrients are presented and give the reader a basis for determining what good, well-balanced diets should contain. The practical application of this basic information is discussed in these chapters and in Chapters 15–18 on feeding horses during various stages of the life cycle.

The important interrelationships between nutrition, disease, and performance are discussed in Chapter 11. The relative value of various feeds for use in horse diets is discussed in Chapter 12. The value of pasture and hay for horses is presented in Chapter 13. Hints on feeding horses for optimal result are discussed in Chapter 14. Chapters 15–18 present information on feeding the foal, growing horse, performance and race horse, and mare and stallion. Chapter 19 discusses purified diets for horses, and Chapter 23 presents information on antibiotics, learning ability, feeding behavior, nutrient toxicity, weight equivalents, weight–unit conversion factors, and the effect of cold weather on horses. Sample diets are given, useful as guides in developing diets for horses. Suggested levels of protein, minerals, and vitamins for use in horse diets are presented. These can be used as guides which can be modified to suit the various feeding situations encountered on horse farms.

This book provides information helpful to those interested in the feeding and nutrition of horses. It is written in a manner designed to be especially valuable to beginners in horse production, established horse owners, and those who are concerned, directly or indirectly, with horse feeding and nutrition. This book will be very helpful to feed manufacturers and dealers and others concerned with producing the many different nutrients, supplements, feeds, and other ingredients used in horse diets. It will also be of value to county agents, farm advisors, consultants, veterinarians, and college and university students and teachers of courses on feeds and feeding, horse production, horse nutrition, and animal nutrition. The text contains basic information for students on these and other courses. It contains many key references for those interested in obtaining further information on a particular subject.

In preparing this book I have had the benefit of suggestions from many eminent scientists in the United States and abroad. Moreover, they supplied me with various publications and photographs for use in this text. I especially want to thank the following who provided the most assistance: Norman K. Dunn, Gerald E. Hackett, and Steven Wickler, California State Polytechnic University, Pomona; Larry M. Slade and James L. Shupe, Utah State University; E. A. Ott, S. Lieb, L. R. McDowell, J. H. Conrad, R. L. Asquith, and D. C. Sharp, University of Florida; G. D. Potter, D. D. Householder, J. L. Kreider, C. H. Bridges, and J. W. Evans, Texas A & M University; H. F. Hintz and H. F. Schryver, Cornell University; Tom N. Meacham and J. P. Fontenot, VPl; R. M. Jordon, University of Minnesota; R. R. Johnson, California State Polytechnic University, San Luis Obispo; J. P. Baker and S. G. Jackson, University of Kentucky; C. G. Depew, Louisiana State University; J. J. Kiser, Iowa State University; M. A. Russell and T. W. Perry, Purdue University; D. R. Topliff and D. W. Freeman, Oklahoma State University; L. M. Lawrence and K. H. Kline, University of Illinois; R. Albaugh, University of California; G. M. Weber, USDA; R. A. Mowrey, Jr., North Carolina State University; M. J. Glade, Northwestern University, N. Comben, England; M. Bradley, University of Missouri; K. Malenowski, Rutgers University; S. Donoghue and S. L. Ralston, University of Pennsylvania.

Tony J. Cunha

1

Horse Industry Future

I. INTRODUCTION

During the last 30 years the horse industry has gone from a minor declining industry to a major and active multibillion dollar industry. During what might be called the horse-power era, from about 1860 until 1915, horses reached a peak of about 27 million head in the U.S. A gradual decline in numbers then occurred as the horse was replaced by the tractor, truck, automobile, and other automation. By 1960, horse numbers had declined to about 3 million head. But, soon thereafter, horse numbers started to increase rapidly as their use for riding, recreation, and sports activities increased (Fig. 1.1).

Unfortunately, there are no exact figures on the number of horses in the U.S. at the present time. Estimates and surveys vary from 5 to 9 million depending on whose figures one uses. Regardless of the approximate number, the horse industry has developed into a big business and all indications are that it will continue as such (1,2).

II. HORSE INDUSTRY FACTS

The American Horse Council (3) 1987 report presented the following equine industry data in their horse industry directory:

1. Horses contribute approximately $15 billion annually to the economy of the U.S.
2. Horse owners account for about $13 billion in yearly investment and maintenance expenditures.
3. Horse sports draw more than 110 million spectators yearly.
4. U.S. race track attendance exceeds 70 million each year. Wagering on horse races surpasses $13 billion yearly. In Canada, an additional 12 million people attend horse races and wager about $1.6 billion yearly.
5. The U.S. Department of Interior estimates that about 27 million people over age 12 ride horses yearly, with about 54% riding on a regular basis and the remainder for occasional recreation.

Fig. 1.1   Youngsters enjoying a ride with horses in Gainesville, Florida area. Horses are an excellent recreational activity for young girls and boys. (Courtesy of T. J. Cunha, University of Florida and Cal Poly University, Pomona.)

6. Horse exports amount to approximately $200 million yearly.

7. Texas and California are the leading horse population states.

8. Breeds with the highest individual population counts are (1) Quarter Horses, (2) Arabian, Anglo-Arab, and Half-Arab, and (3) Thoroughbreds.

9. From 1960 to 1986 yearly registration figures for the major American light horse breeds increased from 70,050 to 306,584 head. These figures indicate the great increase in the number of quality horses that are registered yearly.

The monetary value of all breeds of horses has increased a great deal. As one example, Thoroughbred yearlings in select sales in 1960 averaged about $25,000 compared to $356,000 in 1982 (5).

III. 4-H CLUB HORSE PROJECTS

During 1987, 4-H Club Youth Enrollment in horse and pony organized projects and activities numbered 224,903 (2). This compared to enrollments of 107,417 in swine, 107,376 in beef, 97,939 in sheep, 85,579 in poultry, 70,711 in dairy, and 19,274 in goat 4-H Club projects. These numbers indicate the high degree of interest in the nation's 4-H Club Youth in horses. It's interesting that female enrollment in the horse and pony projects was 152,743 as compared to

72,160 males enrolled. These youth enrollment figures are further evidence that the horse industry will continue to be an important industry. In 1966, 4-H Club horse projects took the lead over 4-H Club beef projects.

The top 10 states with 4-H Club youth enrollment in horse and pony projects in 1987 were as follows (2):

1. Tennessee. 23,402
2. Washington 14,039
3. New York 12,119
4. Texas 10,864
5. Alabama 10,389
6. Ohio 10,139
7. Michigan 9,304
8. Georgia 7,944
9. Oklahoma 7,691
10. North Carolina 6,625

It is estimated that about 50% of the new market for horses each year is due to youth programs. Enjoying horses during their leisure time is one of the most wholesome activities young people can engage in. In 1977 it was estimated that 40% of all horse owners were under the age of 20 and approximately 60% of the horse owners were from families with an annual income under $10,000. These figures indicate the importance of youth in the horse industry. Moreover, it indicates that horses are no longer used primarily by wealthy individuals.

IV. HORSE ACTIVITIES AND USE

As the United States has become more affluent, many who previously could not afford a horse, can now have one. Horses are no longer used primarily by wealthy families. Saddle clubs and 4-H Club projects have grown rapidly. This has resulted in a great increase in people of all ages riding for pleasure and recreation. There has been an increase in riding trails along roads and highways, as well as bridle paths in city parks and other areas. Horse academies and other horse establishments have increased in number and are available to teach people to ride and to train horses. Many of them will board horses for those who do not have facilities for their own animals.

Horses are also used for the game of polo, which is expanding. Riding to the hounds is another area where horses are used. About one-half a million horses are used as cow ponies on the range (Fig. 1.2). Horses are also used in rodeos (Fig. 1.3), racing (Fig. 1.4), as pack animals, for trail rides, mounted patrols, parades, movies, T.V., and other activities. The Forest Service uses horses in some areas. In the laboratory the horse is used to produce certain medicines for

Fig. 1.2 Horses are used for working cattle in a Florida ranch. They can be used in areas where jeeps or trucks cannot go. (Courtesy of T. J. Cunha, University of Florida and Cal Poly University, Pomona.)

Fig. 1.3 Scene at Silver Spurs Rodeo in Kissimmee, Florida. Rodeos are a prominent activity and enjoyed by many. (Courtesy of T. J. Cunha, University of Florida and Cal Poly University, Pomona.)

Fig. 1.4   Horses are used for racing which is the leading spectator sport in attendance in the United States (Courtesy of T. J. Cunha, Cal Poly University, Pomona.)

human use. So the horse industry has established a definite and important role throughout all segments of U.S. society.

V. HORSE CONTRIBUTIONS TO THE PUBLIC

1. Since the early 1930s, horse racing has been America's number one spectator sport. In 1981, horse sports generated $1 billion in revenue from taxes on legalized betting, for federal, state, and local governments (4). In some states this revenue is an important part of their budget. In Florida, for example, where the revenue is divided equally among all its counties, the revenue is a very important part of the budget of the small counties. Without that revenue, many of them would have financial difficulties.

Horse racing, therefore, is an important recreational industry (5,6,7). Moreover, it provides an important revenue source. It is hoped that some of this revenue might be diverted to horse programs in the states where it is generated. For example, in New York, 2% of the revenue is used to support equine research at Cornell University (8). Proper action by the horse industry and state legislature can make this happen elsewhere.

2. People in all walks of life and at all ages like to ride horses for pleasure, recreation, and exercise. Riding is one of the best forms of exercise and is prescribed by many in the medical profession. All organs of the body are exercised by riding.

3. Riding horses for pleasure, work, and recreation saves on energy and helps keep the environment clean since autos, trucks, and other vehicles are not otherwise being used.

4. Horses are used by handicapped people as a source of exercise and for recreation and pleasure. A handicapped person who has the use of his arms can ride (after being helped onto a horse). Some seriously handicapped people ride horses even though they need to be strapped to the horse and the animal is led by a normal rider.

5. Horses provide employment for many industries. It is estimated that from $1000 to $1500 or more is spent on each horse yearly for tack, housing, equipment, feed, health care, and many other items.

6. It is estimated that about one-half million horses are used by U.S. livestock producers on their farms and ranches in working their animals, checking on fences, and doing other chores. They can get to areas the jeep or truck cannot reach. Moreover, they are better adapted to working animals.

VI. RESEARCH AND EDUCATION NEEDS

The science of modern horse production is in its infancy. The technology required for optimum training and performance of the high-level performance horse is likewise just emerging. There is no doubt that less is known about the science and technology of horse production and use than that of other farm animals.

Following are just a few major problems facing the horse industry which indicate the great need for new technology:

1. The average foal crop is 50–60% depending on whose estimate one uses. This means that the majority of mares foal one year and skip the next.

2. It is estimated that only 50% of the thoroughbred horses that start training ever reach the track for racing. Moreover, only 20% of those continue to race throughout the first year. Feet and leg problems are responsible for a large percentage of the drop out. Many of these problems occur because the young horse is pushed to a level of training and performance that is too great and too early, which can result in damage to immature, partially ossified bones, which leads to bone abnormalities.

3. Very little is known about the nutrient requirements of the horse. Even less is known about the nutrient needs of the high-level performance horse. This knowledge is needed to design optimum diets for the specific activity the horse is to be involved in.

4. The field of muscle exercise physiology is in its infancy. As this area is developed, it will provide the knowledge needed to obtain optimum conversion of chemically bound and stored energy into the energy required for optimum muscular contraction and action. The performance of the horse athlete has not improved much over the past 50 years as compared to the human athlete. Un-

locking the secrets of muscle exercise physiology will enable the horse to start improving its performance level.

5. The horse developed as an animal that was used to life in the wild. But stall confinement, access to limited grazing and exercise areas, and intensified production methods used with high-level performance horses has brought on many problems which need solving.

6. Disease and parasite problems with horses are very serious and take a heavy toll yearly, which some people estimate decreases production efficiency 15 to 20%. The objective should be to develop adequate prevention programs rather than just treatment.

7. Studies on buildings, equipment, and automation are needed to reduce labor costs, which are exceedingly high in too many operations. Obtaining high-quality labor is a major problem and indicates the need for training programs designed to improve the labor situation.

VII. NEED FOR MORE UNIVERSITY INVOLVEMENT

During the past 20 years there has been a gradual increase in horse research, teaching, and extension programs in the U.S. (Fig. 1.5), but a greater increase

Fig. 1.5 Universities and colleges are expanding their educational emphasis in the area of horse production. The W. K. Kellogg Arabian Horse Center at Cal Poly University, Pomona, California is one of the finest. (Courtesy of Professor Norman K. Dunn, Cal Poly University, Pomona.)

still is needed in the future. Some universities that do not have an extensive research and/or teaching program have a horse extension program. Some universities have a good teaching and extension program, but little or no research effort. In the last few years the horse program has been curtailed at a few universities. But, overall there has been an increase in horse programs at U.S. universities which is encouraging.

It is important to note that a number of universities have excellent research, teaching, and extension programs with horses. In a few universities the horse programs are outstanding and on a par with their other livestock programs.

An indication of the increased research effort with horses can be obtained by perusing the 1987 "Proceedings of the Tenth Equine Nutrition and Physiology Symposium." There were 100 research reports presented in all areas of horse production. Nine reports were given on teaching and extension work. Moreover, the knowledge presented has been of considerable assistance to the horse and other related industries (Fig. 1.6).

The symposium began in 1968 when the Equine Nutrition and Physiology Society was started by a small group of scientists interested in horse production. Each meeting, which occurs biannually, is held at a different university site. The number and quality of papers has increased with each subsequent symposium.

Fig. 1.6 Attractiveness of a farm is an important part of the total merchandising program. This beautiful scene is of the Westerly Stud Farm, Santa Ynez, California. (Courtesy of Professor Norman K. Dunn, Cal Poly University, Pomona.)

There is no doubt that the symposium has been a major factor in increased horse research, teaching, and extension programs at the university and industry level. In addition to the papers presented at the symposium, other papers are presented in other journals and publications.

These and many other research studies can provide a productive and useful life for the horse. They are also needed to increase the profitability of horse enterprises. Competition is increasing for the recreation and sports dollar. To meet this competition, horse people need to use more scientific information to improve their programs further and thus make them more competitive.

In addition to research, it is important to have a viable teaching program. Students need training in all aspects of the horse industry as well as in related enterprises which supply goods and services for horse people. Without this training, the industry is less able to keep up with enterprises using more sophisticated and up-to-date innovative methods. In addition to undergraduate study, there is a need for graduate training and research in various aspects of the horse industry.

Extension activities are a necessary part of a university program. This includes programs in cooperation with the County Agent or Farm Advisor. It also involves short courses, seminars, field days, clinics, and other activities designed to help the horse and related industries augment their programs.

Teaching, research, and extension programs at universities need to be properly supported and take their rightful place along with other livestock programs. The horse industry is a very important one and needs science and technology to help it survive the future in an increasingly competitive society.

It is important for the horse industries and others involved with them to make their needs known to those in authority to help allocate the needed resources to help them with their problems.

VIII. DRAFT HORSES, MINIATURE HORSES, AND PONIES

A. Draft Horses

Draft horses are still being used in various activities in the U.S. such as logging, heaving–pulling contests, parades, horse shows, circuses, movies, TV, pleasure driving, etc. Many are still being used on small farms. A few religious groups have never abandoned the draft horse on their farms. Most draft horses are found in the north central states and Canada although they are widely distributed in both countries. During recent years there has been an increased interest in the use of draft horses, as evidenced by the number of states that have draft horse organizations.

The first National Research Council (NRC) report in 1949 on the nutrient requirements of the horse gave much more attention to draft horses than light horses but it stated that information on nutrient requirements was very limited (5). Subsequent NRC publications have increasingly involved research information with the light horse breeds since that was the information available. The 1989 NRC report on nutrient requirements of the horse does not have a section on draft or miniature horses, but, with some modification, some of the information presented also has some application to both. Texas A & M studies (9) with purebred Belgian and Percheron horses showed that it takes less feed per unit of body weight to maintain heavy, draft-type horses than lighter horses, an observation that is frequently voiced by draft horse breeders. It is suggested that owners of draft horses, miniature horses, and ponies wanting assistance on diet formulation should visit or call their State Agricultural University and confer with those in charge of the horse program.

B. Miniature Horses

A number of breeders have used small ponies to breed miniature horses. Many are used as pets, in parades, shows, circuses, and other activities. They are in limited supply and the demand is good. They are expensive and sell well. The offspring of some become too large which lessens their demand and sale value.

The American Miniature Horse Association (AMHA) is headquartered in Fort Worth, Texas. It has 2800 registered members with Texas, California, Florida, and Indiana having the most members in the order listed. AMHA members can be found in most states as well as in some foreign countries including England, Japan, and Australia (9,10). To be registered with AMHA, a miniature horse cannot stand taller than 34 inches when measured from the base of the last hairs of the mane to the floor, while the horse is standing squarely. They usually weigh under 300 pounds.

C. Ponies

Ponies are very popular, especially with children. Children given the responsibility to care for a pony develop character and the ability to enjoy an animal and the pleasure it can provide them.

Ponies are used in many research studies on feeding and nutrition. It is easier and less costly to use them in certain studies and the results obtained have application to both the pony and the horse. Much of the information presented in the 1989 NRC publication on nutrient requirements of the horse has application to the pony.

The pony is defined as being under 14 hands, 2 inches by the American Horse Show Association. The small pony is 13 hands and below whereas the large pony ranges from 13 hands to 14 hands, 2 inches. Many pony breed registries define a pony as being under 14 hands.

Ponies are used for pleasure riding by children. Some are used and shown in many events by children and adults. They are used in pony-pulling contests and pony harness racing, with parimutuel betting. They are also used in parades, circuses, movies, TV, and many other events.

IX. CONCLUSIONS

In the last 30 years the horse industry has gone from a minor declining industry to a major one and has an excellent future. Horses contribute approximately $15 billion annually to the U.S. economy. Horse racing is by far the leading spectator sport in the U.S. Horse sports draw more than 110 million spectators yearly. In 1981, horse sports generated $1 billion in revenue for federal, state, and local governments.

In 1987 there were 224,903 4-H Club youth enrolled in horse and pony organized projects and activities which is by far the most popular youth livestock program. It is estimated that 50% of the new market for horses is due to youth programs. The great interest of youth in horse programs speaks well for the future of this industry.

Horses are used for riding, pleasure, recreation, exercise, racing, polo, rodeos, pack animals, parades, movies, TV, mounted patrols, working livestock, and a myriad of other activities. They touch the lives of millions of people.

The horse industry faces many serious production problems which require more university involvement in research, teaching, and extension programs. These programs have been neglected until recent years and as a result less is known about the science and technology of horse production than other farm animals. It is encouraging to note that this situation is being changed and increasing university programs are underway. But, more effort is needed in the future. Horses, ponies, and miniature horses have an excellent future.

REFERENCES

1. Weber, G. M. USDA, Washington, D.C. (personal communication), (1988).
2. USDA. "Annual 4-H Youth Development Enrollment Report." USDA, Washington, D.C., (1987).
3. American Horse Council. "Horse Industry Directory." AHC, Washington, D. C., (1987).
4. American Horse Council. "Horse Industry Directory." AHC, Washington, D. C., (1982).
5. Hintz, H. F., and E. L. Squires. *J. Anim. Sci.* **57**, Suppl. 2, 58 (1983).
6. Hubbard, D. D. Extension Service, USDA, Washington, D. C. (personal communication), (1977).

7. Lochhead, C. *Insight* **4**(51), 44 (1988).
8. Potter, G. D., J. W. Evans, G. W. Webb, and S. P. Webb. *Proc. Equine Nutr. Physiol. Symp., 10th, 1987* p. 133 (1987).
9. Pryor, A. *Calif. Farmer* **270**(3), 48 (1989).
10. Anonymous. The History of the American Miniature Horse. American Miniature Horse Assoc. p. 1, Fort Worth, Texas (1989).

2

Art, Science, and Myths in Feeding Horses

In spite of new developments in horse nutrition, the art and science of feeding are still important in properly feeding a top quality race or high-level performance horse. The science involves knowing the nutritional value of feeds, their limitations, the nutrient requirements of the horse, and the technology of combining this information into well-balanced diets to supply the needed nutrients for the various periods of the horse's life cycle. The art involves knowing how to put all the new technology together and getting the horses to make use of it.

I. ART OF FEEDING

The art consists of knowing how to feed a horse and properly take care of its needs. Each horse is an individual and has peculiar needs and desires. To a certain extent, horses are like humans and vary considerably in likes and dislikes for certain feeds and levels of feed intake. Some prefer more hay, others dislike one hay or concentrate feed as compared to another, while some will want to eat more frequently. Many horses are prima donnas and require a certain degree of pampering to persuade them to consume their feed. Close observation is necessary to determine peculiarities and individual needs. A good horse feeding program thus depends almost as much on the feeder as it does on the diet being used. A well-balanced diet is a necessity. However, a competent feeder is needed to ensure that the diet does the job it is capable of. There is no substitute for a feeder who has integrity, is dependable, alert, a hard worker, a good observer, a student of horses, and who feeds the horses on a regular schedule. A good diet, therefore, is only the beginning of a good feeding program. The art, as practiced by the good feeder, is uniquely important, especially for top quality horses being developed for racing and performance purposes. A good feeder with a good diet means the difference between developing a champion or just another horse (Fig. 2.1).

II. SCIENCE OF FEEDING

The science phase of horse feeding is also very important. It requires a knowledge of feeds, their analyses, their limitations, and how to combine them

Fig. 2.1 To obtain contented horses such as these requires a knowledge of the art and feeding of horses. These are Thoroughbred horses in the Ocala, Florida area. (Courtesy of T. J. Cunha, University of Florida and Cal Poly University, Pomona.)

in a well-balanced diet. Feeds vary considerably in their nutrient content and feeding value, which is often affected by methods of processing and storing. Even though a diet contains the same feed formula, its nutritional value will not always be exactly the same, nor will it give the same results in a feeding program. The grains used, such as corn, oats, barley, and others, and the protein supplements available will vary in protein level as well as other nutrients. Not all the grain and protein supplements (such as soybean meal, cottonseed meal, linseed meal, and others) used during the year will come from the same locality. They will vary depending on where they were produced, stored, processed, and how they were transported. Rolled oats, for example, will vary depending on the feed mill which rolled them and the temperature, moisture, and other conditions involved in the processing. Protein supplements will vary depending on the method of extracting the oil, temperature used, and other factors. In other words, animal feeding is not an exact science. The only thing one is absolutely sure of is that variation in results will occur. However, this variation can be minimized by someone with training and experience in feeds, nutrition, and feeding.

Familiarity with vitamins, minerals, protein, amino acids, fatty acids, carbohydrates, and other nutrients, which are part of a good feed program, is of great importance. Understanding these nutrients, their interrelationships, their level and availability in different feeds, which nutrients need to be added to the diet, and how to accomplish this are areas requiring expertise. One also needs to know the nutrient requirements of the horse during the various stages of its life cycle, including the suckling stage, weanling, yearling, 2 year old, training, gestation, and lactation periods. Diets should be formulated which adequately take care of each stage of the life cycle since the level of nutrients needed in the various diets will vary in order to meet those particular nutritional needs. Development of diets for top quality racing or performance horses requires considerable expertise, especially since only a few of the actual nutritional requirements of the horse have been determined in experimental studies. Extrapolation from experimental work with the pig, cattle, and other animals needs to be done to arrive at tentative nutrient levels for use in horse diets. These tentative levels should be determined by competent nutrition scientists, until such time as more exact information becomes available.

So, in feeding top quality horses, the art and the science have an equally important role. Both are needed and proper attention to the small details involved in each area can mean the difference between developing a champion or just another horse.

III. MYTHS IN FEEDING

More myths surround the feeding of horses than with any other animal. It is partially due to a lack of research information on the feeding and nutrition of horses. Consequently, less is known about their requirements for vitamins, minerals, amino acids, protein, fatty acids, and other nutrients. Horse nutrition is perhaps 15–20 years behind that of other livestock species. As a result, too many horse owners and others are still following folklore, old wives tales, and myths in their feeding programs. The author has worked with many owners or managers in programs to improve the feeding and nutrition of horses. After mutual agreement of the diet to use and the complete feeding program to follow, it was never surprising to occasionally find the person who was actually doing the feeding sprinkling some preparation or special mixture on the diet at feeding time. Their explanation was that the horses would not do well unless they also received these supposed "magic potions." They were actually convinced about the absolute necessity for these additions to the diet. Some horse owners who want to develop a champion are not concerned about the extra cost involved. The overriding factor is their search for some "trade secret" or "potion" that will change the horse into a winner instead of just another horse. Unfortunately, these magic potions for developing champions are not available. The best route to follow is that of feeding a well-balanced diet along with a good breeding and management program. Horse people not knowledgeable in the feeding and nutrition of horses should seek the advice of those who are. Magic potions or other supposed short cuts to producing a champion may do more harm than good.

3

Problems in Supplying Proper Nutrient Supplementation

I. INTRODUCTION

Diet formulation to ensure a proper level of nutrients is more complex than many realize. This becomes more apparent as one acquires experience and knowledge in feeding and nutrition. It becomes even more apparent as one tries to coordinate university and other research findings with those occurring on horse farms under varying conditions of management, environment, stress, disease level, and feeding programs. Nutritional requirements obtained under carefully controlled research laboratory conditions may need modification to provide adequately for the myriad of factors encountered on farms which can alter nutrient needs. Even a proper evaluation of research findings by various research laboratories in the U.S. and abroad will indicate that variation occurs in specific nutrient recommendations made by various scientists whose conditions of study differed. Thus, it requires experience and good judgment to put all the research information together and come up with nutrient levels to use for specific horse programs in different locations and situations (1).

The National Research Council (NRC) nutrient requirements committee on horses is very helpful in developing nutrient requirement levels. The committee consists of six distinguished scientists who have stature and considerable expertise in horse feeding and nutrition. But, even they have differences of opinion and some of their final suggested nutrient levels may be a committee compromise which takes into account the many factors which may affect nutrient needs as well as the varying results available in the scientific literature throughout the world.

II. NUTRIENT REQUIREMENTS OF THE HORSE

As one acquires experience and expertise in horse nutrition, it becomes apparent that there are no exact nutritional requirements (1). At best, one can only recommend an approximation of nutrient needs. It is also apparent that a nutrient requirement obtained under one set of conditions cannot be used to generalize for all conditions and all localities. The NRC-recommended nutrient requirement levels are still the best guideline available and should be used as a starting point by those formulating diets. In some cases the NRC-recommended nutrient levels may be satisfactory and in others a modification may be needed to properly take care of the many factors which may alter nutrient needs. The following discussion in Sections A to V will involve the many factors which can alter nutrient requirements.

A. Variation in Requirements by Animals

Most nutritional requirements are based on the average performance of a group of horses. But, if one produces a deficiency of a specific nutrient in a group of horses, some will show signs of a deficiency quickly, others will take a longer period of time, and some animals will show very little, if any, symptoms. This indicates a difference in their nutrient needs which may be due to many factors including genetic differences. This also shows that only a certain percentage of the horses will first show deficiency signs. They are the indicator animals and are telling the producer something is wrong and needs attention (1).

One such example, even though it is with the pig, is a study at Michigan State University (2). It showed that a level of 4.15 mg of pantothenic acid per pound of feed was adequate for only 5 of the 10 pigs in the experimental lot. These five pigs gained normally and at no time showed any signs of a deficiency. The other five pigs, however, showed typical pantothenic acid deficiency symptoms. This example is with the pig but the same may also occur with the horse.

Thus, individual animals vary in their requirements of nutrients. This means that some margin of safety is desirable in formulating diets and especially with nutrients that are not very stable and may be slowly and gradually destroyed by long storage.

B. Feed Nutrient Levels Vary

Some of the many factors which affect the level of nutrients in feeds are as follows: soil type and level of fertilization; stage of maturity at harvesting; harvesting methods; processing methods; handling and storage methods; exposure to varying temperature, humidity, and other environmental factors; moisture level; rancidity level; variety of feed; time interval between harvesting, processing, storage, and its use; and many other factors.

These and other factors are responsible for feeds varying considerably in feed composition. The 1982 NRC publication on "United States–Canadian Tables of Feed Composition" states that "Organic constituents (e.g., crude protein, cell wall constituents, ether extract, amino acids) can vary as much as plus or minus

15%, the inorganic constituents as much as plus or minus 30%, and the energy values as much as plus or minus 10%" (3). So, feed composition tables are average values which can be used as a guide. But, they need verification with actual analyses of the feeds used by the research scientist and the feed-formulating industry.

C. Variation in Availability of Nutrients in Feeds

There are differences in the availability of nutrients, depending on their form. There is also a difference between the availability of these nutrients as determined by a chemical analysis or a microbiological assay and as determined by the use the horse will make of them. Their use by the horse may be different from the figures published on the nutrient analyses by various assay methods. This does not mean that analytical values are not valuable—they definitely are. It does mean, however, that some degree of reservation should be exercised and that provision should be made for taking care of this possible difference in availability. Just one example of this difference in availability is the Purdue University finding by Dr. W. M. Beeson that zinc in soybean meal protein is less available than that in casein. This is due to the phytic acid in soybean protein forming a complex which makes the zinc less available. A horse-feeding trial is the final criterion for determining whether a certain combination of feeds mixed by a feeder or feed manufacturer is adequate in certain nutrients. As diets vary in their makeup with different feeds, there undoubtedly is some difference in the availability of the nutrients contained therein. This problem requires careful consideration by those concerned with formulating diets.

D. Feed Intake Varies

A 1987 NRC publication entitled "Predicting Feed Intake in Food-Producing Animals" does an excellent job of discussing a large number of social, physiological, environmental, genetic, and management factors that influence feed intake for animals (4). The paper does not include horses but many of the factors influencing feed intake in other animals would also apply to the horse. The variation in feed intake presents problems in proper nutrient supplementation to take care of the many situations encountered with different feeds, localities, environments, genetics, levels of performance and production, and stage of the life cycle in which the horse is being fed. It is obvious that a range in nutrient requirements is needed in order to optimize properly horse performance and efficiency under varying levels of feed intake. Horses may vary considerably in their hay and concentrate intake. Some horses prefer more of the hay or more of the concentrate than other horses. This makes it difficult to provide the same nutrient intake to individual horses since the nutrient supplementation is added to the concentrate mixture and not to the hay.

E. Effect of Higher Level of Performance

The trend is toward higher efficiency, performance, and productivity in horse programs. The program that may be satisfactory for the average producer is usually not adequate for the higher level production unit. The higher producing animals, with increased body metabolic heat from the products they produce, tend to be more susceptible to heat stress. University of Florida studies, for example, have shown that the high-milk-producing dairy cow needs higher levels of potassium than presently recommended by NRC (1). This may also be the case with the heavy milking mare. The Florida studies are a good example of the need to reevaluate nutrient needs continually as the level of performance and productivity increases. The producers of high-level performance horses find that many NRC nutrient levels are not adequate for them once they exceed a certain level of performance and productivity. University and consulting animal nutritionists are of considerable help to these high-level producers by working closely with them on their farms to develop the nutrient levels needed for their conditions (1).

F. Stress Conditions

It is difficult to list all the factors that create stress conditions, since some are not known. Stress can be induced by removing the foal from the mare, moving the mare and foal to a new location, or mixing foals from different mares in a pasture which causes fighting to establish a pecking order. Stress can be reduced by mixing foals of the same size. Changes (especially sudden and big ones) in temperature, moisture, and humidity can also cause stress. Bringing horses together from different areas for strenuous competition is a stress factor. Improper and irregular feeding and management are stress factors. Horse temperaments are different from those of other animals and vary considerably with breed, age, sex, management methods, and individual handling. Improper attention to individual horse needs can cause stress. Poor housing and muddy lots or pastures can be stress factors. University of Missouri scientists think that pigs from certain strains may be more susceptible to stress than others (5). The same may be true for horses. All of these factors, plus other stress conditions, can alter the need for many nutrients. Many scientists, therefore, recommend higher levels of nutrients for moderate or severe stress conditions (6). Higher levels of nutrients should be used with caution, however, since some can cause harmful effects if used at too high a level. Thus, one needs to exercise good judgment when increasing nutrient levels to counteract stress. Well-balanced, nutritious diets should be the horse's first line of defense against stress and infectious diseases.

G. Intestinal Flora

The intestinal flora is affected by the type of diet and nutrients fed. Therefore, the intestinal synthesis of nutrients or the requirements of nutrients by the intestinal microflora can change. This in turn will affect the requirement for certain nutrients in the diet. This accounts for recent reports indicating a need for certain vitamins which normally one would expect the pig to synthesize in sufficient quantities for its needs. Unfavorable intestinal flora can also have detrimental effects on the pig. While these data are for the pig, the horse may react in a similar manner.

H. Compounds in Feeds Can Increase Nutrient Needs

There are certain compounds in feeds that can increase the need for certain nutrients, for example: phytates tie up zinc; oxalates tie up calcium; avidin, streptavidin, and stravidin tie up biotin; thiaminase destroys thiamin; goitrogens increase iodine needs; gossypol increases iron needs; and antimetabolites increase certain nutrient needs. Certain antimicrobial drugs may also increase the need for certain nutrients. Rancidity in feeds may destroy vitamins A, D, E, C, biotin, and possibly other nutrients. There are other factors in feeds which can also influence the nutrients in feeds. Some of them are trypsin inhibitor, saponins, isothiocyanate, molds, salmonella, tannins, and others (1).

Antimetabolites in feeds increase the need for certain nutrients. The antimetabolite may be similar in chemical composition to certain nutrients and thus enter certain enzyme systems in the body and block their action which is detrimental to normal nutrition.

These few examples certainly indicate the complexity of determining exact nutrient requirements. It also indicates the need for a reasonable range in nutrient requirements which can be used to meet varying situations.

I. Nutrient Interrelationships

Experimental information on nutrient interrelationships is still in its early stages. As more of these interrelationships are solved, they will answer some of the unexplained results and differences obtained in nutrition studies. Just a few examples of these interrelationships are: choline and methionine; methionine and cystine; phenylalanine and tyrosine; niacin and tryptophan; calcium, manganese, and copper; zinc, copper, and protein; copper, zinc, and iron; vitamin D, calcium, phosphorus, and magnesium; iron and phosphorus; molybdenum, copper, and sulfur; sodium and potassium; biotin and pantothenic acid; Vitamin B12 and methionine; and vitamin E, selenium, and sulfur amino acids. Nutrition will not be thoroughly understood until all the many interrelationships of nutrients are identified. Neither can one accurately determine the requirement of certain nutrients until their interrelationship with other nutrients are known. The requirements of many nutrients will be modified by the level of other nutrients. Certain nutrients may increase the need for others. A few of these include the following:

excess calcium may increase the need for phosphorus, magnesium, zinc, copper, iron, and total sulfur amino acids; a high fat diet level may increase the requirement for pantothenic acid, certain amino acids, and other nutrients.

This discussion explains why there is some variation in the nutrient requirements reported by different investigators feeding animals different kinds of diets under varied conditions of diet quality, management, and environmental situations (1).

J. Nutrients and Compounds in Water

This is an area which is virtually unexplored and on which considerable emphasis should be placed in the future. Water is a source of minerals and other compounds. Since the horse will usually consume 2–4 pounds of water per pound of feed, the level of minerals in water should be taken into consideration when determining nutrient requirements. Moreover, water intake may be increased depending on the level of activity since the horse can sweat profusely. Nitrites, sulfites, and other chemicals in the water can destroy certain nutrients. It is also possible that high levels of sulfates and other compounds in water may cause diarrhea and other digestive disturbances in the horse (7).

K. Energy Content of Diet

The energy content of the diet will definitely affect nutrient needs. For example, amino acid requirements increase as the caloric density and protein level increases. The need for other nutrients in the diet may be either increased or decreased depending on the level and kind of fat in the diet.

L. Criteria Used to Determine Nutrient Requirements

Some nutrient requirements have been worked out without the use of very complete studies. Many of them involve data primarily on rate of gain and feed efficiency. An example, with the pig, of what effect this may have on nutrient requirements is a Michigan State study by Dr. E. R. Miller on pyridoxine. He showed that a level of 0.5 mg of pyridoxine per kilogram of solids was nearly adequate for good growth and feed conversion in the baby pig. When the blood (hemoglobin, red blood cells, etc.) and urinary xanthurenic acid constituents were taken into consideration, however, the pyridoxine requirement was shown to be higher. It was greater than 0.75 mg, but less than 1.0 mg per kilogram of solids (8). If histopathological data had been obtained, the requirement might have been shown to be even higher. University of Pennsylvania School of Veterinary Medicine data on optimal vitamin A intake for optimal growth averaged 1.4 times the 1978 NRC recommendation; for liver secreted serum constituents 5.4

times; and for red blood cell criteria 10 times the NRC requirements (9). These two examples indicate that some nutrient requirements may be higher when more complete studies involving growth, reproduction, data on blood and other body constituents as well as histopathology are obtained. With such complete studies more accurate data on nutrient requirements are obtained.

M. Variation in Results with Natural versus Purified Diets

Some nutrient requirements are worked out using purified diets. Sometimes this information may be applied directly to natural or practical diets as fed on the farm. However, as compared to purified diets, there is some difference in the availability of nutrients in natural diets. Many of the nutrients, such as vitamins and minerals, are added to purified diets in relatively pure form. In natural diets, these vitamins and mineral elements are in their natural state. In most cases, they are in different forms from when they are fed in purified or laboratory diets. Moreover, the two different diets may have different effects on the intestinal synthesis of certain nutrients which subsequently affects their need in the diet. Thus, there is some difference in requirements as worked out with purified and natural diets. Two examples of this difference in diets, with the pig, are the Purdue finding by Dr. W. M. Beeson that zinc in soybean protein is less available than that in purified casein and a Michigan State University study by Dr. E. R. Miller showing that vitamin D needs were higher with a soybean protein diet than with a purified casein protein diet. The zinc requirements were 18 ppm with the casein diet and 50 ppm with a soybean diet. The vitamin D requirements were 45 IU per pound of feed with a casein diet and 227 IU per lb. of feed with a soybean protein diet.

There is apt to be some difference in amino acid requirements as worked out with synthetic amino acids as compared to their need supplied in natural or practical diets. The digestibility of protein varies depending on the feed used. Thus, amino acid requirements worked out with one diet may not necessarily apply to another kind of diet. This means some care must be used in applying results obtained with purified diets directly to natural diets. This does not impair the value of using purified diets in nutritional studies but rather indicates these same nutrients also need to be studied with natural diets. The data obtained with purified diets serve as a valuable guide for studying problems encountered in natural diets fed on the farm.

N. Variation in Deficiency Symptoms

Single nutrient deficiencies are seldom encountered under farm conditions. In most cases, multiple nutrient deficiencies occur. As a result, a complex deficiency will arise; this may be a combination of symptoms described for various single nutrients, or it may be something entirely different. Conditions such as reduced appetite and growth or unthriftiness are common to malnutrition in general. The nature of the deficiency may be detected only by careful review of the dietary history of the animal and by close observation of the symptoms obtained. A fruitful field is open for studies involving multiple deficiencies such as occur on the farm under varying conditions of feeding. Almost no experimental information is available on multiple deficiencies with the horse.

Nutritional deficiencies may exist without the appearance of definite deficiency symptoms. These may be called borderline deficiencies. This results in horses which look normal in appearance but which are growing or performing at a lower and inefficient rate. In fact, many deficiency symptoms do not appear in an average- or low-producing group of horses until the farmer introduces superior breeding stock or improves feeding and management which causes performance to increase beyond a certain level. The nutritional deficiency may first show itself only by slight tissue depletion which may have very little, if any, effect on the performance of the animal. As the deficiency becomes more severe, however, it will affect chemical processes in the body and will eventually result in symptoms which can be observed by looking at the horse.

O. Variation in Treatment of Deficiencies

Many nutritional deficiencies can be treated by supplying the missing nutrient. If treatment begins early enough, most, if not all, of the symptoms can usually be cured by supplying the missing substance. This will not always be the case, however, if the deficiency is long standing and certain changes have occurred in the body which cannot be repaired by feeding the missing nutrient. This must be borne in mind when giving a group of deficient horses a highly fortified diet to cure nutritional deficiencies. If the horses have been deficient for too long it may be too late to cure them, although those that are the least deficient may respond to treatment and recover well. In some cases, they may not be depended upon for use as herd replacement animals, however. Their reproductive systems and possibly other body organs may not be normal enough to produce a healthy, viable foal. Severe deficiencies are detrimental to the horse and, in many cases, cause permanent harm. Thus, a good feeding program should eliminate periods in which horses are allowed inadequate feed intake or improperly balanced diets and thus become deficient in certain nutrients.

P. Will the Growing Foal Be Kept for Reproduction?

Most studies on nutrient needs during growth have not been followed up with gestation and lactation studies to determine if the nutrient level was adequate to develop a normal reproductive tract during the growing period. Until this is done, in some cases the level of a nutrient that is adequate for growth may not be adequate for horses kept for breeding purposes later on. Some examples to illustrate this are the studies performed at Wisconsin, Washington, and Purdue which showed that the diet fed during growth influenced the ability of pigs to conceive, reproduce, and lactate many months later (10). In many cases, the diets which gave the best results during growth gave the poorest results later on during gestation and lactation. This indicates that studies are needed which involve the entire life cycle of the horse and which take into account the specific nutrient needs during each phase and how they affect requirements during the other phases. These kinds of studies may be very beneficial in raising the re-production rate of mares since most mares foal one year and skip the next.

Q. Immunity Effect on Nutrient Needs

A number of recent studies indicate that nutrient levels which are adequate for growth, feed efficiency, gestation, and lactation may not be adequate for normal immunity and for maximizing the animal's resistance to disease. Magnesium, phosphorus, sodium, chloride, zinc, copper, iron, and selenium have been shown to improve an animals ability to cope with infection (11). Deficiencies of thiamin, riboflavin, niacin, pantothenic acid, pyridoxine, folacin, choline, and vitamin C decrease immune response in swine. Protein and/or amino acid levels appears to be involved in immune response also. Recent University of Wisconsin studies show that the dietary methionine levels that are adequate for growth may be inadequate for maintaining the chick's ability to mount an immune defense to challenging organisms (6).

Proper feeding of the mare will ensure that she will produce enough colostrum to feed the foal. The antibodies in colostrum are absorbed by the foal for about 12 to 36 hr after birth. They are very important in protecting the foal from infectious agents shortly after birth since the foal may require 7 to 14 days or more to manufacture antibodies against antigens.

It is apparent that a proper level of nutrient intake is necessary for an adequate health care system designed to prevent diseases which can decrease animal productivity about 15–20% in the U.S. and 30–40% in developing countries (12). Genetics may also be involved in the variation observed in the ability of immune systems of individual horses to ward off diseases (13).

R. Subclinical Disease Level

This refers to a disease condition that occurs at a low level, even though nothing seems to be wrong with the horse's appearance. However, it causes the horse to perform below the level it should. A good example of subclinical disease level is the response to antibiotics by the pig. When hogs are placed in a new house, they respond very little, if any, to antibiotics. The longer they are kept in the house, however, the greater the response to antibiotics. For lack of a better explanation, this seems to be due to a buildup of microorganisms in the house which causes a subclinical disease level in the pigs. Muddy lots, changing climate, and a lack of sanitation all have an effect on the subclinical disease level encountered on the farm. This can vary considerably and accounts for the variability in the response obtained in research laboratories and on farms to antibiotics and other antimicrobial compounds. Kentucky studies showed that the response to antibiotics in field tests can be twice as great as that under university experimental conditions which are cleaner and in which smaller, less thrifty pigs are used less frequently (14).

Much information is needed on the effects of subclinical disease levels on performance and productivity in the horse. Undoubtedly high-level performance horses competing in horse shows, racing, and other performance activities in different locations encounter some subclinical and clinical disease level and stress problems. Moreover, horses are now being moved from one country to another for sales, breeding, and competitive events. This has resulted in the spread of certain diseases, and more stringent regulations are needed to prevent this spread (15).

Diseases can destroy red blood cells, tissues, vital organs, and other parts of the body system. This increases the need for the nutrients required for the repair and restoration of damaged cells. So extra nutritional supplementation may be helpful to the animal that is sick and in poor condition.

A nutritional deficiency can cause abnormality of the body's epithelial tissues and subsequent penetration of organisms. Properly fed animals are usually more resistant to most diseases, particularly parasitic and certain bacterial diseases (6, 16). Moreover, proper nutrition enhances their recovery from all diseases and increases efficiency, performance, productivity and profitability.

S. Molds in Feeds

Molds in feeds may affect the availability of biotin. Streptavidin and stravidin isolated from streptomyces are biotin-binding substances (17). They act similarly to avidin in raw egg white in typing up biotin (18). Streptomyces are molds found in soil, moldy feeds, manure, and litter. There are some indications that molds in feeds may influence the need for biotin, folacin, vitamin K, and other nutrients in the pig. Whether a similar effect occurs with the horse is not known. A report from England, however, has shown that biotin supplementation is beneficial for hoof integrity in the horse (19).

T. Performing as an Athlete

The high-level performance horse is expected to compete as an athlete at race meets, horse shows, trail rides, fairs, clinics, rodeos, polo, and other events.

This requires considerable training and physical conditioning to compete successfully. This in turn requires proper nutrition and veterinary care to prevent and treat, if necessary, the many skeletal, muscular, and behavioral problems not encountered with other classes of livestock. In many cases, higher levels of certain nutrients are required for the competing horse athlete depending on the level of activity involved. This area of nutritional need is comparatively new and awaits further research for more accurate determination of the nutritional requirements of the horse athlete.

U. Selection for Higher Performance

As horses are selected for faster growth rates, higher level of reproduction, higher milk production, and improved performance as an athlete, there is no doubt that the required level of certain nutrients will increase. Without proper nutrient intake, the horse may not be able to realize its full genetic potential.

V. Other Factors

There are a number of other factors which may affect nutrient needs. Some of these include: environmental temperature and humidity; destruction of nutrients by rancidity, light or irradiation; hormones; feed additives; toxins; enzymes; management procedures; and others.

III. CONCLUSIONS

When dealing with horses and other animals, the only sure thing is that variation in response will occur. This also applies to nutrient requirements since many factors can influence their need. Therefore, there are no exact nutritional requirements but only approximate requirements. The NRC-recommended nutritional requirements are still the best guide available. But, they may need some modification to meet the myriad of conditions under which they are applied under many and varied farm situations.

REFERENCES

1. Cunha, T. J. *Feedstuffs* **59**(42), 1, 42–44, 47 (1987).
2. McMillen, W. N. *Mich. Agric. Ext. Serv., Bull.* **299** (1949).
3. Conrad, J. H., C. W. Deyoe, L. E. Harris, P. W. Moe, R. L. Preston, and P. J. Van Soest. *N.A.S.-N.R.C., Publ.* (1982)
4. Fox, D. G., C. A. Baile, H. R. Conrad, R. Ewan, L. J. Koong, G. L. Rumsey, and P. W. Waldroup. *N.A.S.-N.R.C., Publ.* (1987).

5. Ellersieck, M. R., T. L. Verum, and T. L. Durham. *J. Anim. Sci.* **48**(3), 453 (1979).

6. Cunha, T. J. *Feedstuffs* **57**(41) 37, 38, 40, 42 (1985).

7. Shirley, R. L., C. H. Hill, J. T. Maletic, O. E. Olson, and W. H. Pfander. *N.A.S.-N.R.C., Publ.* (1974).

8. Miller, E. R., D. A. Schmidt, J. A. Hoefer, and R. W. Luecke. *J. Nutr.* **62,** 407 (1957).

9. Donoghue, S., D. S. Kronfeld, S. J. Berkowitz, and R. L. Copp. *J. Nutr.* **111,** 365 (1981).

10. Cunha, T. J. "Swine Feeding and Nutrition." Academic Press, New York, (1977).

11. Miller, E. R. *J. Anim. Sci.* **60**(6), 1500–1507 (1985).

12. Cunha, T. J., K. H. Shapiro, J. M. Fransen, H. J. Hodgson, J. E. Johnston, W. H. Morriss, R. R. Oltjen, W. R. Pritchard, R. R. Spitzer, and N. L. Van Demark. *N.A.S.-N.R.C., Publ.* pp. 141–250 Vol. 1 (1977).

13. Smith, A. T. *J. Anim. Sci.* **51,** 1087 (1980).

14. Cromwell, G. L. *Anim. Health Nutr.* **38**(4), 18 (1983).

15. Powell, D. G. *Univ. Ky. Coll. Agric., Coop. Ext. Serv., Bull.* Equine Data Line, Sept., pp. 1–3 (1985).

16. Cunha, T. J. *Squibb Int. Swine Update Rep.* **3**(1), 1, 4–7 (1984).

17. Cunha, T. J. *Feed Manage.* **35**(35), 14–24 (1984).

18. Ghaiet, L., and F. J. Wolf. *Arch. Biochem. Biophys.* **106,** 1 (1964).

19. Comben, N., R. J. Clark, and D. J. B. Sutherland. *Vet. Rec.* **115, ** 642 (1984).

4

The Digestive Tract

I. DIGESTIVE SYSTEM

The digestive system, or the alimentary canal, in the horse is about 100 ft long from the mouth to the anus (Fig. 4.1). The entire tract can be divided into two functional parts: the foregut and the hindgut (Table 4.1). The foregut functions in a manner similar to that of simple stomached animals such as the pig. The hindgut has some similarity to ruminants in that microbial action is not as great as in the ruminant nor is absorption of the nutrients produced there as efficient, because the cecum is located posterior to the small intestine. The ruminant (beef, dairy, sheep, goats, and others) has four compartments in its stomach. Moreover, it has a large rumen where considerably more microbial action occurs and which is located anterior to the small and large intestines which favors greater absorption of the nutrients produced in the rumen. Therefore, the horse is somewhere between the pig and the ruminant in its utilization of feeds.

The digestive system of the horse is involved in feed consumption, chewing, mixing, digesting, absorbing feed nutrients, and eliminating the undigested portion as solid waste. It takes about 65 to 75 hr for feed to proceed from the mouth to the anus.

II. FUNCTION OF DIGESTIVE SYSTEM

A. Mouth

The function of the mouth is to grasp and take in food, moisten it with saliva, reduce particle size by chewing and thus increase surface area for digestive enzymes. Saliva contains amylase, an enzyme involved in carbohydrate breakdown. Horses produce considerable amounts of saliva while eating.

B. Pharynx

It is located in the upper back part of the horse's mouth and serves to guide feed into the esophagus. Feed or water cannot return to the mouth from the esophagus because the soft palate blocks its return. The epiglotis also closes and prevents passage of the feed into the lungs.

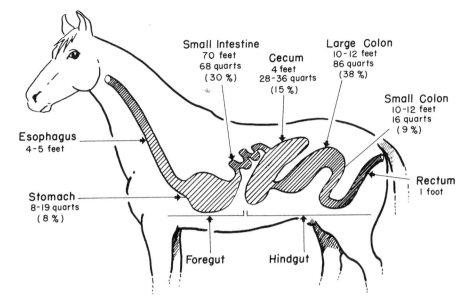

Fig. 4.1 Digestive tract of the horse. (Courtesy of D. D. Householder and G. D. Potter, Texas A & M University.)

C. Esophagus

The esophagus is a 4–5 ft muscular tube which conveys feed and water to the stomach via waves of muscular restrictions called peristalsis. These waves are usually irreversible.

TABLE 4.1

Digestive System

Foregut (similar to pig)	Hindgut (some similarity to ruminant)
Mouth	Cecum[a]
Pharynx	Large colon[a]
Esophagus	Small colon[a]
Stomach	Rectum[a]
Small intestines	Anus

[a]Some refer to these as the large intestines. Some will divide the large intestine into cecum, ventral and dorsal colons, transverse colon, small colon, and rectum.

D. Stomach

A muscle called the sphincter regulates the opening of the esophagus into the stomach. The horse's stomach is small compared to that of other animals. The stomach comprises 8–10% of the total digestive tract and cannot tolerate too much feed at one time.

On a relative basis, the stomach of a mature cow has the capacity for over 10 times as much feed as the stomach of a mature horse. The stomach of the horse, therefore, requires frequent consumption of small quantities of feed, rather than very large amounts at one time. Thus, if a horse is fed too much feed or roughage, it may result in labored breathing and rapid fatigue. Severe overeating can cause colic, a ruptured stomach, or founder. Overfeeding should be avoided, and horses should be fed two or three times a day depending on their level of activity.

Soon after eating, the feed is rapidly passed through the stomach and into the small intestine. Sometimes the feed consumed at the beginning of the meal passes on to the small intestine before the meal is completed. Therefore, if too much feed is consumed at one time, it may have only limited contact with the stomach gastric secretions, which aid in digestion. The stomach will empty itself in about one-third the time it takes a ruminant.

Feeds begin to breakdown due to enzymatic and microbial digestion in the stomach. The stomach has a microbial population which allows a small amount of digestion to take place. Gastric juices contain hydrochloric acid and the enzymes pepsin and gastric lipase. Pepsin helps digest protein into smaller fragments called peptides and lipase helps digest fat.

E. Small Intestine

The small intestine is about 70 ft long and connects the stomach to the large intestine. It makes up about 30% of the digestive tract. The small intestine and its nearby organs, the liver and pancreas, provide most of the enzymes for digestion. Peristalsis of the intestinal wall causes efficient mixture of its contents.

The pancreas produces the enzymes trypsin, pancreatic lipase, and amylase. The pancreatic amylase breaks down starch. Trypsin breaks down protein and peptides into amino acids. Pancreatic lipase hydrolyzes fats to fatty acids and glycerol. The liver secrets bile which promotes emulsification of fats. The horse does not have a gallbladder to store bile.

Fatty acids, simple sugars, amino acids, vitamins, and minerals are absorbed from the small intestine. The stomach and small intestine serve as the first sites for digestion and absorption of nutrients.

F. Large Intestine

The large intestine consists of the cecum, large colon, small colon, and rectum. This area makes up about 60–62% of the digestive tract. Feed and nutrients not digested in the stomach and small intestine flow to the large intestine.

The horse has a much larger cecum than the cow. It makes up 38–40% of the capacity of the digestive tract. The cecum is the horse's fermentation vat. It contains a microbial population which is somewhat similar to that of ruminants (1). Most of the digestion of forage fiber (hay and pasture) occurs in the cecum. Therefore, the horse cannot handle as much roughage in the diet. For example, the horse digests the fiber of average-quality grass hay with about two-thirds of the efficiency of the cow (2). One can state the cecum is not as valuable to the horse as the rumen is to the cow.

A large portion of the forage consumed is digested by the bacteria and protozoa in the cecum and to some extent in the remainder of the hindgut area. Microbial action also produces vitamins, volatile fatty acids, and amino acids. The volatile fatty acids may supply about one-fourth of the energy used by the pony (3,43). It is not known how much of the horse's requirement for vitamins is produced and absorbed from the hindgut area. The level of certain vitamins may not be adequate for the high-level performance horse. One report indicates that certain B-complex vitamins (e.g., thiamin) may not be synthesized or absorbed in great enough quantities to meet the requirements of hard working horses (4). These horses may need supplemental thiamin added to their diets (4). This was also indicated in early work by Carroll et al. (5). They found that the horse synthesized thiamin, riboflavin, niacin, pyridoxine, folacin, and biotin primarily in the hindgut area. They also felt that inadequate thiamin was synthesized.

How much amino acid absorption occurs in the hindgut area is also not known. Much of the protein and its degradation products not utilized and absorbed in the small intestine are broken down to ammonia in the hindgut area. Microbial action produces amino acids from the ammonia but one report indicated that no appreciable amino acid absorption occurs in the hindgut area (6). This would indicate that the horse, unlike the ruminant, cannot convert low-quality protein to high-quality protein by microbial action in the hindgut and then absorb the indispensable amino acids which are synthesized (7). Therefore, the horse needs high-quality protein in the diet to provide its indispensable amino acid requirements.

The cecum serves as the main site of water absorption. This results in relatively dry feces.

The foal has a small digestive tract and the cecum does not become fully functional until it is about 15–24 months old. So, foals and young horses are limited in their ability to use much forage. It is best to use high-quality hay and/or pasture for them. Moreover, they need some supplemental concentrate in order to meet nutritional requirements and for adequate growth and development.

III. DIGESTION AND ABSORPTION

There is still a great deal to learn about digestion and absorption in the horse and what can be done to improve it. The information available is discussed below.

A. Protein Digestion

Protein digestion starts in the stomach and then continues in the small intestine, which is the main site for conversion to amino acids and where the majority of the amino acids are absorbed. Amino acids of microbial origin are absorbed from the cecum and large intestine. However, it is thought that amino acids synthesized by the microbes in the cecum and large intestine are not utilized too efficiently by the horse (6, 8–11).

The young horse is largely dependent on the balance and level of amino acids in the feed and cannot depend on microbial synthesis of amino acids in the cecum to take care of its indispensable amino acid needs. Protein digestion in the young horse is similar to that of the pig because the cecum is not fully developed. This accounts for the fact that lysine supplementation improves the growth of young horses but not that of mature horses (12, 13).

A Texas study (14) showed that the role of the small intestine in digestion of hay protein appears to be rather small, and the role of the large intestine in digestion of hay nitrogen can be comparatively very high. They suggested that a highly digestible source of amino acids should be provided in the diet of young, growing horses, even if they are fed high-quality hay.

Horses are not fed nonprotein nitrogen compounds such as urea since the amount utilized by the horse is very small (15).

B. Carbohydrate Digestion

The easily digested carbohydrates are broken down and absorbed largely in the small intestine. Here they are absorbed as glucose and other simple sugars. Some of these carbohydrates, however, reach the cecum and large intestine where they are digested by microorganisms and end up as volatile fatty acids.

These volatile fatty acids (VFA), which consist mainly of acetate, proprionate, and butyrate, are absorbed in the large intestine and are used as a source of energy.

The complex carbohydrates, or the fibrous part of the diet (such as hay and pasture), are digested primarily in the cecum and large intestine. The horse, however, is only about two-thirds as efficient as the ruminant in the digestion of average-quality grass hay (2, 16). The higher the quality of the forage, the better it is digested by the horse. The horse comes close to the ruminant in the digestion of a high-quality alfalfa hay. The principal end products of fiber digestion are volatile fatty acids.

A Maryland study (17) showed that cellulolytic bacteria numbers per gram of ingesta were similar in the pony's cecum and the steer's rumen, whether or not oats was included in the diet. But, none of the types of protozoa found in ponies were found in steers and vice versa. Cecal fistulas have been developed which allow convenient access to the cecum (18) and large intestine (19).

Work in Kentucky showed that distillers' dried-grain solubles and dehydrated alfalfa meal increased cellulose digestibility by equine cecal organisms in *in vitro* digestion trials (20).

A Texas study (21) showed that prececal apparent starch digestibility was similar for two, three, and four meals per day. They felt the results of the study indicate that high-concentrate diets can be safely fed to horses in two meals per day when horses are fed a maintenance level of energy.

C. Fat Digestion

The horse does not have a gallbladder, but this does not appear to affect the digestion of fat. Unfortunately, there is still very little information available on fat digestion in the horse. The fat consumed in the diet appears to be digested and absorbed in the small intestine. The composition of body fat in the horse is similar to that of dietary fat since the fatty acids are absorbed from the small intestine before they can be altered by microorganisms in the large intestine (22). Mature horses have been fed 15% beef tallow in the diet with good results. Therefore, the horse can tolerate a fair amount of fat in the diet. Studies in Virginia (23) showed that ponies can consume 20% corn oil in the diet with good results. The addition of the 20% corn oil did not affect the apparent digestibility of crude protein.

A Texas study showed that a high-energy diet containing 10% fat supported rapid weight gains in yearling horses (24). There were no radiographic indications of skeletal abnormalities in the horses. In another study the Texas group showed that adding 10% fat to the diet had a sparing effect on muscle glycogen reserves (25).

D. Minerals

Ponies fed a conventional diet absorb 96% of the sodium and chloride and 75% of the soluble potassium and phosphate entering the large bowel from the ileum.

A Cornell study (26) showed that the major sites of net phosphorus absorption from all feed sources were the dorsal large colon and the small colon. The calcium content or the type of feedstuff did not affect the site of absorption. Another Cornell study (27) showed that the upper part of the small intestine appeared to have the greatest calcium absorptive potential and to be the major effective site of net calcium absorption. However, the lower portion of the small intestine may also be a site of significant calcium absorption. Relatively little calcium was absorbed from the large intestine. The fact that phosphorus, but not calcium, is absorbed from the large intestine of the horse may explain why excess dietary phosphorus greatly decreases calcium absorption, but excess calcium has a less dramatic effect on phosphorus absorption (28). Much of the excess calcium is absorbed from the small intestine and so is not available to hinder phosphorus absorption from the large intestine.

Magnesium is not efficiently absorbed from the large intestine of the horse (29). There is some indication that the small intestine is the primary site of zinc and copper absorption in the horse (15).

E. Factors Affecting Digestion

Horses and ponies are quite similar in their digestion of feeds (30). Therefore, some of the information obtained with ponies is cited in this book with the assumption that it will approximate what would occur with the horse. The pelleting of roughage decreases fiber digestion. This is probably due to the fine grinding of the roughage (before it is pelleted) which causes it to pass faster through the digestive tract (31, 32). All grains should be cracked, crimped, or rolled for foals and mature horses with poor teeth. The digestion of wheat and milo is improved when the kernel is rolled or cracked. Processing of corn or oats does not seem to significantly increase their digestibility (33). The digestibility of protein, dry matter, and fiber of diets containing 40% dry-rolled or steam-processed flaked milo was similar (34). Most horse owners prefer the corn cracked or rolled, and the oats rolled.

The digestibility of diets containing roughage and grain may be decreased with increased dietary intake, although the level of intake of a diet consisting entirely of forages may not affect digestibility (35). The kind and quality of forage (hay or pasture), its level in the diet, and the kind and level of grain used may have some influence on digestibility, although experimental information is not adequate on this. Alfalfa was fed as cubes, pellets, chopped, or as loose hay. No difference in digestibility of protein, dry matter, or grass energy occurred

between the different forms of hay although the intake of cubes was the greatest
(36). Another study (37) found no differences in the digestibility of chopped or
long timothy hay.

Studies at Cornell (33) with ponies fed a pelleted diet showed that giving them
water before or after eating had no effect on digestibility of the diet. Feed intake
was decreased, however, when the ponies were given water only after they had
eaten. This observation is in agreement with most horse owners who prefer that
their horses have access to water before eating. In fact, some horses will not eat
unless they first drink some water.

There is still very little information available on the effect of frequency of
feeding on digestibility of the diet. Studies at Cornell (33) showed no apparent
difference in digestibility of a pelleted diet fed to ponies either once, twice, or six
times a day (Table 4.2). It is usually recommended, however, that horses be fed
two or three times a day because they have a relatively small stomach. The
frequency of feeding would depend on the amount fed each time and the level of
activity of the horse.

Yeast culture supplementation of yearling horses allowed free-choice exercise
and fed mixed hay–grain diets significantly increased the apparent digestibilities
of neutral detergent fiber, hemicellulose, acid detergent fiber, cellulose, and
nitrogen (38). A prior study with mature horses showed no benefit from yeast
culture supplementation (39).

A New Jersey study (40) showed that the major difference among forages in
digestible organic matter and energy was the result of different carbohydrate
compositions of the forages. The legume forages with the greater portion of
soluble carbohydrate were the most digestible. The higher soluble carbohydrate
content along with the high apparent digestibility of soluble carbohydrate is the
distinct advantage legumes have over grasses (40).

There is one report suggesting that light exercise may improve diet digestibility but heavy work may decrease it (41).

TABLE 4.2

Effect of Feeding Frequency on Digestibility of Pony Diet[a]

Feeding frequency	Percent digestibility of diet constituents			
	Dry matter	Crude protein	Neutral detergent fiber	Acid detergent fiber
Once daily	71.5	82.3	45.9	28.2
Twice daily	71.0	80.6	44.6	27.6
Six times daily	72.0	79.5	44.2	28.1

[a]Data from Hintz and Schryver (33).

IV. DIGESTIVE SYSTEM DISORDERS

The digestive system of the horse is subject to twisting and impaction. The natural response of a horse with a digestive disorder is to roll, and this action may cause a section of the intestine to become twisted. Impaction may occur when a section of the intestine is damaged by parasites or toxins. There are two locations in the hindgut which are susceptible to impaction because of their small diameters. One of these areas is at the junction of the small intestine and cecum (the illioceal orifice). The other area is located between the large and small colon at the pelvic flexure (42).

In addition to impaction, colic may result from the formation of gaseous products of microbial digestion. Major causes of founder are also related to by-products of microbial digestion. If large amounts of soluble starch (like that in grains) are available to microbes in the hindgut area, large quantities of gaseous products are produced. These compounds may not be absorbed or disposed of as quickly as they are produced, so gaseous swelling of the hindgut may occur. This swelling may frequently cause the horse to develop colic (42).

The anatomy of the digestive tract requires some bulkiness in the diet to help prevent founder and colic. Therefore horses require a certain amount of forage in their diet. However, poor-quality hay high in fiber fed without adequate water intake may sometimes cause impaction and colic. The use of high-quality hay and/or pasture eliminates or minimizes the occurrence of impaction and colic.

Horses are usually fed less hay than are ruminants. This is because the cecum has only about one-fifth the capacity of the rumen. It is though that the minimum daily roughage needs of the horse are at least 0.5 lb of hay per 100 lb bodyweight. Many horse owners, however, prefer to feed twice this level of roughage. The hay intake, therefore, will vary with its quality, the remainder of the diet, the stage of the life cycle, the individual horse, and how it responds to the hay level.

V. CONCLUSIONS

The horse is somewhere between the pig and the cow in its ability to digest feeds. The horse has a small stomach and requires frequent consumption of small quantities of feed. If fed too much feed at one time, labored breathing, rapid fatigue, colic, founder, and a ruptured stomach may occur. The horse has a large cecum which makes up about 40% of its digestive tract. The cecum, the horse's fermentation vat, has a microbial population which is similar to that of the cow. Most of the digestion of forages occurs in the cecum. Microbial action in the cecum produces volatile fatty acids, vitamins, and amino acids. The volatile fatty acids produced may supply about one-fourth of the horse's energy needs. Some of the vitamins produced are absorbed but it is not known to what degree they provide for the needs of the high-level performance horse. Amino acid absorption is thought to be low from the hindgut area. Young horses, therefore, cannot depend on amino acid synthesis in the cecum to supply their indispensable amino acid needs. Therefore, the diet fed must supply the indispensable amino acids. With a lack of adequate high-quality protein sources in the diet, supplementation with lysine has benefited the young horse.

REFERENCES

1. Baker, J. P., T. M. Leonard, and W. J. Hudson. *Proc. Conf.—Distill. Feed Res. Counc.* **28,** 19 (1973).
2. Hintz, H. F. *Veterinarian* **6,** 45 (1969).
3. Argenzio, R. A., J. E. Lowe, D. W. Peckard, and C. E. Stevens. *Am. J. Physiol.* **226,** 1035 (1974).
4. Topliff, D. R., G. D. Potter, J. L. Kreider, and C. R. Cregar. *Proc. Equine Nutr. Physiol. Symp., 7th,* p.167 (1981).
5. Carroll, F. D., H. Gross, and C. E. Howell. *J. Anim. Sci.* **8,** 290 (1949).
6. Potter, G. D., and J. P. Baker. *Proc. Amer. Soc. Anim. Sci. Meet.,* Penn State University (1970).
7. Householder, D. D., and G. D. Potter. *Tex. A&M Univ. Mimeo Rep. Horse Feed. Nutr.* (1988).
8. Wysocki, A. A., and J. P. Baker. *Proc. Equine Nutr. Physiol. Symp., 4th,* p. 21 (1975).
9. Reitnour, C. M., and R. L. Salsbury. *Br. Vet. J.* **131,** 466 (1975).
10. Slade, L. M., D. W. Robinson, and K. E. Casey. *J. Anim. Sci.* **30,** 753 (1970).
11. Reitnour, C. M., and R. L. Salsbury. *J. Anim. Sci.* **35,** 1190 (1973).
12. Hintz, H. F., H. F. Schryver, and J. E. Lowe, Jr. *J. Anim. Sci.* **33,** 1274 (1971).
13. Brewer, L. H., and J. D. Word. *Fed. Proc., Fed. Am. Soc. Exp. Biol.* **27,** 730 (abstr.) (1968).
14. Gibbs, P. G., G. D. Potter, G. T. Schelling, J. L. Kreider, and C. L. Boyd. *J. Anim. Sci.* **66,** 400 (1988).
15. Hintz, H. F., H. F. Schryver, and C. E. Stevens. *J. Anim. Sci.* **46,** 1803 (1978).
16. Evans, J. W., A. Borton, H. F. Hintz, and L. D. Van Vleck. "The Horse." Freeman, San Francisco, California, 1977.
17. Kern, D. L., L. L. Slyter, J. M. Weaver, E. C. Leffel, and G. Samuelson. *J. Anim. Sci.* **37,** 463 (1973).
18. Teeter, S. M., W. E. Nelson, and M. C. Stillions. *J. Anim. Sci.* **27,** 394 (1968).
19. Baker, J. P., H. H. Sutton, B. H. Crawford, Jr., and S. Lieb. *J. Anim. Sci.* **29,** 916 (1969).
20. Baker, J. P., and R. E. Pulse. *J. Anim. Sci.* **37,** 285 (1973).
21. Massey, K. J., G. D. Potter, G. T. Schelling, and W. L. Jenkins. *Proc. Equine Nutr. Physiol. Symp., 9th,* p.42 (1985).
22. Garton, G. A. *Nutr. Abstr. Rev.* **30,** 1 (1960).
23. Bowman, V. A., J. P. Fontenot, K. E. Webb, Jr., and T. N. Meacham. *Va. Polytech. Inst., Livest. Res. Rep.* **172,** 72 (1977).
24. Scott, B. D., G. D. Potter, J. W. Evans, J. C. Regor, G. W. Webb, and S. P. Webb. *Proc. Equine Nutr. Physiol. Symp., 10th,* p.101 (1987).
25. Meyers, M. C., G. D. Potter, L. W. Greene, S. F. Crouse, and J. W. Evans. *Proc. Equine Nutr. Physiol. Symp., 10th,* p.107 (1987).
26. Schryver, H. F., H. F. Hintz, P. H. Craig, D. E. Hogue, and J. E. Lowe. *J. Nutr.* **102,** 143 (1972).

27. Schryver, H. F., P. H. Craig, H. F. Hintz, D. E. Hogue, and J. E. Lowe. *J. Nutr.* **100,** 1127 (1970).
28. Schryver, H. F., H. F. Hintz, and J. E. Lowe. *Cornell Vet.* **64,** 493 (1974).
29. Hintz, H. F., and H. F. Schryver. *J. Anim. Sci.* **35,** 755 (1972).
30. Slade, L. M., and H. F. Hintz. *J. Anim. Sci.* **28,** 842 (1969).
31. Haenlein, G. F. W., R. G. Smith, and Y. M. Yoon. *J. Anim. Sci.* **25,** 1091 (1966).
32. Hintz, H. F., and R. G. Loy. *J. Anim. Sci.* **25,** 1059 (1973).
33. Hintz, H. F., and H. F. Schryver. *Proc. Cornell Nutr. Conf.* pp. 108–11 (1973).
34. Schurg, W. A., and P. D. Kigin. *J. Anim. Sci.* **57,** Suppl. 1, 270 (1983).
35. Reid, J. T., and H. F. Tyrrell. *Proc. Cornell Nutr. Conf.* p.25 (1964).
36. Todd, L., W. C. Sauer, and R. J. Coleman. *J. Anim. Sci.* **57,** Suppl. 1, 273 (1983).
37. Gallagher, J. R., H. F. Hintz, and H. F. Schryver. *Animal Prod. Aust.* **15,** 349 (1984).
38. Glade, M. J. Norwestern University, Evanston, Illinois (personal communication), 1988.
39. Webb, S. P., G. D. Potter, and K. J. Massey. *Proc. Equine Nutr. Physiol. Symp., 9th,* p.64 (1985).
40. Fonnesbeck, P. V. *J. Anim. Sci.* **27,** 1336 (1968).
41. Olsson, N., and A. Ruudvere. *Nutr. Abstr. Rev.* **25,** 1 (1955).
42. Freeman, D. W., and D. R. Topliff. *Okla. State Univ., Agric. Ext. Facts* No. 3973 (1985).
43. Argenzio, R. A., and C. E. Stevens. *Am. J. Physiol.* **228,** 1224 (1975).

5

Vitamin Requirements

I. INTRODUCTION

Most of the work on vitamins has been done since 1911. At that time, Casimir Funk (1) who was working at the Lister Institute in London coined the term *vitamine*. Later the letter e was dropped, and the present term *vitamin* was adopted.

Vitamins are organic compounds. All are different in structure and function. They are not related to each other as are proteins, fats, and carbohydrates. Vitamins are needed in only very small amounts. They perform many important functions in the animal's body. Without vitamins a horse cannot grow, reproduce, lactate, work, or perform. Thus, it is very important that all vitamins be supplied in adequate amounts in horse diets. Unfortunately, there is a lack of experimental information on the level of vitamins to use as well as which of the vitamins are needed by the horse. Moreover, it is not known which vitamins need to be added to well-balanced horse diets.

II. LISTING OF VITAMINS

A number of vitamins are water soluble and some are fat soluble. This means the vitamins will dissolve in either water or liquid fat. *Water-soluble vitamins* include vitamin C and the B-complex vitamins, thiamin, riboflavin, niacin, pantothenic acid, choline, biotin, vitamin B6, vitamin B12, folacin, myo-inositol, and *p*-aminobenzoic acid. *Fat-soluble vitamins* include vitamin A, vitamin D, vitamin E, and vitamin K.

The list includes 11 B-complex vitamins, 4 fat-soluble vitamins, and vitamin C. Later in the chapter, information will be given on different forms of some of the vitamins and their relative activity. The vitamins are listed in Table 5.1. Some scientists do not feel that myo-inositol and *p*-aminobenzoic acid should be classified as vitamins but rather as vitamin-like compounds. Some also feel that choline does not qualify as a true vitamin since it is required at levels higher than other vitamins. Moreover, choline is not known to participate in enzyme systems. But, all are included in this chapter until their role in nutrition is better clarified.

TABLE 5.1

List of Vitamins

Fat-soluble Vitamins	Water-soluble Vitamins	Water-soluble Vitamins[a]
A	C	Myo-inositol
D	*B-complex vitamins*	p-Aminobenzoic acid
E	Thiamin	
K	Riboflavin	
	Niacin	
	Pantothenic acid	
	B6	
	Choline[b]	
	Biotin	
	Folacin	
	B12	

[a]Some scientists do not consider myo-inositol and p-aminobenzoic acid to be classified as vitamins. Some feel they should be classified as vitamin-like substances.

[b]There are also some who state that choline is included as a vitamin but it does not qualify as a true vitamin since it is required at far greater levels than other vitamins and is not known to participate in any enzyme system.

The fat-soluble vitamins contain only carbon, hydrogen, and oxygen, But some of the water-soluble vitamins also contain nitrogen, cobalt, and sulfur.

The fat-soluble vitamins are primarily excreted in the feces and the water-soluble vitamins primarily in the urine. The water-soluble vitamins are relatively nontoxic when used at reasonable levels. Vitamins A and D and choline (as choline chloride) can be harmful, however, if used at excessive levels. Considerable information on vitamin tolerance in animals is available in a 1987 NRC publication (2).

The fat-soluble vitamins are absorbed along with dietary fats via mechanisms that are probably similar to those of fat absorption. Conditions favorable for fat absorption evidently favor the absorption of fat-soluble vitamins. Diseases or metabolic abnormalities that interfere with fat absorption also adversely affect absorption of these vitamins (3). Oral administration of fats that are not absorbed, such as mineral oil, may result in undesirable loss of fat-soluble vitamins in the feces (3).

III. STATUS OF VITAMIN KNOWLEDGE

Some of the vitamins are synthesized by the horse. The level synthesized will vary with the vitamin and the kind of diet fed. The cecum serves as the area where a good deal of vitamin synthesis occurs. It is not known, however, how much of the vitamins synthesized in the cecum are absorbed in the large intestine. Indications are that the level absorbed may be low. As one cannot depend on the horse to synthesize all of the B vitamins it needs, many horse owners use B-vitamin supplementation of diets for the young horse and those being developed for racing or high-level performance purposes. If the vitamins are supplemented at reasonable levels, the cost is low and the benefits derived could pay for the cost many times over. In spite of the tremendous amount of information on vitamins, there is very little experimental information available on the vitamin needs of the horse. Moreover, it is not known which vitamins need to be added to well-balanced diets and under which conditions during the various stages of the life cycle of the horse. Therefore, scientists making up diets, or vitamin supplements, need to extrapolate and use information obtained with other animals as a guide in deciding how much to feed the horse. These estimates can be used until research provides more exact information.

Although the author would prefer not to make recommendations without research data obtained with the horse, the horse people need some guidance now and cannot wait 10, 20, or more years before sufficient research information becomes available to give them the exact help they need. It is felt the level of vitamins recommended in this chapter will be better than some of those presently being used by many horse owners, which in many cases are extremely high and could be causing harmful effects. It should be pointed out that the author has based his recommendations on information obtained with horse owners who were willing to use similar vitamin levels in cooperative studies with him during the past 40 years. Therefore, the vitamin levels recommended have some basis for being used.

IV. THE EXISTENCE OF BORDERLINE DEFICIENCIES

A borderline deficiency of any vitamin may exist without the horse showing any symptoms. When this occurs, growth, reproduction, and performance may be affected. These borderline deficiencies are difficult to detect and are costly to the horse producer in terms of reduced productivity. The fact that a horse does not show overt deficiency symptoms is not an adequate reason for continuing to feed poor-quality diets. So a deficiency of a group of vitamins may exist on the farm without the horse showing deficiency signs such as have been described for single vitamin deficiencies. This is especially so if the nutritional level is so low that growth and performance is poor.

V. SINGLE VITAMIN DEFICIENCIES RARELY FOUND

Under farm conditions, one will not usually find a single vitamin deficiency. In almost every case, a multiple vitamin deficiency will exist. In other words, the deficiency symptoms may be a combination of symptoms described for the various single vitamins or they may be something entirely different. Conditions such as unthriftiness, reduced appetite, poor growth and performance are common to malnutrition in general. Studies are needed to determine the symptoms and performance obtained with horses that have multiple deficiencies of the type that may be encountered on the farm under varying conditions of feeding and management.

VI. VITAMIN NEEDS BECOMING MORE CRITICAL

In recent years, the need for vitamins has become more critical since the trend is toward keeping horses in close or near total confinement. Horses are not kept on high-quality pastures as much now as in the past. Most of the horses used for pleasure riding, and sometimes for racing or high-level performance, are now kept in or near the larger cities. High-quality pasture in adequate amounts is seldom available for these horses. In some cases, what is called a pasture has very little vegetation and is essentially an exercise area. Moreover, the quality (or greenness) of the hay now used is, in many cases, not good. Therefore, few horses are receiving a high level of vitamin intake from a lush, green pasture or from a high-quality, leafy, green hay. This is a major reason for the increasing vitamin supplementation of diets. There are also other reasons, some of which follow below.

1. Increased selection for improved performance and racing ability which increases nutrient needs.
2. Genetic differences between horses, which can result in different nutrient needs.
3. The depletion of certain nutrients in soils, which can affect nutrient level of the feeds grown thereon.
4. Newer methods of handling and processing feeds, which can affect nutrient level and its availability for the animal.
5. Certain nutrient interrelationships, which can affect vitamin needs.
6. Changing environmental conditions in horse units, which can increase nutrient needs.
7. Increased stress and subclinical disease level conditions resulting from closer contact between horses in close confinement at home and at various performance events and locations. This can also increase certain nutrient needs.
8. Molds in feeds, which can increase the need for certain vitamins.
9. Presence of antimetabolites in feeds, which can increase certain vitamin needs.

10. Trend toward earlier weaning which increases the need for vitamins in diets fed shortly thereafter.

These reasons, as well as others, have resulted in an increased need for adding vitamins to horse diets. Supplementation with vitamins, however, needs to be done carefully. One needs to avoid using excessively high levels in order to prevent harmful effects which can occur from the misuse of vitamins.

Diets are being supplemented with vitamins even though some research studies may not indicate a need for them. Most feed formulators use vitamin supplementation as a precaution to take care of stress factors, subclinical disease level, and other conditions on the average farm which may increase vitamin needs. Moreover, they want to provide a small safety margin against vitamin losses which can occur in the feed during storage and which may be affected by temperature, humidity, rancidity, moisture level, length of storage, and other factors. Moreover, many farmers buy commercial protein supplements (fortified with vitamins and minerals) to mix with their grain. Unfortunately many of them attempt to save money and use less than the recommended protein supplement level, and thus a lack of vitamins and other nutrients may occur.

VII. UNIDENTIFIED NUTRIENT FACTORS

There is some indication that unidentified nutrient factors still exist for livestock and poultry. It is possible that a vitamin could be involved. Responses in growth and reproduction to sources of unidentified factors have been variable and in some cases no response is obtained. Most of this evidence is available for poultry and swine; none is available for the horse. To what extent one or more nutrient factors are involved is not clear. For poultry it appears that several factors are involved. Some scientists feel that there are no unidentified factors, whereas others feel that there are. Some feel that correction of an imbalance of nutrients could be involved. Feeds such as green forages, alfalfa meal, fish solubles, distiller's solubles, yeast, whey, fish meal, meat meal, liver, certain fermentation residues, and other feeds have been reported as sources of these factors.

Many horse people find that placing hard-to-breed mares in a lush, green pasture is very helpful in getting them to conceive. High-quality pasture and dehydrated alfalfa meal have been shown to be sources of unidentified factors for other animals. It is possible that unidentified factors could be involved in the low foal crop. At least an open mind should be kept on this possibility until more experimental information is obtained. In the meantime, it is recommended that as much use as possible be made of high-quality, lush, green pastures in horse feeding programs. This should especially be the case for young, growing horses and breeding mares.

VIII. PASTURE WILL DECREASE VITAMIN NEEDS

The use of short, lush, green leafy pastures will minimize the vitamin deficiencies that can occur in horses. Unfortunately, what many consider a pasture is,
in many instances, no more than an exercise area. Pasture conditions must be
considered when formulating diets. In areas where the pasture season is of short
duration, green, leafy alfalfa meal is an excellent pasture substitute.

IX. SUGGESTED VITAMIN LEVELS

Table 5.2 gives a suggested vitamin premix to use. This premix can be added
to the diet fed each day at the level shown in Table 5.3, or it can be mixed in the
diet at the levels shown in Table 15.9.

Table 5.4 gives the approximate daily level of the vitamins provided by the
premix during the various stages of the horse's life cycle. The levels shown in
Table 5.3 can be changed if conditions warrant doing so. If the horses are being
pushed hard for racing or high-level performance purposes the levels might be
increased somewhat. If the horses have an excellent pasture program, the levels
could be decreased, since high-quality pasture is a good source of vitamins. The
key to changes should be how well the horses look and perform with these

TABLE 5.2

**Suggested Vitamin Premix for Racing or
Performance Horses**[a]

Vitamin	Per ounce of premix
Vitamin A (IU)	40,000
Vitamin D (IU)	4,000
Vitamin E (IU)	160
Vitamin K (mg)	20
Thiamin (mg)	24
Riboflavin (mg)	40
Niacin (mg)	120
Pyridoxine (mg)	12
Pantothenic acid (mg)	48
Choline (mg)	600
Vitamin B12 (μg)	120
Folacin (mg)	12
Biotin (mg)	1

[a]Recommendations on myo-inositol, p-aminobenzoic
acid, and vitamin C will be given later in this chapter in the
discussion on each of these vitamins.

TABLE 5.3

**Suggested Level of Vitamin Premix To Be Fed Daily to
Horses Used for Racing or Performance Purposes**

Horse		Daily rations (oz.)[a]
Foals		
(a)	From birth to 2 months of age	$\frac{1}{8}$
(b)	From 2 months of weaning	$\frac{1}{4}$
(c)	From weaning to training	$\frac{1}{2}$
(d)	During training	1
Mares		
(a)	During gestation	1
(b)	During lactation	1
(c)	Barren mares	$1\frac{1}{2}$
Stallions		
(a)	During breeding season	$1\frac{1}{2}$
(b)	When idle	1

[a]These levels might be varied some depending on stress factors and other
conditions which might justify it.

vitamin levels. If they do well, it is an indication the vitamin levels are suitable
and should be continued as is.

The vitamin levels shown in Table 5.4 are most likely higher than they should
be in many cases. Moreover, some of the vitamins may not need to be added to a
well-balanced diet. Most horse owners with valuable racing or high-level performance horses, however, are adding them anyway as insurance against a possible
need. For those who wish to add vitamins, the recommendations given in Tables
5.2, 5.3, 5.4, and 15.9 can be used as a guide. These levels can be safely used
and have been proved successful by horse owners who have used them. These
levels can be used as a guide until more exact research information becomes
available.

The vitamin levels recommended are based on the assumption that horses will
consume about 2% of their body weight as total feed daily. The level of feed
intake will also vary from 2% (up and down) depending on the stage of the life
cycle. However, the 2% figures is a good rule of thumb.

The levels shown in Table 5.4 indicate that the highest level of vitamins per
pound of feed should be concentrated on the very young horse, horses in training, breeding stallions, and barren mares. These are stages in the life cycle where
it was felt that the highest concentration of vitamins would do the most good. It
was also felt that using higher levels of vitamins with barren mares might help to
get them to breed. It should be emphasized that the vitamin levels recommended
should be adequate for all stages of the life cycle unless some unusual condition
occurs to raise the requirement or to destroy some of the vitamins in the diet

TABLE 5.4

Vitamin Levels Supplied Daily per Animal in Total Diet[a]

Vitamins	Foals				Mares			Stallions	
	Birth to 2 months	2 Months to weaning	Weaning to training	During training	Gestation	Lactation	Barren	Idle	Being bred
Vitamin A (IU)	5,000	10,000	20,000	40,000	40,000	40,000	60,000	40,000	60,000
Vitamin D (IU)	750	1,500	3,000	6,000	6,000	6,000	9,000	6,000	9,000
Vitamin E (IU)	20	40	80	160	160	160	240	160	240
Vitamin K (mg)	2.5	5	10	20	20	20	30	20	30
Thiamin (mg)	3	6	12	24	24	24	36	24	36
Riboflavin (mg)	5	10	20	40	40	40	60	40	60
Niacin (mg)	15	30	60	120	120	120	180	120	180
Pyridoxine (mg)	1.5	3	6	12	12	12	18	12	18
Pantothenic acid (mg)	6	12	24	48	48	48	72	48	72
Choline (mg)	75	150	300	600	600	600	900	600	900
Vitamin B12 (µg)	15	30	60	120	120	120	180	120	180
Folacin (mg)	1.5	3	6	12	12	12	18	12	18
Biotin (µg)	125	250	500	1,000	1,000	1,000	1,500	1,000	1,500

[a]Anyone wishing to add these vitamins in the feed can do so by estimating the average number of pounds of total feed the animals will consume daily during the various phases of the life cycle and then divide this into the figures shown above. The result will be the amount which needs to be added per pound of total feed. It should be remembered that horses, on the average, will consume about one-half concentrates and one-half hay and/or pasture. Therefore, the vitamin premix needs to be added to the concentrate mix at twice the level indicated since the other half of the diet will be hay and/or pasture to which no vitamins are added.

(such as rancidity). The level of B vitamins recommended is formulated using information available for the horse and also using research data available for the pig and chick. The vitamin needs of the pig and chick are not too far different (per pound of feed) and were used as a guide when inadequate or no information was available for the horse. In some cases, data for other animals were also used as a guide. The recommended vitamin levels are reasonable even though some extrapolation from other data was used to determine the level. These levels can be changed when research data indicate a need to do so. In the meantime, they are a safe guide to follow.

The remainder of this chapter will discuss each of the vitamins separately.

Table 15.9 gives a vitamin premix which can be added to concentrate diets. The level added varies depending on the stage of the horse's life cycle and the suggested concentrate diets shown in Chapters 15–18. It gives the option of the B-complex vitamins being left out. It also provides the option of whether vitamins E and K are to be included or left out. Vitamins A and D are included in all four options.

X. VITAMIN A

A. Names Used Previously

Antiinfective vitamin, ophthalamin, retinol, biosterol, and fat-soluble A are names that have been applied to this vitamin. Vitamin A should be used for all β-ionone derivatives exhibiting qualitatively the biological activity of all-*trans*-retinol. At least 12 major forms or derivations of retinol occur in nature (3). Phrases such as "vitamin A activity" and "vitamin A deficiency" are preferred usage.

B. Units and Forms

Vitamin A is expressed in international units (IU). There are three different esters of vitamin A which are called vitamin A alcohol, vitamin A acetate, and vitamin A palmitate. On an IU basis, they all have the same biological or feeding value for the horse. One IU provides the vitamin A activity of 0.3 μg of all-*trans*-retinol. The 1989 NRC publication on nutrient requirements of the horse indicates that 1 mg of carotene is equivalent to 400 IU of vitamin A (3).

C. Deficiency in Horse Diets

Many horse diets lack vitamin A activity. This is apt to occur when horses are fed hay which is not green in color or kept on dry pastures for long periods of time. Green color in hay or pasture is indicative of vitamin A activity. Actually, vitamin A *per se* does not occur as such in hay or pasture; instead carotene is present, which is changed to vitamin A in the intestinal wall of the horse. There is a possibility that some may be converted to vitamin A in the liver and other organs. The conversion of carotene to vitamin A by horses is not very efficient, especially carotene from grass forages (4). Thus, the use of carotene values as presented in the literature may exaggerate the value of carotene as a source of vitamin A (5). Anything that affects the integrity of the intestinal tract, such as a parasitism or a nutritional deficiency, can decrease carotene conversion to vitamin A. In certain situations, high nitrate levels in the forage or water will also interfere with carotene conversion. Therefore, just because horses are kept on green pasture is not absolute assurance that they are receiving an adequate intake of vitamin A activity. Beef cattle, for example, have been benefited by vitamin A supplementation when they have been fed on excellent quality green pasture or when fed corn silage, which should have supplied many times their carotene needs.

D. Effects of Deficiency

A deficiency of vitamin A is characterized by anorexia (lack of appetite), poor growth, night blindness, lacrimation, joint lesions, keratinization of the eye cornea, skin and respiratory symptoms, abscess of the sublingual glands, impaired conception, elevated cerebrospinal fluid pressure, convulsive seizures, reproductive problems, and progressive weakness (5–11). A severe deficiency may result in a weak, blind, dead, or deformed foal. A decline in plasma, liver, and kidney vitamin A concentrations also occur.

E. Function

Vitamin A is required for normal eye function, cell differentiation, and maintenance of tissues in reproductive, nerve, and urinary tracts. It is also needed for growth, reproduction, and lactation. An adequate amount of vitamin A may provide some resistance to infection (10). Bone remodeling in the growing animal is also modulated by vitamin A (3).

F. Requirements

The 1989 NRC publication (3) recommends a level of 650–1680 IU of vitamin A per pound of feed. The level varies depending on the stage of the life cycle of the horse. These levels are shown in Tables 6.5 and 6.6 The highest requirements occur in pregnant and lactating mares and working horses. Tables 5.2, 5.3, 5.4, and 15.9 give suggested levels of vitamin A to use in horse diets.

A Pennsylvania study (12) reported that calculated estimates of optimal vitamin A intake for maximal growth average 1.4 times the 1978 NRC requirements, for liver secreted serum constituents 5.4 times, and for red blood criteria 10 times the NRC recommendations. They stated that the recommended vitamin A intake of 16.4 IU per pound bodyweight for weanling horses appeared less than optimal and that hematologic criteria and serum biochemistries may be more sensitive indicators of vitamin A nutriture than growth. They suggested that a daily intake of 1.5–5 times the 1978 NRC vitamin A requirements may represent an optimal range for growing horses. Two reports from Finland (13, 14) stated that dried hay and stored oats are not sufficient to provide adequate circulating levels of vitamins A and E in pregnant mares and weanlings if the horses are kept in stables for long periods of time without green forage or vitamin supplementation. One study (14) agreed with the Pennsylvania report (12) that the 1978 NRC vitamin A levels were too low for the winter supplementation period.

G. General Information

Both vitamin A and carotene are stored in the liver and fatty tissues of the body. This store of vitamin A, if large enough, can last for months, and is helpful in providing the horse with a reserve which can be drawn upon when requirements are higher than the level provided in the diet. The most accurate method of determining the vitamin A status is its level in the liver since it is indicative of vitamin A storage. However, liver samples are not ordinarily obtained. Blood samples are often used to determine the level of vitamin A. These samples can indicate if the horse has sufficient vitamin A or if a deficiency is present. It is not, however, an exact measure of vitamin A status. The blood vitamin A level usually becomes low only when the liver vitamin A store is almost depleted. Therefore, the blood may show an adequate level of vitamin A, and the liver may just be getting to the point of becoming deficient in vitamin A. Shortly thereafter the blood vitamin A level will become low indicating that the liver store is also very low. Liver values vary from about 6 to 2000 IU of vitamin A per gram of liver. This indicates that liver levels are a good indicator of vitamin A intake and storage. Vitamin A values vary from about 20 to 90 IU per 100 ml of blood. When liver vitamin A stores are depleted, blood levels drop quickly to levels between 5 and 20 μg/100 ml. This variation is not as great as in the liver stores of vitamin A.

The cost of vitamin A has decreased to the point where horse owners should not have to run liver or blood analyses to determine if their horses need it. Instead, enough should be added to the diet to be sure the supply is adequate. This is especially the case with horses used for racing or high-level performance. Vitamin A can be supplied in a number of ways, such as addition to mineral mixtures, in the diet via special vitamin or mineral supplements, or it can be injected. Injecting would be the least desirable, unless the animal was sick or not eating properly. Because it is difficult to know whether vitamin A supplementation is needed on a specific horse farm, injecting one-half the horses and leaving the other half as controls is a good way to determine if supplementation will help. This is important since vitamin A needs vary a great deal depending on the breeding of the horses, the diet used, the level of performance required, the kind and quality of hay and pasture used, and many other factors. The best way to find out is to try it and see if it helps.

Studies at Rutgers University (4, 15, 16) showed that the carotene in certain species of forage may be more available to the horse and more efficiently converted into vitamin A, since with certain hay species the carotene is utilized very inefficiently. In their study, a lower intake of carotene from alfalfa hay maintained the blood plasma vitamin A level in the horse more effectively than carotene from the grass hays.

The carotene content of hays varies considerably. Stage of maturity, season of the year, species of forage, variation in the exposure to sunlight, rain, and the drying conditions during harvesting are major cause of the wide differences in the carotene content of forages (17).

At the American Association of Equine Practitioner's Convention in New Orleans in December 1967, Professor H. D. Stowe of the University of Kentucky presented a paper on the effect of vitamin A and E supplementation on mares that had been barren for about 3 years. He reported that the daily feeding or injection of either 100,000 IU of vitamin A, 100 IU of vitamin E, or both to barren mares resulted in substantial improvement in conception rate. These levels are somewhat similar to those suggested in Table 5.4.

Excess vitamin A should be avoided. Prolonged feeding of excess vitamin A may cause lethargy, colic, hair loss, depression, anorexia, anemia, bone fragility, hyperostosis, and exfoliated epithelium (5, 11, 12). The presumed safe upper level is about 7272 IU per pound of feed, which is about 4–10 times the nutritional requirement (2). The mechanism for vitamin A toxicity is still not known. The large capacity of the liver to store vitamin A gives considerable protection against vitamin A toxicity as well as against a deficiency occurring. This storage also makes it difficult to determine vitamin A requirements.

Vitamin A in the form of alcohol is referred to as retinol whereas retinal is the aldehyde form and retinoic acid is the acid form. There are some European data which indicate that β-carotene may be required for normal reproduction in cows (18, 19). Some scientists feel this is due to the high concentration of β-carotene in the corpus luteum which may reflect its need for reproduction at a high rate. Whether the dairy cow needs both carotene and vitamin A is not definitely established, and many dairy scientists in the U.S. and other countries have not been able to verify the European data. More studies are needed to clarify this possible role of carotene. It is mentioned since many horse producers wonder whether or not it is needed along with vitamin A. Data are not available to verify a need for β-carotene if vitamin A intake is adequate for the horse.

A Michigan State study (20) showed that peak vitamin A activity in milk occurred 1 day postpartum and preceded by 3 days the maximum vitamin A activity in foal serum and the lowest vitamin A activity in the mare serum. Their data confirmed the importance of colostrum as a vitamin A source for the neonate and the relatively low serum vitamin A level of newborn foals.

An NRC report (21) recommended that the safe upper limit of nitrate nitrogen is 100 mg/liter of water and nitrite nitrogen in 10 mg/liter in drinking water for livestock and poultry. Nitrates are not harmful *per se* but they can be converted into nitrites by microorganisms in the digestive tract and may increase vitamin A needs.

Grain in storage loses carotene activity and thus decreases the level of vitamin A in the diet.

XI. VITAMIN D

A. Names Used Previously

The antirachitic vitamin, sunshine vitamin, rachitasterol, and rachitamin are terms previously used for vitamin D. The term *vitamin D* should be used for all steroids exhibiting qualitatively the biological activity of cholecalciferol. Phrases such as "vitamin D activity" and "vitamin D deficiency" are preferred.

B. Units and Forms

Vitamin D activity is exhibited by at least 10 sterol derivatives. Vitamin D2 and D3, however, are the most important. Vitamin D2 is the form found in plant products such as hay and in irradiated yeast. Vitamin D3 is the animal form found in fish oils, in irradiated milk, and in the horse's skin after exposure to the sun. Vitamin D2 (or calciferol) is obtained from ergosterol by irradiation. Vitamin D3 (or activated 7-dehydrocholesterol) is obtained from 7-dehydrocholesterol on irradiation.

One IU of vitamin D is defined as the biological activity of 0.025 μg of crystalline vitamin D3. New developments show that vitamin D3 is metabolized in the liver into 25-hydroxy-D3 (25-OHD3) and into 1,25-dihydroxy-D3 (1,25-$(OH)_2$D3) in the kidneys. These products are much more active than D3 in bone calcification and formation. Eventually, these breakdown forms of vitamin D will become possible alternatives for use in providing vitamin D for the horse. Both 25-OHD3 and 1,25$(OH)_2$D3 are natural compounds found in the body. NRC (2) states that 25-OHD3 is 2–5 times as potent as vitamins D2 and D3 and 1,25$(OH)_2$D3 (calcitriol) is 5–10 times as potent as vitamins D2 and D3.

C. Deficiency in Horse Diets

Almost all feeds are very low in vitamin D activity. The horse obtains vitamin D from sunlight (sunlight changes sterols in the skin to vitamin D), sun-cured hay, or from its addition to the diet. Horses probably get all the vitamin D they need from sunlight activity during the summer months if they are kept outside for at least a few hours a day. In the winter this may not be the case, since a good many horses are outside only a part of the time. Moreover, in many areas there are few sunny days, and the sunlight available is less effective than in the summer. In many locations, therefore, it is unsafe to rely entirely on winter sunlight to take care of vitamin D needs. This causes many horse owners and feed manufacturers to add vitamin D to horse diets. Moreover, vitamin D is so low in price that its use in the diet is low-cost insurance against a possible deficiency.

The action of sunlight in producing vitamin D in the skin is more potent in the summer than in the winter, more effective at noon than in the morning or evening, and more potent at high altitudes where the sun is closest to the earth. Sunlight which comes through ordinary window glass is ineffective in producing vitamin D in the skin since the glass does not allow penetration of ultraviolet rays.

D. Effects of Deficiency

A lack of vitamin D results in the failure of bones to calcify normally. This is known as rickets in the young and osteomalacia in the adult. A deficiency of vitamin D results in symptoms similar to those of a lack of calcium or phosphorus or both. This is because all three are concerned with proper bone formation. Vitamin D deficiency symptoms include reduced bone calcification, stiff and swollen joints, stiffness of gait, softness of bones, bone deformities, frequent cases of fractures, and reduction in blood serum calcium and phosphorus, and unthriftiness. Some Studies in Florida (22) showed that young ponies deprived of sunlight and vitamin D developed a loss of appetite, slower growth, as well as decreased bone ash, cross-sectional cortical area, density and breaking strength of metacarpal bones. A deficiency of calcium, phosphorus, or vitamin D can cause bone deformities caused by the weight of the animal and the pull of the body muscles on weak, porous bones.

E. Requirements

The 1989 NRC publication on nutrient requirements of the horse (3) recommends 136–364 IU of vitamin D per pound of feed. The level varies depending on the stage of the life cycle (Table 6.14). Tables 5.2, 5.3, 5.4, and 15.9 give suggested levels of vitamin D to use in horse diets. These levels provide the needed safety factor to eliminate conditions caused by a vitamin D deficiency.

More vitamin D is needed if the level of calcium and phosphorus is low or if the ratio between them is not correct. The greater the imbalance of calcium to phosphorus, the greater the benefit derived from vitamin D. However, if the ratio becomes too imbalanced it can reach a point where vitamin D can no longer be beneficial in correcting the imbalance effects. No amount of vitamin D will compensate for severe deficiencies of either calcium or phosphorus in the diet. Therefore, one should not depend on vitamin D being able to make up for the wrong level of calcium and phosphorus or a very wide ratio of one to the other.

Some recent studies indicate that certain molds in feeds can interfere with vitamin D3 (23). The mold *Fusarium roseum* in corn interferes with the absorption of vitamin D3 in the chick. Other molds may also be involved. The result is bone disorders. A number of flocks have been successfully treated by adding water-dispersable forms of vitamin D to the drinking water at 3–5 times the normal recommended level. It is possible that molds in horse feeds could also interfere with vitamin D absorption. If so, this could account for some unexplained leg and bone problems in the horse such as have occurred with the chick.

F. General Information

Excess vitamin D intake should be avoided since it will cause weight loss, calcification of the blood vessels, heart, and other body soft tissues (Fig. 5.1). It will also cause bone abnormalities and kidney damage (24–26). The presumed safe upper level is 4–10 times the recognized dietary requirement. Vitamin D3 is 10–20 times more toxic than vitamin D2 (2, 25). NRC (2) has set 1000 IU of vitamin D3 per pound of feed as a maximum safe level for long-term feeding (more than 60 days).

Vitamin D is needed for both the absorption and utilization of calcium and phosphorus in proper bone formation. Therefore, an adequate level of vitamin D is needed to develop and maintain sound bone that is so essential for the horse. Feet and leg problems occur frequently with horses. Therefore, horse diets containing an adequate level of vitamin D is very important.

Attention should be paid to horses that may be suffering from certain diseases or parasites which cause intestinal, liver, and kidney damage, which may interfere with the conversion of vitamin D to $1,25(OH)_2D3$. For example, humans with kidney damage cannot synthesize $1,25(OH)_2D3$ and the administration of $1,25(OH)_2D3$ has been found to alleviate the painful and serious bone disease which goes with chronic renal failure (27).

Vitamin D is stored in the body for a long time, but not to the same extent as vitamin A. Thus, the horse can resist a deficiency of vitamin D for a considerable time.

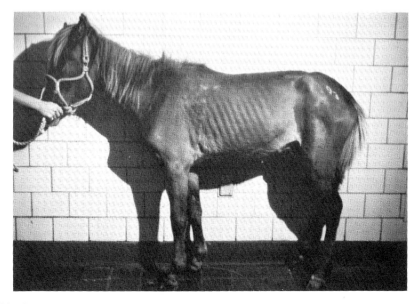

Fig. 5.1   Note effect of excess vitamin D in the horse. The weight loss occurs because of difficulty in eating because the tongue is partially calcified. (Courtesy of H. F. Hintz, Cornell University.)

Vitamin D3 is more effective than D2 for the chick and this may be true for livestock since recent studies, which need more verification, indicate D3 to be more active than D2 for livestock.

The metabolite $1,25(OH)_2D3$ functions with the parathyroid hormone (PTH) to bring about blood and calcium homeostasis. Also, $1,25(OH)_2D3$ acting with PTH mediates the resorption of bone with the release of calcium and phosphorus (2).

Certain plants have toxic levels of vitamin D-like compounds. One study (28) described hypercalcemia and calcinosis in horses consuming jasmine while grazing Florida pastures. The active calcitropic principle appears to be 1,25-dihydroxy-D3 glycoside (29) which results in hyperabsorption of calcium by animals eating plants containing it.

XII. VITAMIN E

A. Names Used Previously

The term vitamin E should be used for all tocol and tocotrienol derivatives exhibiting qualitatively the biological activity of α-tocopherol. Vitamin E has been called the antisterility vitamin and factor X. Phrases such as "vitamin E activity" or "vitamin E deficiency" are the preferred usage.

B. Units and Forms

There are eight forms of vitamin E which are found in nature in plant feeds. There are four tocopherols (α, β, γ, and δ) and four tocotrienols (α, β, γ, and δ) (27). Tocopherol is the preferred form of vitamin E since it has the greatest nutritional value. For example, if α-tocopherol is given a value of 100, the other forms have less vitamin E activity with the γ and δ forms having practically no vitamin E activity (27). Therefore one can be misled by a total tocopherol analysis of feeds. The most important thing to know is the α-tocopherol value of feeds.

On an equivalent IU basis 1 mg of d-α-tocopherol acetate is equal to 1.36 mg of dl-α-tocopherol acetate (30). The dl-α-tocopherol acetate is accepted as the International Standard having a defined activity of 1 IU per mg. Therefore, 1 IU and 1 mg of dl-α-tocopherol acetate are used interchangeably. Synthetic free tocopherol, dl-α-tocopherol, has the potency of 1.1 IU or 1.1 mg.

C. Deficiency in Horse Diets

There is increasing evidence that vitamin E supplementation may be needed for the horse in certain situations. What is not known, however, is when the supplementation should be provided.

The importance of vitamin E and its value in animal feeding is becoming increasingly recognized. About 20 years ago it was thought that vitamin E was of no value in supplementing animal diets. Research since then, however, has definitely shown a need for vitamin E supplementation in many situations. Part of the increased interest in vitamin E is due to the finding that selenium is deficient in 44 states as well as in many foreign countries. A lack of selenium causes considerable loss to all animals. It was also found there is an important interrelationship between vitamin E and selenium. Vitamin E can substitute to a certain extent for selenium and selenium can likewise substitute for some of the vitamin E, but neither one can substitute entirely for the other. In certain situations it is difficult to determine which deficiency symptoms are due to vitamin E and which are due to selenium. Therefore, most scientists call them vitamin E and/or selenium deficiencies. Present evidence shows that both vitamin E and selenium are needed in many situations. Both are needed nutritionally and both have a metabolic role as well as an antioxidant effect in the body. Therefore, vitamin E and selenium are very important in animal diets, and deficiencies are occurring widely in the Unites States and throughout the world. Some of the reasons why this is occurring are listed below.

1. Heating and pelleting feeds lowers their vitamin E value. As more pelleted feeds are used, the more one needs to consider the possible need for E supplementation.

2. Rancidity destroys vitamin E. The increasing use of fats or feeds with unsaturated fatty acids, which are susceptible to rancidity, results in the possibility of rancidity developing, as does storing feeds in hot or damp areas, or in areas with poor air circulation. Antioxidants should be used to protect against rancidity.

3. More animals are being kept in close confinement, and less green pasture and green alfalfa or other green hays are being used. Green pastures and green hay are excellent sources of vitamin E. As the green color in pasture or hay decreases so does vitamin E level.

4. The vitamin E level in feeds has been overestimated in the past. Vitamin E is a tocopherol of which there are many forms. Only α-tocopherol has maximum vitamin E activity.

5. More selenium deficiencies are occurring and are now being recognized. This increases the need for vitamin E since it can substitute for part of the selenium requirement and in certain cases is needed along with selenium.

6. It is now known that selenium in feeds has low and variable availability to the animal. An analysis of selenium in feeds may overestimate its value.

7. Horses are growing faster today and more is demanded of them at an early age. This could increase the need for vitamin E as well as other nutrients, including selenium.

8. There may be differences in vitamin E needs between breeds or strains of horses. This should always be kept in mind as one evaluates the adequacy of diets for horses. Some horses may have much higher requirements for certain nutrients.

9. Some horses require less feed than others. This means they may need to be fed more vitamin E, or other nutrients, per pound of feed depending on their level of total feed intake.

These and other factors account for the increased level of vitamin E and/or selenium deficiencies being observed. It is no longer safe to say that a well-balanced diet for horses contains all the vitamin E a horse needs. This is especially so for the horse being trained and pushed for racing and high-level performance at an early age. It is also important that these top performing horses be able to reproduce so they can perpetuate themselves. This makes it doubly important that horses be supplied with the proper level of vitamin E.

A report from Finland (31) indicated that relatively low blood serum concentrations of α-tocopherol and 25-hydroxyvitamin D suggest that during the winter season a deficiency of these two vitamins may occur in northern countries such as Finland. Significant differences were observed between horses from different stables, especially for serum α-tocopherol levels. Other Finnish studies (13, 14) indicated that dried hay and stored oats are not sufficient to ensure adequate levels of vitamin E in pregnant mares in winter and that a deficiency may develop within a few months if the horses are kept in stables for long periods of time without fresh green forage or vitamin supplementation.

D. Effects of Deficiency

One study (32) reported a degenerative myelopathy in six horses which were from 13 months to 14 years of age. Ataxia was evident in all, including uncoordination of the hind limbs, and an abnormally wide-based gait and stance. Histological examination revealed degeneration of the neural processes in the ventral and lateral funiculi.

Signs of a possible deficiency of both vitamin E and selenium have been described in the foal (33, 34). Myodegeneration is common, with pale diffuse or linear areas in skeletal and cardiac muscle. Histological examination revealed hyaline and granular degeneration, as well as swelling and fragmentation of muscle fibers from several sites, including the tongue. Subcutaneous and intramuscular edema, pulmonary congestion, and occasionally, steatitis were also observed.

A 1987 study (35) described a vitamin E-responsive degenerative myeloencephalopathy in Standardbred and Paso Fino horses, from 3 to 36 months old. Symmetric ataxia and paresis, along with laryngeal adductor, cervicofacial, local cervical, and cutaneous trunci hyporeflexia, were characteristic. No clinical signs were observed before 3 months of age. The onset of gait abnormality was usually abrupt. Clinical signs then remained static or progressed for weeks or months. Severely affected animals often fell while running. On the basis of genetic studies, this disorder appeared to have a familial disposition.

Combinations of vitamin E and selenium have been used in the treatment of "tying-up" in horses (5). Tying-up is characterized by lameness and rigidity of the muscles of the loin area. The urine may be coffee-colored because of the myoglobin released from the damaged muscle cells. Many veterinarians have reported that injections of a combination of vitamin E and selenium are beneficial for this condition. Dr. H. E. Hill (36) reported that tying-up appears in horses of all ages and affects approximately 2–5% of the horses at race tracks. Two-thirds of the cases occur after 1–2 days of rest from a rigid training schedule. Dr. Hill stated that the horses could be divided into three groups: (1) those that tied-up regularly and frequently, (2) those that tied-up occasionally, and (3) those that had muscle soreness or stiffness as the predominant symptom. Dr. Hill used an injectable preparation which contained 2.5 mg of selenium as sodium selenite and 25 mg of d-α-tocopherol acetate per milliliter.

As discussed earlier for vitamin A, Dr. H. D. Stowe, of the University of Kentucky, reported that the feeding or injection of either 100,000 IU of vitamin A, 100 IU of vitamin E, or both daily to barren mares resulted in substantial improvement in conception rates. Canadian veterinarians (37–40) have shown

that the breeding performance of mares and stallions improved under the continuous administration of vitamin E. They used daily doses of 1000–2000 IU of
vitamin E. More studies are still needed to verify more adequately the role of
vitamin E in reproduction. Many practicing veterinarians as well as horse owners
and trainers feel that vitamin E is helpful in reproduction. Unfortunately, there is
still a lack of enough experimental information to substantiate the experience
obtained in the field. Horse owners can try utilizing vitamin E supplements on
horses to determine if it has a beneficial effect on reproduction problems.

There is some indication that the use of vitamin E may improve racing
performance (37, 41). Vitamin E supplementation was begun 3–4 months before
and continued throughout the racing season. Levels of 2000–5000 IU of vitamin
E were fed per horse daily. The investigators felt that stamina was increased, less
time was spent for recuperation, and the horses were still running well late in the
season. They also reported that the horses became easier to manage and train,
especially those that were nervous and highly strung. Unfortunately there is a
lack of research data obtained at race tracks by university scientists on the effect
of vitamin E on racing performance. It would also be difficult to obtain since
horse trainers are reluctant to allow anyone access to their training procedures.
Therefore, it is difficult to make a definite recommendation on the use of vitamin
E at high levels for improving racing performance until adequate research studies
are conducted.

Dr. L. M. Slade of Utah State University (42) has recently shown that supplemental vitamin E is beneficial in the diet of endurance horses, especially when
diets high in unsaturated oils are used. The combination of adding 12% vegetable
oil and 1000 IU of vitamin E resulted in the most desirable response in blood
packed cell volume, hemoglobin, and glucose after exercise. He feels the endurance horse requires approximately 4000–4500 IU of vitamin E in the diet
daily.

E. Requirements

The 1989 NRC publication (3) recommends from 22.7 to 36.4 IU of vitamin
per pound of diet. The level varies with the stage of the life cycle (see Table
6.14). This level is considerably higher than that recommended by the 1978 NRC
report (5). A University of Kentucky study showed that it took 1.2 mg of vitamin
E (given intramuscularly) and 10.5 mg of vitamin E (given orally) per 100 lb
bodyweight to maintain erythrocyte (red blood cell) stability (43).

Dr. John B. Herrick of Iowa State University recommends the following
levels of vitamin E supplementation for horses (in IU) per day: foals, 10; growing, 20; mature (idle), 20; breeding season, 50–100; training season, 50–100;
and racing season, 1000–2000 (44.). Tables 5.2, 5.3, 5.4, and 15.9 give suggested levels of vitamin E to use.

A number of factors affect the requirement for vitamin E. These include the selenium, sulfur amino acids, and protein levels and the level of rancidity, unsaturated fatty acids, copper, iron, antioxidants, and possibly others. Until these interrelationships are better understood, it will be difficult to state precise vitamin E requirements.

F. General Information

There is no experimental information on the levels of vitamin E which are toxic for the horse. One should be careful, however, and not use the vitamin in excess. Evidently, vitamin E is less toxic than other vitamins. The 1989 NRC (3) indicates that a conservative maximum tolerable level of vitamin E is 454 IU per pound of dry diet.

Vitamin E functions as an antioxidant at the cellular membrane level with a structural role in cell membranes. It functions in the transfer of hydrogen for the reduction of free radicals within the cell.

Vitamin E is stored throughout all body tissues with liver having the highest level.

The stability of all naturally occurring tocopherols is poor and large losses occur during processing, manufacturing of feeds, and storage. The losses may be as high as 80% or more of the original tocopherols in the feed (27). This may account for more interest in supplementing vitamin E in horse diets, especially for high-level performance animals.

During the commercial synthesis of dl-α-tocopherol it is esterified to acetate as a means of stabilizing it. The acetate form is also available with gelatin coating. Both products are quite stable (27).

XIII. VITAMIN K

A. Names Used Previously

Antihemorrhagic vitamin, coagulation vitamin, prothrombin factor, phylloquinones, and 2-methyl-1,4-naphthoquinone are names previously given to vitamin K. The term vitamin K should be used for 2-methyl-1,4-naphthoquinone and all derivatives exhibiting qualitatively the biological activity of phytylmenaquinone (phylloquinone). Thus, phrases such as "vitamin K activity" and "vitamin K deficiency" are the preferred usage.

B. Units and Forms

There are a number of compounds that are similar in structure and that all have vitamin K activity. Vitamin K exists in three series: the phylloquinone (K1), the menaquinone (K2), and the menadione (K3). There are two natural forms, K1 and K2. Vitamin K1 occurs naturally in green plants, and vitamin K2 is present in microorganisms and is formed by intestinal microorganisms. In addition, several synthetic compounds have been prepared that have vitamin K activity. One of them is 2-methyl-1,4-naphthoquinone, which is called menadione. Some of the synthetic vitamin K compounds are water soluble, in contrast to the natural products (K1 and K2), which are fat soluble. The major water-soluble forms are menadione sodium bisulfite (MSB), menadionine sodium bisulfite complex (MSBC), and menadionine pyrimidinol bisulfite (MPB). Their vitamin K activity depends upon the menadione content and their water solubility. Sometimes MSB is coated with gelatin to increase stability.

C. Deficiency in Horse Diets

Vitamin K deficiencies are now observable with other farm animals where previously a need was not thought possible. Vitamin K supplementation is now in wide practice. It is not known whether horse diets should be supplemented with vitamin K. On the basis of experiences with the pig, the possible reasons for vitamin K supplementation will be given just in case some of them might be applicable to the horse.

1. A mycotoxin produced by certain molds may be in the feed and thus increase vitamin K needs.

2. An antimetabolite may be in the feed and thus increase vitamin K requirements. The presence of dicoumarol, an antagonist of vitamin K, will cause vitamin K deficiency.

3. As confinement feeding has increased, horses have less access to high-quality green pasture which is a good source of vitamin K.

4. Less high-quality green hay is available, a good source of vitamin K, and so diets contain less vitamin K.

5. Horses normally synthesize vitamin K in the intestinal tract. Factors in the diet or medicinal treatments could decrease vitamin K synthesis.

6. Horses are growing faster today and more is demanded of them at an early age. This could be increasing vitamin K requirements.

7. The trend toward using solvent-extracted soybean meal and other plant protein sources has decreased their content of vitamin K.

8. Any decrease in the opportunity for coprophagy (eating of feces) lessens vitamin K intake since vitamin K synthesized in the intestinal tract cannot be consumed via this route.

9. There may be differences in vitamin K needs between breeds or strains of horses.

10. Some horses require less feed than others. Therefore, they may need more vitamin K per pound of feed to satisfy their total daily needs.

Undoubtedly other possibilities presently exist for vitamin K deficiencies showing up, but those listed are the main causes. Therefore, horse owners should be alert to possible symptoms of vitamin K deficiency. On the basis of experience with other animals, only some of the animals may show a need for vitamin K supplementation.

D. Effect of Deficiency

Clinical signs of vitamin K deficiency are low prothrombin levels, increased clotting time, and hemorrhaging. Vitamin K is needed for prothrombin formation in the liver. The prothrombin is necessary for blood clotting. Thus, a lack of vitamin K will lead to hemorrhages because the blood will not clot. The adequacy of vitamin K in the diet can be measured by determining the clotting time of the blood. If the clotting time is increased, it means there is either a decreased synthesis or utilization of vitamin K. Blood clotting time is very important if horses are cut, injured (causing a rupture of the blood vessels), or if an operation is necessary. One should always determine blood clotting time on any valuable horse before operating on it.

E. Requirements

The 1989 NRC report (3) states that vitamin K requirements of the horse have not been determined. The vitamin K in pasture and good-quality hay and intestinal synthesis of vitamin K presumably meet the vitamin K requirements of the horse in all but the most unusual circumstances. Whether any special conditions will decrease vitamin K synthesis and produce a deficiency in the horse is not known. It is logical to assume that the young foal may not synthesize enough vitamin K until the microbial flora in its digestive tract becomes established. How long this takes is not known, but it could take from 1 to 2 weeks after birth, or possibly longer. Therefore, it is a good policy to run blood clotting time on any foal that needs surgery within a few weeks after birth. If it is not normal, vitamin K should be supplied until it is.

There is no evidence to indicate that it is necessary to add vitamin K to horse diets. Some horse owners, however, are adding it to make sure no deficiency exists. There also is no information available on the vitamin K requirement of horses. The chick has a requirement of 0.24 mg of vitamin K1 per pound of feed. The pig has a requirement of 0.23 mg of menadione per pound of diet. Tables 5.2, 5.3, 5.4, and 15.9 give suggested levels for diets. If symptoms of a deficiency are occurring on the farm, higher levels of vitamin K may be needed.

Menadione can be ingested at levels as high as 1000 times the dietary requirement with no adverse effects (2). When administered parenterally, doses of 0.95 to 3.77 mg/lb of bodyweight caused acute renal failure in the horse (45).

XIV. VITAMIN C

A. Names Used Previously

Vitamin C has been called ascorbic acid, cevitamic acid, antiscorbutic vitamin, scorbutamin, and hexuronic acid. The term *vitamin C* should be used for all compounds exhibiting qualitatively the biological activity of ascorbic acid. Thus, phrases such as "vitamin C activity" and "vitamin C deficiency" are preferred.

B. Units and Forms

One IU of vitamin C is the activity contained in 0.05 mg of the vitamin. Thus, 1 mg of vitamin C is the same as 20 IU of vitamin C. The activity of vitamin C is usually expressed in milligrams of vitamin C.

Vitamin C occurs in two forms. One is ascorbic acid (reduced form) and dehydroascorbic acid (oxidized form). Both forms are biologically active, but the majority of vitamin C exists as ascorbic acid. Only the L-isomer of ascorbic acid has vitamin activity, the D-form being inactive.

C. Deficiency in Horse Diets

The experimental information on the need to supplement horse diets with vitamin C is very small. There are many horse owners and trainers who feel vitamin C supplementation is beneficial. Many scientists feel that the horse synthesizes all the vitamin C it needs (5, 46). The unanswered question is whether there might be special conditions that either alter the synthesis of vitamin C or increase its need beyond the level synthesized. The pig, for example, is occasionally benefited by vitamin C supplementation, even though it is not supposed to require vitamin C in the diet. Some feel vitamin C may be helpful for the pig under high stress conditions. The same could be true for the horse, although this information is not yet available.

Two studies (47, 48) showed that 20 g of oral ascorbic acid/day resulted in increased plasma ascorbic acid in horses in training. They also suggested that ascorbyl palmitate administered orally produced higher plasma ascorbic acid concentrations than ascorbic acid. Other studies showed that ascorbic acid absorption from oral doses is very low (3).

D. Effect of Deficiency

A deficiency of vitamin C results in symptoms of scurvy, which are bleeding, swollen, and ulcerated gums, weak bones, loosening of the teeth, and fragility of the capillaries that result in hemorrhages throughout the body. Ascorbic acid is stored to only a limited extent in the body, thus, it needs to be supplied regularly. The level of vitamin C in the blood plasma is a good indicator of its intake.

There was an early report that showed that vitamin C supplementation increased fertility of stallions and mares (49). Another report showed that vitamin C is often used in the treatment of epistaxis (bleeding from the nose) and sometimes even seems to alleviate the condition (50). Unfortunately, there is still very little experimental information on which to base recommendations on the use of vitamin C with the horse.

There is one report in Europe indicating that the conception rate of sows was 20% higher when the boars were given 2 g of ascorbic acid per day during the hot summer months when conception rate is lower. High temperatures increase vitamin C needs in hens, and egg shell quality is improved by vitamin C supplementation during hot weather. Evidently vitamin C requirement may be increased during periods of high temperatures. There have been some reports indicating that in animals on vitamin A-deficient diets, vitamin C in the blood may become low. This is not well established, however. It emphasizes the importance, though, of making sure that horses have enough vitamin A in their diets.

E. Requirements

There is no experimental information available on the requirement for vitamin C by the horse. The writer has talked to a number of horse owners who feel that the daily use of 1000 mg (1 g) of vitamin C helped in getting hard-to-breed mares to conceive. However, this is only an observation, and the work was not controlled to make sure that the vitamin C treatment itself actually did the job. The observation is being presented, however, since the author has heard it frequently.

Conditions under which vitamin C might be beneficial are (1) during hot weather, (2) under stress conditions, (3) during rapid growth or high level of performance, and (4) when something interferes with vitamin C synthesis. There may be a relationship between vitamin C synthesis and a lack of energy, vitamin E, and selenium. Extrapolating from work with other species, not all animals will respond to vitamin C supplementation. Therefore, one needs to observe individual horse response, and not group or lot averages, to determine if vitamin C is beneficial.

The 1989 NRC report (3) states that a dietary concentration of ascorbic acid of 1 g per kg of diet appears to pose no hazard to chickens, pigs, dogs, cats, and, probably, horses.

F. General Information

Ascorbic acid shows fairly good stability in unpelleted feeds. However, with pelleting and the stress of heat and moisture, stability is not as good. So ascorbic acid coated with ethylcellulose which is much more stable than untreated ascorbic acid should be used (51). Ohio State workers found ethylcellulose-coated vitamin C to have 97, 93, and 87% of the initial level at 7, 28, and 57 days after mixing with swine feed (51). As more stable vitamin C products are developed they should be used in diet mixtures. Adding extra vitamin C may compensate for that which is destroyed.

Vitamin C is essential for the hydroxylation of proline and lysine, which are important constituents of collagen. Collagen is essential for growth of cartilage and bone. Vitamin C enhances the formation of intercellular material, bone matrix, and tooth dentin in the pig (52).

Vitamin C is interrelated with iron, thiamin, pantothenic acid, riboflavin, biotin, folacin, and vitamin B12. When vitamin C is deficient, the utilization of folacin and vitamin B12 is impaired, resulting in anemia (27). These interrelationships are not yet entirely understood and may be involved in determining vitamin C requirements.

A Canadian study using the pig showed that a greater response to vitamin C occurred at a low energy intake than at an intermediate or high energy intake. It appears that a low energy intake restricts the level of free glucose available for vitamin C synthesis (53).

XV. THIAMIN

A. Names Used Previously

Vitamin B1, oryzamin, anti-beriberi vitamin, antineuritic vitamin, torulin, polyneuramin, and aneurin are names previously applied to this vitamin. It is now called *thiamin*. Its use in phrases such as "thiamin activity" and "thiamin deficiency" is acceptable.

B. Deficiency in Horse Diets

Horses fed poor-quality hay have been shown to develop a thiamin deficiency (54). Horses poisoned by yellow star thistle (*Centaurea solstitialia*) causing glossopharyngeal (throat) paralysis recovered following 5–7 days administration of 1 g of thiamin daily (55). Supplemental thiamin is beneficial in the treatment of thiamin deficiencies resulting from bracken fern poisoning (56). A recent study (57) indicated that incoordination, staggering, and muscular tremors were observed in 27 mules, 2 months after introduction to bracken fern-infested pasture. Eight died and the rest recovered after removal from the pasture and injection with 100 mg of thiamin. At necropsy, generalized congestion, pulmonary edema, and serosal and mucosal hemorrhages were noted.

C. Effect of Deficiency

Experimentally produced thiamin deficiency causes anorexia (loss of appetite), loss of weight, incoordination (especially in the hind legs), lower blood thiamin, elevated blood pyruvic acid, and a dilated and hypertrophied heart (5, 54, 58, 59). A study involving the use of amprolium (a thiamin antimetabolite) to produce a thiamin deficiency resulted in bradycardia and dropped heart beats, ataxia, muscular fasciculations, and periodic hypothermia of peripheral parts (hooves, ears, and muzzle). Some of the horses exhibited blindness, diarrhea, and body weight loss (60).

D. Requirements

The 1989 NRC report (3) states that although research data are limited, on the basis of research with horses and other species the thiamin requirement appears to be no more than 1.36 mg per pound of diet dry matter for maintenance, growth, and reproduction, unless high levels of antithiamin compounds are consumed. For performance horses it may be prudent to ensure that their diets contain 2.47 mg per pound of diet dry matter. A Texas A & M study (61) suggested that the NRC-recommended allowance of 1.36 mg of thiamin per pound of diet may not be adequate for performing horses. Thiamin is synthesized by the horse and it is estimated that 25% of the free thiamin in the cecum is absorbed by the horse (5, 62). The exact thiamin requirement of the horse is not known. Tables 5.2, 5.3, 5.4, and 15.9 give suggested levels of thiamin to use in horse diets. In certain situations, such as the heavy stress of training, racing, or performance, thiamin supplementation may be beneficial.

E. General Information

There is limited storage of thiamin in the body, which indicates the horse needs a regular supply in the diet. Thiamin functions as a constituent of enzyme systems and is essential in the utilization of carbohydrates and protein. Thiamin plays a very important role in glucose metabolism. Since the breakdown of carbohydrates is increased during racing or performance, it is important that thiamin be available in sufficient quantity.

Thiamin toxicity in the horse has not been reported. This could be due to excess thiamin being rapidly excreted in the urine; however, excessive levels of thiamin in the diet should not be used. NRC (2) states that dietary intakes of thiamin up to 1000 times the requirement level are apparently safe for most animal species. So, oral toxicity in horses is very unlikely for thiamin (3).

Thiamin is heat labile, so excess heat or autoclaving can reduce the thiamin level in feeds.

Disease conditions as well as diarrhea and malabsorption may increase thiamin requirement. Endoparasites such as Strongylids and Coccidia compete with the host for thiamin contained in the feed (27).

Thiamin antimetabolites and thiaminases can cause a thiamin deficiency. Thiaminase splits the thiamin molecule and makes it inactive.

XVI. RIBOFLAVIN

A. Names Used Previously

Vitamin G, vitamin B2, lactoflavin, riboflavine, ovoflavin, uroflavin are previously used names for *riboflavin*. Its use in phrases such as "riboflavin activity" and "riboflavin deficiency" is acceptable.

B. Deficiency in Horse Diets

A number of studies show that riboflavin synthesis occurs in the intestinal tract of the horse. It is not known whether riboflavin supplementation will benefit horse diets, however. Early work indicated that periodic ophthalmia (moon blindness) responded to 40 mg of riboflavin daily even though the horses were on green pasture or hay, which are good sources of the vitamin (63). Many scientists, feel, however, that other factors may be involved with periodic ophthalmia. Some think it is probably due to an immunological response to other disease conditions (see Chapter 20).

C. Effect of Deficiency

A lack of riboflavin causes a decreased growth rate and lowered feed utilization. Riboflavin is an essential constituent of enzyme systems needed for the utilization of fats, carbohydrates, and proteins in the body. It is important in the release of feed energy and in the utilization of nutrients in the diet. It is not definitely known how much riboflavin is involved with periodic ophthalmia (equine uveitis). Periodic ophthalmia can cause severe eye damage, cataracts, and blindness. Invasion of the cornea by leptospira (64) or microfilaria (*Onchocerca cervicalis*) have been implicated in periodic ophthalmia (65).

D. Requirements

The dietary requirement of riboflavin for the horse is probably no more than 0.9 mg per pound of diet dry matter (3). Tables 5.2, 5.3, 5.4, and 15.9 give suggested levels of riboflavin to use for racing or performance horses.

Studies with the pig indicate that the riboflavin requirement is higher at lower temperatures.

E. General Information

The amount of riboflavin excreted in the urine is closely correlated with its intake (66). Rapid excretion of excess riboflavin may account for its lack of toxicity. Even so, it is best not to feed excessive levels of the vitamin in the diet.

The trend toward intensive horse production and the use of less pasture in horse feeding has decreased the level of riboflavin in horse diets. Pasture and high-quality leafy green hays are excellent riboflavin sources.

Riboflavin is more toxic when administered parenterally than when given orally. The toxic level with the horse is not known but data obtained with other animals indicate that dietary levels between 10 and 20 times the requirement can be tolerated safely (2). Riboflavin is not absorbed well and is excreted rapidly in the urine which means it needs to be supplied regularly since little storage occurs in the body (2).

XVII. VITAMIN B12

A. Names Used Previously

The names animal protein factor, zoopherin, erythrotin, factor X, and physin have been used for vitamin B12. The term *vitamin B12* should be used for all compounds exhibiting the biological activity of cyanocobalamin, hydroxycobalamin, and nitrocobalamin. Thus, phrases such as "vitamin B12 activity" or "vitamin B12 deficiency" are preferred usage.

B. Effects of Deficiency

Vitamin B12 functions in the utilization of carbohydrates, fat, and protein. Therefore, it is very important in feed utilization. In animals other than the horse, a deficiency of B12 results in poor growth and reproduction, anemia (normocytic, normochromic), posterior incoordination, unsteadiness in gait, poor appetite, hyperirritability, and a rough hair coat. A vitamin B12 deficiency has not been produced in the horse. Some of the deficiency symptoms obtained with other animals, however, may occur with the horse.

C. Requirements

There is no experimental information reported on the vitamin B12 requirements of the horse beyond that supplied by intestinal synthesis (3). Tables 5.2,

5.3, 5.4, and 15.9 give suggested levels of vitamin B12 for racing and performance horses. The estimated requirements of most animals range from about 4 to 10 μg per pound of diet.

D. General Information

A vitamin B12 deficiency has never been produced experimentally in horses: however, severely debilitated, anemic, heavily parasitized horses appear to respond to vitamin B12 injections (67). Parasitism is a severe problem with horses. In certain warm, humid areas, horses are treated monthly for parasites. In other areas, they are treated every 2 months. Parasitized horses usually bleed internally and thus more blood regeneration is needed. It is reasonable to assume that the horse's need for vitamin B12, other B-complex vitamins, iron, copper, other minerals, protein, and other nutrients concerned with blood and hemoglobin formation would increase due to parasitism. Many horse owners and trainers are in agreement and therefore add vitamin B12 and other B vitamins to their horses' diets. They feel it is good insurance against a possible need. This is especially so with horses used for racing and high-level performance purposes where a high hemoglobin level is necessary for high oxygen-carrying capacity to all tissues of the body. Most horse trainers like to have levels of hemoglobin at 16% or more.

The horse synthesizes vitamin B12 from cobalt in the diet. This means the diet of the horse must provide enough cobalt for microorganism vitamin B12 synthesis. The level of cobalt needed is quite low, however, since horses have remained in good health while grazing pastures so low in cobalt that ruminants confined to them died (68). Vitamin B12 contains about 4% cobalt. Vitamin B12 is absorbed from the lower digestive tract (large intestine) of the horse (5, 67, 69–71).

A study with a purified diet showed that mature horses receiving 6 μg of vitamin B12 daily for 11 months failed to show any signs of a deficiency (70). This indicates that the horse has considerable capability to synthesize this vitamin, as has been shown by a number of scientists.

No toxic effects have been reported from using vitamin B12 at excessive levels with the horse. However, excessive levels are best avoided. Mouse data suggest that dietary levels of at least several hundred times the requirement level are safe (2). An Ohio study involved feeding 100 times the vitamin B12 requirement to sows and it benefited reproduction (72).

Feed ingredients of plant origin are devoid of vitamin B12. Traces may be present, however, due to synthesis by microorganisms which adhere to the plants. Vitamin B12 is unique in that it is synthesized in nature only by microorganisms. Animal product feeds are good sources of vitamin B12, however.

Vitamin B12 is necessary for proprionic acid utilization. Therefore, a diet involving increased volatile fatty acid production (of which proprionic acid is one) would require vitamin B12 adequacy.

Vitamin B12 is metabolically related to other essential nutrients such as methionine, choline, and folacin. It is stored principally in the liver.

Absorption of vitamin B12 is mainly in the ileum and is facilitated by the presence of the intrinsic factor which is released in gastric juice. Whether this occurs in the horse is not known.

XVIII. NIACIN

A. Names Used Previously

Pellagra-preventive factor, nicotinic acid, PP factor, pellagramine, vitamin PP, and niamid are names previously used for *niacin*. Phrases such "niacin activity" and "niacin deficiency" are acceptable usage. Niacin is a generic term for two compounds which have vitamin activity, nicotinic acid and nicotinamide.

B. Effect of Deficiency

Niacin functions as a component of enzyme systems and is required by all living cells, making it a very important vitamin. These enzyme systems play a role in cell respiration and in the digestion of carbohydrates, protein, and fat.

A niacin deficiency has never been produced in the horse. A deficiency of niacin in the pig results in slow growth, poor appetite, occasional vomiting, dermatitis, loss of hair, rough hair coat, normocytic anemia, diarrhea, and a high incidence of necrotic lesions in the colon and cecum. Whether some of these would occur in the horse is not known.

Very little storage of niacin occurs in the body.

C. Requirements

There is no experimental evidence reported on the niacin requirement of the horse (3). Studies have shown, however, that the horse can synthesize niacin (62, 73, 74). Tables 5.2, 5.3, 5.4, and 15.9 give suggested levels of niacin for racing and high-level performance horses.

The niacin requirements of most animals varies from 5 to 20 mg per pound of diet.

D. General Information

Research studies have shown that the niacin in corn, oats, grain sorghums, wheat, and other cereals (and their by-product feeds) is in a bound form, which is largely unavailable to the pig. Whether this is the case for the horse is not known. This is an important question since data obtained with the rat and poultry (which are nonruminants) indicate that the niacin in the cereal grains, and their by-products, are largely unavailable to them. One study (75) indicated that about 40% of the niacin in oilseeds is in a bound form. Milk, meats, and fish contain no bound forms of niacin.

The pig, rat, and chick can convert the amino acid tryptophan into niacin. The horse can also synthesize niacin from tryptophan (73). Niacin, however, cannot be converted back to tryptophan. However, conversion from tryptophan is a very inefficient process. In the pig, it takes 50 mg of tryptophan in excess of the tryptophan requirement to yield 1 mg of niacin (76). Unfortunately, most diets for nonruminants (such as the pig) do not contain excess tryptophan. Some recent studies indicate that some amino acids are very important in horse diets. This is especially the case with lysine for the young, growing horse. Therefore, it is important to make sure that the diet supplies enough niacin to the horse so that it will not need to use tryptophan to make niacin. An important consideration is the fact that niacin is very low in price, whereas currently tryptophan is quite expensive.

The need for niacin by the horse will be influenced by the amount of tryptophan in the diet. The tryptophan level is influenced by the level of total protein in the diet and the level of tryptophan in the protein feeds used. Therefore, a low-protein diet could be low in tryptophan. This would decrease the amount of tryptophan available to the horse from which to synthesize niacin. If the tryptophan level is too low, it would result in a tryptophan deficiency. This should be avoided especially with young, developing horses and those being trained for racing and high-level performance purposes. The possibility of a niacin deficiency or a tryptophan deficiency would be greater if the niacin in the cereal grains, and their by-product feeds, is largely unavailable to the horse, as is the case for the pig, poultry, and the rat. This information causes many horse owners and trainers to add niacin to their diets as insurance against a possible niacin need. Vitamin B6 is needed in the synthesis of niacin from tryptophan. Therefore, an adequate level of vitamin B6 is also required in the diet.

No toxic effects from using excess niacin in the diet of the horse have been reported. It is best, however, not to feed niacin at excessive levels. Limited research with laboratory animals indicate that nicotinic acid is toxic at levels greater than 159 mg per pound bodyweight per day (2). Nicotinic acid can be tolerated somewhat better than nicotinamide. Limits for parenteral administration could be lower than those for oral intake (3).

XIX. PANTOTHENIC ACID

A. Names Used Previously

The names liver filtrate factor, yeast filtrate factor, and antidermatitis factor have been used for *pantothenic acid*. Its use in phrases such "pantothenic acid activity" and "pantothenic acid deficiency" is acceptable.

B. Form

Pantothenic acid is available as the calcium salt (calcium pantothenate), and because of its stability and crystalline nature, the salt is commonly used by the feed industry to add to horse diets. Calcium pantothenate is frequently marketed as the racemic mixture (both D and L forms) of DL-calcium pantothenate. Only the D isomer of pantothenic acid has vitamin activity. Therefore, the following amounts should be used: (1) 1000 mg D-calcium pantothenate = 920 mg pantothenic acid; (2) 1000 mg DL-calcium pantothenate = 460 mg pantothenic acid.

C. Effect of Deficiency

Pantothenic acid appears to be synthesized in the intestinal tract of adult horses (58) and ponies (62). A pantothenic acid deficiency has not been described for the horse (3). In animals, other than the horse, pantothenic acid deficiency symptoms include growth and reproductive failure, skin and hair lesions, gastrointestinal symptoms, and nervous system lesions. The pig shows a "high-stepping" or "goose-stepping" condition with its hind legs. It is not known which of these symptoms would also occur in the horse.

Pantothenic acid is a component of important coenzymes and is concerned with carbohydrate, fat, and protein digestion. It also has many other important functions in the body. Therefore, a deficiency of this vitamin results in a wide variety of deficiency symptoms, and it is very important to make sure it is not lacking in the diet.

D. Requirements

The 1978 NRC publication on nutrient requirements of the horse suggests a level of 6.8 mg of pantothenic acid per pound of feed for growth and for maintenance of mature horses (5). This level is fairly close to that required by the pig. The 1989 NRC report (3) states that a pantothenic acid requirement has not yet been established for the horse.

Tables 5.2, 5.3, 5.4, and 15.9 give suggested levels of pantothenic acid to use in supplementing diets for racing and high-level performance horses.

E. General Information

The amount of pantothenic acid excreted in the urine is influenced by the level consumed (77). No toxicity has been reported from using excessive levels. However, it is best not to feed the vitamin at excessive levels. Pantothenic acid can be administered orally or in the diet at an intake of 4.55 g per pound bodyweight with no adverse effects (2). The acute LD_{50} for calcium pantothenate administered parenterally is about 1 g/kg body weight (3).

The inclusion of biotin in the diet of a pantothenic acid-deficient pig was effective in prolonging the life of the pig, but caused pantothenic acid deficiency symptoms to appear in half the time. This may be due to some interrelationship of biotin and pantothenic acid (78). Whether this relationship exists in the horse is not known.

Pantothenic acid is not stored in appreciable amounts in the body. Some of the pantothenic acid in feeds occurs in a bound form which lessens its availability to the animal but little information is available on the degree of availability.

XX. VITAMIN B6

A. Names Used Previously

Factor Y, yeast eluate factor, vitamin H, adermia, antiacrodynia rat factor, and antidermatitis rat factor all have been used as names for *vitamin B6*. Its use in phrases such as "vitamin B6 activity" and "vitamin B6 deficiency" is acceptable.

B. Form

Vitamin B6 includes three compounds: pyridoxine, pyridoxal, and pyridoxamine. There may also be other forms of pyridoxine. Pyridoxine, pyridoxal, and pyridoxamine are equal in activity for animals under many conditions. Under others, however, pyridoxal and pyridoxamine may show slightly less activity than pyridoxine. The three forms, however, show very different activities for many microorganisms. In yeast, glandular organs, and meats, most of the vitamin B6 is present as pyridoxal and pyridoxamine, with only traces of pyridoxine. Pyridoxine is the main form in plant products. Thus, vitamin B6 studies must consider the form of the vitamin present in the feed as well as the effectiveness of each form in the animal or microorganism response. No information is available with regard to these forms of vitamin B6 and their relative effectiveness with the horse.

C. Effect of Deficiency

Vitamin B6 functions as part of enzyme systems and is, therefore, a very important vitamin. It is concerned with protein, carbohydrate, and fat metabolism. In the absence of vitamin B6, the amino acid tryptophan cannot be utilized by the animal.

In animals other than the horse, a deficiency of vitamin B6 causes poor growth, anemia, dermatitis, epileptic-like fits or convulsions, nerve degeneration, reproductive failure, spastic gait, and impaired vision.

D. Requirements

No experimental information is available on the vitamin B6 requirements of the horse (3). Tables 5.2, 5.3, 5.4, and 15.9 give suggested levels of vitamin B6 to use.

E. General Information

It has been shown that the horse synthesizes vitamin B6 in the intestinal tract. The largest amount was found in the cecum and large intestine (58). No data are available on toxicity resulting from using excessive vitamin B6 levels with the horse. NRC suggests that dietary levels of at least 50 times nutritional requirements are safe for most species (2).

A Georgia study (79) with the pig showed that increasing the fat level (corn oil) in the diet decreased the vitamin B6 requirement of 3-week-old pigs. When high protein diets are fed, the requirement for vitamin B6 is increased since it is involved in most reactions of amino acid metabolism.

Only small quantities of vitamin B6 are stored in the body. The level of the vitamin in feeds may be decreased by improper heat processing and storage. Riboflavin and niacin are needed to convert pyridoxine to its active form in the body. Urinary xanthurenic acid, which is excreted in the urine, is a sensitive indicator of pyridoxine deficiency. The synthetic form of pyridoxine hydrochloride is generally used in diet supplementation.

XXI. CHOLINE

A. Names Used Previously

Bilineurine and lipotropic factor are names used previously for choline. Its use in phrases such as "choline activity" and "choline deficiency" is acceptable.

B. Effect of Deficiency

Choline has many important functions in the body. One of its roles is in fat metabolism. It is involved in the transport of excess fat from the liver. In animals other than the horse, a lack of choline causes fat to accumulate in the liver, which results in fatty livers. Other deficiency symptoms include unthriftiness, poor body conformation, lack of coordination in movements, poor reproduction, and mortality in the young at birth. Since a choline deficiency has not been produced it is not known which of these symptoms would occur with the horse.

C. Requirements

There is no experimental information available on the choline requirements of the horse (3). Tables 5.2, 5.3, 5.4, and 15.9 give suggested levels of choline to use.

D. General Information

There is an interrelationship between choline and methionine. Methionine provides methyl groups (CH_3) which can be used for choline synthesis. This phenomenon is called *transmethylation*. Vitamin B12 is, in some way, concerned with transmethylation. This interrelationship indicates that the diet requirement for choline depends on the level of methionine. If the methionine level is high enough, it decreases the need for choline. Many diets, however, are only adequate in supplying methionine. This points up the importance of making sure that the diet is adequate in choline so that methionine will not be depleted for choline synthesis. For swine, it is estimated that when methionine is fed at levels in excess of that required for rates of protein synthesis, 4.3 mg of methionine was equal to 1 mg of choline in providing methylating capacity (52). Choline, however, cannot be used to synthesize methionine. Extra choline would only spare methionine which might otherwise be used to synthesize choline.

At a conference in Indiana, the author was told by a practicing veterinarian that the use of 6–8 g of choline daily (a 50% choline product) was effective against heaves in the horse. While this fact has not been confirmed by research, it is provided for anyone who might wish to try choline for this condition. This has been confirmed by a number of other practicing veterinarians.

Data obtained with other animals show that studying the need for choline is a complicated matter and one must be sure the diet is adequate in protein, methionine, folacin, vitamin B12, and possibly other nutrients.

Excess choline should be avoided since there is not a large safety factor between the required level and that which may be harmful (2). No data are available for the horse.

The form most commonly used in supplementing diets is choline hydrochloride which may be in solid, liquid, or deliquescent form.

XXII. FOLACIN

A. Names Used Previously

Folic acid, pteroylglutamic acid, vitamin Bc, citrovorum factor, vitamin M, *L. casei* factor, norite eluate factor, SLR factor, and factor U have all previously been used to designate this vitamin. The term *folacin* should be used for folic acid and related compounds exhibiting qualitatively the biological activity of folic acid. Thus, phrases such as "folacin activity" and "folacin deficiency" are preferred usage.

B. Effect of Deficiency

A folacin deficiency has not been described in the horse (3).

In animals other than the horse, a folacin deficiency results in reduced growth rate and anemia. In the horse, one study reported poor performance associated with low blood serum folacin levels of malnourished horses. They also showed that the horses responded to the administration of 20 mg of folacin, which suggested that horses fed in confinement may benefit from folacin supplementation (80). Green forage is a good source of folacin. It may be that folacin may be helpful when green grazing and high-quality, leafy green is not available. A number of studies have indicated that horses on pasture have higher serum folate levels (80–82).

C. Requirements

There is no experimental information on the folacin requirements of the horse (3). Tables 5.2, 5.3, 5.4, and 15.9 give suggested levels of folacin to use in horse diets for those interested in doing so. No adverse responses to ingestion of excess folacin have been reported in any animal species (2).

D. General Information

It has been shown that the horse synthesizes folacin in the intestinal tract. The largest amount was found in the cecum and large intestine (58). There is no information showing toxicity from using folacin at excessive levels. It is a good policy, however, to avoid excessive levels.

The folacin molecule contains p-aminobenzoic acid (PABA). It is thought by some investigators that PABA functions as a precursor of folacin. Under some conditions this is true, but it is not known if this is always the case.

In feeding trials with swine where moldy corn was used, folacin supplementation increased growth rate up to 15% and feed efficiency up to 9%. However, folacin supplementation was of no benefit when corn without mold was fed (83). This indicates molds may increase folacin requirement and this possibility may exist for the horse.

Folacin functions as a coenzyme for several enzymes involved in body metabolism. Folacin is nutritionally interrelated with other vitamins such as vitamin B12, choline, riboflavin, ascorbic acid, and possibly other nutrients. The interrelationships increase the complexity of determining folacin requirement.

The major portion of folacin is stored in the liver. Certain sulfa drugs increase the need for folacin as has been demonstrated with the pig and the chick.

Roberts (82) concluded that if folacin supplements were to be given, daily oral administration would be the method of choice. In another study (84), it was found that orally administered folic acid was absorbed poorly in the horse. This worker noted that natural folacin compounds are polyglutamates rather than monoglutamates, which may be absorbed by a different mechanism.

XXIII. BIOTIN

A. Names Used Previously

The names vitamin H, factor X, coenzyme R, factor W, bios II, bios IIB, and antiegg white injury factor have been used for *biotin*. Its use in phrases such as "biotin activity" and "biotin deficiency" is acceptable.

B. Effect of Deficiency

Biotin has many important functions in the body. It functions as a coenzyme in carbohydrate, fat, and protein metabolism. A deficiency in animals other than the horse causes growth retardation, loss of hair, dermatitis, nervous system disturbance, spasticity of the hind legs, tongue lesions, and cracked hooves.

A 1983 and 1984 report by Comben *et al.* (85, 86) indicated that the addition of biotin to the diet at a high level was beneficial for defective hoof horn formation in horses. Biotin-deficient cases were characterized by a hoof horn that tended to crumble at the lower edges of the hoof walls, and that was generally prone to poor conformation and damage to the walls, soles, and the white line junction (see Fig. 5.2 and 5.3).

The optimum level of biotin supplementation depended on the body weight of the horse and has yet to be refined. In their study, the following levels, all in mg in biotin per animal daily, seemed to be appropriate: riding horses, 15; heavy horses, up to 30; donkeys and ponies, 5–10. They found it was essential to provide an adequate level of biotin and that it be continued for a sufficient period of time. They stated that improvement in hoof horn strength may occur within 3–5 months but biotin supplementation should be continued for 9–12 months or even longer in severe cases and for very large horses. As hoof horn strength improved, the conformation of the hooves also improved. Once the condition of the hooves was corrected, the horses were then able to maintain their hooves in good condition after high-level biotin supplementation was terminated. They mentioned that horses which were previously tender on their feet lose their

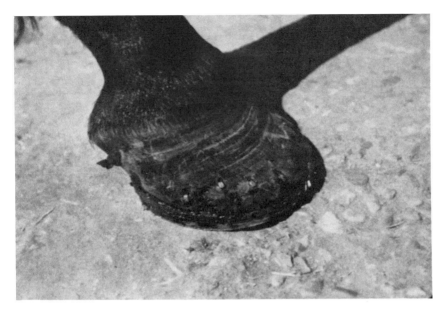

Fig. 5.2 Right fore hoof of a 5-year-old gelding hack prior to biotin supplementation. The horse was extremely tender on its feet, could not be worked, and was losing shoes every few days.

Fig. 5.3 Same right fore hoof as in Fig. 5.2 after 5 months of 15 mg of biotin supplementation daily. The horse no longer lost shoes and had no signs of feet tenderness. Biotin improved the integrity of the hoof horn. (Figs. 5.2 and 5.3 are courtesy of N. Comben, R. J. Clark, and D. B. Sutherland and permission of *Veterinary Record* **115**, 642, 1984.)

apprehension and become more confident, moving with a free and better style. They recommended, however, that a lower, nutritional requirement level of biotin of 2–3 mg of biotin daily should then be used in the diet to maintain healthy hooves in the horse.

They stressed that biotin supplementation should only be expected to maintain or to improve the condition of the surface keratin tissues: the periople, hoof walls, sole, frog, and the white line junctions. No effect on any of the other structures of the foot may be expected. Therefore, biotin supplementation should not be regarded as a panacea for all foot ills.

This is a preliminary report and more studies are needed to confirm these findings. But, the authors of the article also discussed a number of field trials in other locations, including Australia, where biotin supplementation was also found to be beneficial for the horse. Kempson (87) suggested that structural defects in the stratum externum of the hoof horn could be remedied by biotin alone, but poor attachment of the horn squames required biotin plus supplemental calcium and protein.

Biotin supplementation may be quite complex, since it interrelates with so many other nutrients, such as pyridoxine, pantothenic acid, folacin, thiamin, vitamin B12, and ascorbic acid. Therefore, anyone wishing to use a biotin supplement will need to experiment with different levels, since beneficial results may depend on having the correct balance between biotin and other nutrients. Rancidity in feeds can affect biotin requirement since biotin is easily destroyed by this condition (88). Another factor affecting the requirement for biotin is molds in the feed, which can affect biotin availability. Streptomyces are molds found in soil and possibly in bedding and moldy feeds. Streptavidin and stravidin, which are isolated from streptomyces, are biotin-binding substances. They tie up biotin similarly to avidin in raw egg white (89). Therefore, if present in the feed they could tie up biotin and cause a deficiency of this vitamin. Whether this occurs on the farm with horses is not known.

C. Requirements

There are no experimental data available to indicate the biotin requirements of the horse (3). The horse, like other animals, has been shown to synthesize some of its biotin requirements in its intestinal tract (58). On the basis of work with the pig, the level of biotin needed for growth ranges from 23 to 36 μg per pound of total diet. For reproduction, a level of 90 μg per pound of feed is recommended by NRC (52). These levels can be used as a guide for those wishing to try biotin supplements for the horse.

Tables 5.2, 5.3, 5.4, and 15.9 give suggested levels of biotin for the horse.

D. General Information

Biotin is an essential coenzyme involved in carbohydrate, protein, and fat metabolism. It contains sulfur in its molecule. The D-isomer of biotin is the biologically active form whereas the L-isomer is inactive.

Biotin exists in natural materials in both bound and free forms. Much of the bound biotin has low availability. For example, some reports indicate the following percentage of biotin availability in grains: corn, 100; oats, 32; milo, 20–60; barley, 20 or less; and wheat, 10 or less. Whether the same availability occurs for the horse is not known.

When biotin deficiencies are encountered in a sow herd, only 10–20% of the sows may show signs of deficiency. Sometimes a smaller percentage of the sows may exhibit visible symptoms and not all the symptoms may be present. This means the detection of a marginal biotin deficiency is difficult. This is especially so when the usefulness of blood plasma biotin as an indicator of biotin status is still open to question.

Studies with poultry and swine show they can tolerate levels of 4–10 times their nutritional requirements. Since biotin is not well retained in the body, the maximum tolerable level may be much higher (2). How much biotin the horse can tolerate in the diet is not known.

With swine it was found that long-term studies (two to four litters) were needed to determine the value of biotin supplementation for reproduction. One reproductive cycle was not sufficient to show a need for biotin in diets that were marginally lacking in the vitamin. So, long-term studies with horses might be needed.

XXIV. MYO-INOSITOL

A. Names Used Previously

The names inositol, *i*-inositol, *meso*-inositol, inosite, nucite, and dambose have been used. Today it is called *myo-inositol*. Its use in phrases such as "myo-inositol activity" and "myo-inositol deficiency" is acceptable.

B. Effect of Deficiency and Requirements

Myo-inositol has an effect in preventing certain types of fatty livers. Its lipotropic activity is usually synergistic with that of choline. A deficiency of inositol also causes growth retardation and loss of hair. Presently there is no information on myo-inositol deficiency symptoms in the horse. There is also no information on the need to add myo-inositol to horse diets. Inositol has been used for swine at levels of 0.1–0.3% of the diet. This level can be used as a guide for anyone wishing to try adding inositol to horse diets.

There are some scientists who question whether myo-inositol should be considered a vitamin.

XXV. *p*-AMINOBENZOIC ACID (PABA)

A. Names Used Previously

Anti-gray hair factor, BX factor, vitamin Bx, chromotrichia factor, and trichochromogenic factor have been used as names for this vitamin. Its use in phrases such as "*p*-aminobenzoic acid activity" and "*p*-aminobenzoic acid deficiency" is acceptable.

B. Effect of Deficiency and Requirements

There are a few scientists who question whether PABA should be considered as a vitamin. It is a part of the molecule or structure of folacin. Some feel, therefore, that if the diet is adequate in folacin, there is no need to add PABA to the diet. The University of Kentucky published a report showing that the addition of PABA to a purified diet increased the rate of gain in foals (90). Whether this same effect would also occur with a practical-type diet used on the farm is not known. In the study 100 mg of PABA was fed per foal per day. This level can be used as a guide for those who wish to try PABA supplements to horse diets.

XXVI. CONCLUSIONS

Some of the vitamins needed by the horse are provided by a well-balanced high-quality diet. Others are synthesized primarily in the cecum and hindgut area. How adequate both sources of vitamins are in meeting the horse's requirement is not definitely known. It appears that some vitamin supplementation is needed under certain conditions, especially for the high-level performance horse. Unfortunately, there is very little research information regarding which vitamins may be needed and the levels of use.

Borderline deficiencies of vitamins may occur without observable symptoms. When this happens, poor growth, reproduction, and performance may occur. Vitamin requirements are becoming more critical as intensified horse production increases and less high-quality green pasture and hays, which are good sources of vitamins, are available and used.

Vitamin supplementation requires some expertise and care to avoid feeding excessive levels which can cause harmful effects. Suggested levels of vitamins to use in supplementation are given in Tables. 5.2, 5.3, 5.4, and 15.9. Vitamin premixes shown in Table 15.9 include the options of using all vitamins or of not using the B-complex vitamins. Another option is addition of vitamins A, D, E, and K or deletion of either E or K or both from the premix. Thus, the degree of vitamin supplementation will depend on the feeding program, the degree of performance involved, and other factors.

Since the research information on the vitamin needs of the horse is limited, reference is made to results obtained with other animals, as each vitamin is discussed in this chapter.

REFERENCES

1. Funk, C. J. Physiol. (London) **43**, 395 (1911).
2. Combs, G. F. Jr., R. Blair, J. W. Hilton, R. L. Horst, G. E. Mitchel, Jr., and J. W. Suttie. "Vitamin Tolerance in Animals." NAS-NRC, Washington, D. C., (1987).
3. Ott, E. A., J. P. Baker, H. F. Hintz, G. D. Potter, H. D. Stowe, and D. E. Ullrey. "Nutrient Requirement of Horses." NAS-NRC, Washington, D. C., (1989).
4. Fonnesbeck, P. V., and L. D. Symons. J. Anim. Sci. **26**, 1030 (1967).
5. Hintz, H. F., J. P. Baker, R. M. Jordon, E. A. Ott, G. D. Potter, and L. M. Slade. "Nutrient Requirements of Horses." N.A.S.–N.R.C., Washington, D.C. (1978).
6. Howell, C. E., G. H. Hart, and N. R. Ittner. Am. J. Vet. Res. **2**, 60 (1941).
7. Hart, G. H., H. Goss, and H. R. Guilbert. Am. J. Vet. Res. **4**, 162 (1943).
8. Stowe, H. D. Am. J. Clin. Nutr. **21**, 135 (1968).
9. Anderson, A. C., and G. H. Hart. J. Vet. Res. **4**, 307 (1943).
10. Cunha, T. J. Horse Care Fall, p. 66 (1988).
11. Donoghue, S., and D. S. Kronfeld. Compend. Contin. Educ. Pract. Vet. **2**(8), S121 (1980).
12. Donoghue, S., D. S. Kronfeld, S. J. Berkowitz, and R. L. Capp. J. Nutr. **111**, 365 (1981).
13. Maenpaa, P. H., T. Koskinen, and E. Koskinen. J. Anim. Sci. **66**, 1418 (1988).
14. Maenpaa, P. H., A. Pirhonen, and E. Kaskinew. J. Anim. Sci. **66**, 1424 (1988).
15. Fonnesbeck, P. V., and G. W. Vander Noot. J. Anim. Sci. **23**, 1232 (1964).
16. Fonnesbeck, P. V., and G. W. Vander Noot. J. Anim. Sci. **25**, 891 (1964).
17. Garton, C. V., G. W. Vander Noot, and P. V. Fonnesbeck. J. Anim. Sci. **23**, 1233 (1964).
18. Brief, S., and B. P. Chew. J. Anim. Sci. **60**, 998 (1985).
19. Chew, B. P., H. Rasmussen, M. H. Pubols, and R. L. Preston. Theriogenology **18**, 643 (1982).
20. Stowe, H. D. J. Anim. Sci. **54**, 76 (1982).
21. Shirley, R. L., C. H. Hill, J. T. Maletic, O. E. Olson, and W. H. Pfander. "Nutrients and Toxic Substances in Water for Livestock and Poultry." NAS-NRC, Washington, D. C., (1974).
22. El Shorafa, W. M., J. P. Feaster, E. A. Ott, and R. L. Asquith. J. Anim. Sci. **48**, 882 (1979).
23. Naber, E. C., Feed Manage. **39**, 13 (1975).
24. Hintz, H. F., H. F. Schryver, J. E. Lowe, J. King, and L. Krook. J. Anim. Sci. **37**, 282 (1973).
25. Harrington, D. D., and E. H. Page. J. Am. Vet. Med. Assoc. **182**, 1358 (1983).
26. Bille, N. Nord. Veterinaermed. **22**, 218 (1970).
27. McDowell, L. R. "Vitamins in Animal Nutrition." Academic Press, San Diego, California, (1989).
28. Krook, L. P., R. H. Wasserman, J. N. Shively, A. H. Tashjian, Jr., T. D. Brokken, and J. F. Morton. Cornell Vet. **65**, 26 (1975).

29. Wasserman, R. H., J. D. Henion, M. R. Haussler, and T. A. McCain. *Science* **194,** 853 (1976).
30. Marusich, W. L., G. Ackerman, W. C. Reese, and J. C. Bauernfeind. *J. Anim. Sci.* **27,** 58 (1968).
31. Maenpaa, P. H., R. Lappetelainen, and J. Virkkunen. *Equine Vet. J.* **19**(3), 237 (1987).
32. Liu, S. K., E. P. Dolensek, C. R. Adams, and J. P. Tappe. *J. Am. Vet. Med. Assoc.* **183,** 1266 (1983).
33. Schougaard, H., A. Basse, G. Gessel-Nielson, and M. G. Simesen. *Nord. Veterinaermed.* **24,** 67 (1972).
34. Wilson, T. M., H. A. Morrison, N. C. Palmer, G. C. Fenley, and A. A. Van Dreumel. *J. Am. Vet. Med. Assoc.* **169,** 213 (1976).
35. Mayhew, I. G., C. M. Brown, H. D. Stowe, A. L. Trapp, F. J. Derksen, and S. F. Clement. *J. Vet. Intern. Med.* **1,** 45 (1987).
36. Hill, H. E. *Mod. Vet. Pract.* **43,** 66 (1962).
37. Darlington, F. G., and J. B. Chassels. *Summary* **9,** 64 (1957).
38. Darlington, F. G., and J. B. Chassels. *Summary* **12,** 52 (1960).
39. Darlington, F. G., and J. B. Chassels. *Summary* **8,** 1, 52 (1956).
40. Darlington, F. G., and J. B. Chassels. *Summary* **9,** 50 (1957).
41. Darlington, F. G., and J. B. Chassels. *Summary* **8,** 10, 71 (1956).
42. Slade, L. M. Utah Agricultural Experiment Station Logan (unpublished data), (1979).
43. Stowe, H. D. *Am. J. Clin. Nutr.* **21,** 135 (1968).
44. Herrick, J. B. *VM/SAC, Vet. Med. Small. Anim. Clin.* **66,** 1064, 1069 (1971).
45. Rebhun, W. C., B. C. Tennant, S. G. Dill, and J. M. King. *J. Am. Vet. Med. Assoc.* **184,** 1237 (1984).
46. Stillions, M. C., S. M. Teeter, and W. E. Nelson. *J. Anim. Sci.* **32,** 249 (1971).
47. Snow, D. H., and M. Frigg. *Proc. Equine Nutr. Physiol. Symp., 10th,* p. 55 (1987).
48. Snow, D. H., and M. Frigg. *Proc. Equine Nutr. Physiol. Symp., 10th,* p. 617 (1987).
49. Davis, G. K., and C. L. Cole. *J. Anim. Sci.* **26,** 1030 (1943).
50. Johnson, J. H., H. E. Gainer, D. P. Hutchison, and J. G. Merriam. *Proc. Am. Assoc. Equine Pract.* p. 115 (1973).
51. Mahan, D. C., and L. J. Saif. *J. Anim. Sci.* **56,** 631 (1983).
52. Speer, V. C., F. Ahrens, G. L. Allee, G. L. Cromwell, D. W. Friend, and A. J. Lewis. "Nutrient Requirements of Swine." NAS-NRC, Washington D.C., (1987).
53. Brown, R. G., J. G. Buchanan-Smith, and V. D. Sharma. *Can. J. Anim. Sci.* **55,** 353 (1975).
54. Carroll, F. D. *J. Anim. Sci.* **8,** 290 (1950).
55. Martin, A. A. *Nutr. Abstr. Rev.* **45,** 85 (1975).
56. Lott, D. G. *Can. J. Comp. Med. Vet. Sci.* **15,** 274 (1951).
57. Diniz, J. M., J. R. Bashe, and N. J. deCamargo. *Arq. Bras. Med. Vet. Zootech.* **36,** 512 (1984).
58. Carroll, F. D., H. Goss, and C. E. Howell. *J. Anim. Sci.* **8,** 290 (1949).
59. Bertone, J. J., H. F. Hintz, and H. F. Schryver. *Nutr. Rep. Int.* **30,** 281 (1984).
60. Cymbaluk, N. F., P. B. Fretz, and F. M. Loew. *Am. J. Vet. Res.* **39,** 255 (1978).
61. Topliff, D. R., G. D. Potter, J. L. Kreider, and C. R. Creagor. *Proc. Equine Nutr. Physiol. Symp., 7th,* p. 167 (1981).
62. Linerode, P. A., J. S. Mehring, and W. J. Tyznik. *J. Anim. Sci.* **25,** 1259 (1966).
63. Jones, T. C. *Am. J. Vet. Res.* **3,** 45 (1942).
64. Roberts, S. J. *J. Am. Vet. Med. Assoc.* **133,** 189 (1955).
65. Cello, R. M. *Proc. 8th Annu. Conv. Am. Assoc. Equine Pract.* p. 39 (1962).
66. Pearson, P. B., M. K. Sheybani, and H. Schmidt. *Arch. Biochem.* **3,** 467 (1944).
67. Evans, J. W., A. Borton, H. F. Hintz, and L. D. VanVleck. "The Horse." Freeman, San Francisco, California, (1977).
68. Filmer, J. F. *Aust. Vet. J.* **9,** 163 (1933).

69. Davies, M.E. *J. Appl. Bacteriol.* **31,** 286 (1968).
70. Stillions, M. C., S. M. Teeter, and W. E. Nelson. *J. Anim. Sci.* **32,** 252 (1971).
71. Salminen, K. *Acta Vet. Scand.* **16,** 84 (1975).
72. Teague, H. S., and A. P. Grifo, Jr. *J. Anim. Sci.* **23,** 894 (1964).
73. Schweigert, B. S., P. B. Pearson, and M. C. Wilkening. *Arch. Biochem.* **12,** 139 (1947).
74. Pearson, P. B., and R. W. Leucke. *Arch. Biochem.* **6,** 63 (1944).
75. Ghosh, H. P., P. K. Sarkar, and B. C. Guha. *J. Nutr.* **79,** 451 (1963).
76. Firth, J., and B. C. Johnson. *J. Nutr.* **59,** 223 (1956).
77. Pearson, P. B., and H. Schmidt. *J. Anim. Sci.* **7,** 78 (1948).
78. Colby, R. W., T. J. Cunha, C. E. Lindley, D. R. Cordy, and M. E. Ensminger. *J. Am. Vet. Med. Assoc.* **113,** 589 (1948).
79. Sewell, R. F., and M. M. Nugara. *J. Anim. Sci.* **20,** 951 (1961).
80. Seckington, I. M., R. G. Huntsman, and G. C. Jenkins. *Vet. Rec.* **81,** 158 (1967).
81. Allen, B. V. *Vet. Rec.* **103,** 257 (1978).
82. Roberts, M. C. *Aust. Vet. J.* **60,** 106 (1983).
83. Purser, K. *Anim. Nutr. Health* April, p. 38 (1981).
84. Allen, B. V. *Proc. Assoc. Vet. Clin. Pharmacol. Ther.* **8,** 118 (1984).
85. Comben, N., R. J. Clark, and D. J. B. Sutherland. *Ann. Congr. Br. Equine Vet. Assoc.* p. 1 (1983).
86. Comben, N., R. J. Clark, and D. J. B. Sutherland. *Vet. Rec.* **115,** 642 (1984).
87. Kempson, S. A. *Vet. Rec.* **120,** 568 (1987).
88. Pavcek, P. L., and G. M. Shull. *J. Biol. Chem.* **146,** 351 (1942).
89. Ghaiet, L., and F. J. Wolf. *Arch. Biochem. Biophys.* **106,** 1 (1964).
90. Stowe, H. D. *J. Anim. Sci.* **25,** 895 (1966).

6

Mineral Requirements

I. INTRODUCTION

The old saying "no feet and legs—no horse" is still as true today as it ever was. In meeting this problem, it should be stressed that horse breeders need to select against poor feet and legs in breeding stock. This is easier said than done when high-priced animals or special pedigreed stock are involved. Many horse owners are reluctant to dispose of certain animals for sentimental or other reasons. If one studies the results obtained with other classes of livestock, it is obvious that many feet and leg problems are hereditary and strict selection against them has been effective. Therefore, poor feet and legs will not be entirely eliminated until horse owners start putting more emphasis on selection against the use of mares and stallions that have a hereditary weakness for leg abnormalities. It is true, however, that heredity is not the sole cause and that improper nutrition and management are responsible for many feet and leg problems in horses. The problem, however, will not be solved solely by proper nutrition and management, although it will help a great deal. Mares do not foal every year. This has made it more difficult to put as much selection pressure on leg weakness as one should. The average foal crop in the United States is between 50 and 65%. Therefore, it is difficult to eliminate every mare or stallion with feet and leg problems. This problem has to be met sometime, and the sooner started the better. The breeders who eliminate the hereditary aspects of feet and leg problems will eventually be in a favored position. This will take time and will require no compromise or exceptions in selecting and culling breeding stock where it is apparent that heredity is involved.

Good bone, feet, and legs are not due solely to proper calcium and phosphorus nutrition. Other minerals including trace minerals, protein, vitamins, and other nutrients are also involved. Therefore, a well-balanced diet supplying all nutrients is needed to ensure that the horse is receiving all the necessary nutrients. Some of the bone abnormalities in the young foal might trace back to inadequate development *in utero*. If the mare does not supply the foal with all the necessary nutrients *in utero*, the result may be harmful. Corrections of these effects after birth may not be possible. This means a horse breeder who is developing high-level performance horses must pay attention to all stages of the

horse's life cycle and make sure that proper nutrition is supplied in each one. A failure to do so will eventually manifest itself in foals with poor bone structure which causes subsequent feet and leg abnormalities. Usually the fastest growing horses are those experiencing the most bone problems. This is due to the fast growth increasing the nutrient needs of the horse. The diet, therefore, may be adequate for a slow-growing horse, but deficiencies start to show up as growth rate or level of performance increases.

II. DECLINE OF FERTILITY IN SOILS

There is no doubt that soils in the United States are declining in fertility. A high percentage of farmers fertilize at levels below those recommended by university soil scientists. Trace mineral additions to soils are being neglected. As a result, many pastures, hays, grains, and other feeds are lower in mineral content. Trace mineral deficiencies are occurring more frequently throughout the country. Many horse ailments, which are on the increase, may be due in part to this decline in soil fertility. Horse owners frequently note that hay does not have the "strength" or feeding value it had in the past.

As this decline in soil fertility continues, horse owners will have to supply not only the minerals now known to be deficient in the diet but also other minerals that may eventually become low in the soil and in the crops grown thereon. Therefore, proper mineral supplementation of horse diets is a necessity. One must also think in terms of adding minerals back to the soil at the proper levels. This will increase the mineral content of the crops and also the tonnage yield as well as other nutrients in the crop.

III. SOIL, PLANT, AND ANIMAL INTERRELATIONSHIPS

The soil, plant, and animal interrelationships are very complex. Plants do not need selenium or iodine and can grow normally and produce optimum yields even though they contain an insufficient amount of these two minerals to meet animal requirements. Of the trace elements needed by animals, only iron, copper, zinc, manganese, and molybdenum are needed by plants. Cobalt is needed only by nitrogen-fixing microorganisms in the nodules of roots of legume plants. Certain plants grow normally and produce optimum yields even though they contain less iron, zinc, manganese, copper, and cobalt than required by animals. Moreover, normally growing plants may contain levels of molybdenum and cooper that are excessive for animals. In certain areas, where selenium-accumulator plants exist, the levels of selenium in these plants are high enough to cause harmful effects to horses, cattle, sheep, and swine.

Sodium is essential only for certain plants. It is rare for plants to suffer from a deficiency of sodium and chlorine. The commonly used feeds do not contain sufficient sodium to meet animal body needs. As a result, supplementation with sodium is needed and sodium chloride is usually the source of sodium provided in animal feeding. Most concentrates and some roughages are low in chloride. Studies at Cornell University showed that a chloride deficiency occurred in dairy cows fed a diet adequate in sodium.

It is apparent, therefore, that mineral requirements and tolerance to them by plants is different from those of animals. Another example is that boron is an important and essential mineral element for plants and is widely used in fertilizers. But on the basis of present knowledge, animals do not require boron.

This discussion indicates that an excellent looking pasture may be lacking in a number of minerals needed by the horse. So, looks can be deceiving with regard to the nutritional value.

IV. NEW TRACE MINERAL NEEDS

Additional problems occur as soils decline further in fertility after long periods of use. A need for additional trace mineral elements in animal diets may occur. Nickel, vanadium, chromium, tin, silicon, and molybdenum are some of the new mineral elements that may become deficient in the future. This need will be increased more rapidly by: the production of higher producing and performance animals; higher forage and crop yields; new methods of processing and refining feeds; new fertilizer processing methods; and other factors. Each mineral element may eventually be added to the present list of trace mineral elements now being supplemented in animal diets (namely copper, iron, cobalt, iodine, manganese, zinc, and selenium). But, how long before these six new mineral elements become a problem in practical horse diets is not yet known.

A chromium (Cr) deficiency results in impaired glucose tolerance, and molybdenum (Mo) has been shown to be an important part of the enzyme, xanthine oxidase, as well as other enzyme systems. Fluorine (F) prevents dental caries, may be helpful in osteoporoses in aging women, and is needed for anemia prevention. Nickel (Ni) is needed for normal reproduction. Silicon (Si) is needed for growth and proper bone development. Growth rate is decreased as a result of a deficiency of tin (Sn) or vanadium (V). Reduced wing and tail feather growth in chicks occurs as a result of a deficiency of vanadium. All of these deficiencies have been produced with highly purified special diets.

Eventually deficiencies of these trace elements may occur under specialized conditions. Therefore, one needs to keep an open mind regarding additional trace mineral elements being required in animal supplements in the future. Already, molybdenum supplementation, in a few areas where it is very low in forages, would be helpful in counteracting copper toxicity with sheep. Without the molybdenum, copper levels normally used in sheep diets become toxic. Without enough molybdenum, copper accumulates in the liver and causes toxicity.

Molybdenum has recently been found to be part of several enzyme systems. Thus, it might be classified as an essential mineral element. In a highly purified diet, molybdenum has been shown to be essential for lambs, chicks, and turkey poults. It does not need to be added to practical diets, although that may be a possibility in the future. Molybdenum is best known for its toxicity in areas where it occurs in excess in certain locations in the United States and in other countries of the world. Horses are very tolerant to high levels of molybdenum and fail to show any signs of toxicity in these pastures that severely affect cattle. But, their suckling foals may be affected by the high molybdenum levels.

Fluorine might also be included in the list of essential mineral elements. At high levels it is harmful, but at low levels it decreases tooth decay; this means that it affects teeth beneficially in some manner. There is no need to consider adding fluorine to horse diets. However, excess fluorine in the diet needs to be avoided as will be discussed later in this chapter.

V. GRAZING VERSUS HARVESTED FORAGE

Another factor influencing soil and plant mineral levels is whether the forage is harvested for hay or whether it is grazed by horses. Many studies indicate that 60–80% of the minerals in the forage are recycled back to the soil via fecal and urine matter when animals graze an area. Moreover, organic matter is added back to the soil. If the forage and crops are harvested and fed elsewhere, then the soils require either manure or fertilizer application to replace some of what the crops took from the soil. It is known, however, that most farmers do not fertilize their soils at the levels that their State Agriculture Experiment Stations recommend as being practical and economical. Moreover, whenever economic conditions adversely affect the farmer, there is decreased fertilizer application or decreased supplemental mineral feeding of animals as a means of cutting back on expenses. As a result, a number of studies show a gradual decrease in mineral elements contained in many crops grown in the U.S. This has increased the need for mineral supplementation in horse diets.

VI. PLANTS DIFFER IN MINERAL LEVELS

Plants differ in mineral needs and levels. For example, legumes contain higher levels of cobalt than grasses. Wheat and oats are higher in manganese than corn, grain sorghums, or barley. Different varieties of the same plant may show considerable differences in the levels of certain trace minerals. Therefore, the kind of forage grown, grazed, harvested, and fed may vary in mineral composition.

Moreover, the NRC report (1) states that the minerals in the same feeds may vary plus or minus 30%. This variation is considerable and accounts for feed companies analyzing different loads of the same kind of feed to determine their mineral and other nutrient composition. They use nutrient supplementation to keep the nutrient content at about the same level in their manufactured feeds. The use of the computer makes this easy to do once the feed analyses are obtained. This is called quality control in diet formulation.

VII. PRODUCTIVITY AND CONFINEMENT INCREASES SUPPLEMENTATION NEEDS

Horses are being pushed to race and perform at an early age. This increases the need for a well-balanced diet to supply all the nutrients required at their proper levels. Diets that are adequate for the average horse may be inadequate for the racing or high-level performance horse. This needs to be taken into consideration as one develops diets. The animal being kept for occasional pleasure riding does not need as high a level of nutrients as an animal being trained to race as a 2-year-old, the horse that is performing frequently, or the horse that is working in various capacities daily. The level of productivity has considerable effect on the level of nutrition needed.

Horses are being kept more in confinement than previously. High-quality, green pasture is being used less because most horses are kept near or in the larger cities. In these urban areas, land is high priced and scarce for use by horses. Therefore, the land available is used for buildings, exercise areas, and other facilities for handling the horses. Very little, if any, land is available for developing high-quality pastures. Therefore, the horses are being handled under confined or semi-confined conditions. This means that almost all, if not all, of their feed is purchased. The lack of a good-quality pasture puts more stress on the nutritional adequacy of the diet. Pasture is an excellent source of minerals, vitamins, protein, and other nutrients. A good-quality pasture covers up many sins or omissions in a diet and can compensate for a lack of certain nutrients. Therefore, horses not having access to a high-quality pasture need a very well-balanced diet supplying all their nutritional needs. Declining soil fertility, increased productivity, and confined conditions increase the need to supply adequate minerals in the diet.

VIII. REQUIREMENTS VERSUS APPROXIMATE NEEDS

It is very difficult to give an exact answer on specific mineral requirements. This is due to the countless number of factors that affect mineral needs.

At best, therefore, present requirement data should be considered an approximation which may vary depending on many factors. This accounts for the "safety factor" used by those applying NRC requirement recommendations to diets which vary considerably, and are used under varied environmental conditions with feeds of varying quality and with animals having different productivity and stress levels. So, considerable expertise and proper judgement is needed as NRC recommendations are applied.

Minerals and other nutrients, such as vitamins, amino acids, and calories, are interrelated to some extent. This means there is a correct level for each nutrient in relation to the level of all other nutrients to obtain the best response. Many of these interrelationships are known.

But there are others not yet recognized. Some may be discovered as additional research or more sophisticated laboratory procedures are used. It is possible, however, that we may never have an exact nutrient requirement level to meet all animal, feed, farm, and other conditions. But we should develop nutrient level recommendations that satisfy them to the best extent possible.

IX. MINERAL INTERRELATIONSHIPS

Mineral interrelationships are very important in determining mineral requirements, and a few examples of the many interrelationships follow.

Calcium, phosphorus, and vitamin D are interrelated. All three are needed for proper calcium and phosphorus utilization. A deficiency of any one gives similar deficiency symptoms.

Magnesium is also related to calcium and phosphorus requirements. Bone contains about 0.75% magnesium, but scientists are still not sure how magnesium functions in bone. Magnesium is also related to iron level needs.

Excess calcium increase zinc needs by reducing zinc absorption. All plant protein feeds contain phytate, which combines with zinc in the intestinal tract and decreases its absorption. Zinc requirements in the pig can vary from 18 ppm to as high as 150 ppm (with excess calcium) in a plant protein (grain–soybean mean) diet. Excess calcium may also increase the need for copper, manganese, iron, iodine, phosphorus, magnesium, and other minerals.

Excess copper can increase the zinc and iron needs of the pig. There is an interrelationship between molybdenum, copper, and sulfur in a complex mechanism not yet entirely understood in ruminants. Excess molybdenum can increase copper needs in ruminant animals by 2–3 times. A deficiency of molybdenum can result in copper toxicity in sheep in a few isolated areas where the low molybdenum level causes copper to accumulate in the liver of sheep, and harmful effects occur.

Recent studies indicate that high levels of potassium, which are needed for high milk production, increase the sodium needs of the dairy cow. Other recent studies indicate a sodium and potassium interrelationship as well as a potassium–magnesium relationship.

An example of a mineral and vitamin interrelationship is that of selenium and vitamin E. Selenium can substitute to a certain extent for vitamin E, and vice versa, but neither one can substitute entirely for the other. An example of a vitamin an amino acid interrelationship is that of niacin and tryptophan. The pig, chick, horse, and bacteria can use the indispensable amino acid, tryptophan, and synthesize niacin from it. But they cannot convert niacin back to tryptophan.

These few examples of mineral and other nutrient interrelationships indicate the complexity of determining exact mineral requirements for animals. Until these and other interrelationships are well understood, only an approximation of mineral requirements can be made. This means that nutrition specialists need to establish a safe range for each mineral element to be used in meeting varying farm conditions, different diets, and different uses for animals with varying levels of productivity.

X. MINERAL AVAILABILITY

The availability of minerals varies substantially. Determining the exact availability of most minerals is difficult. With some minerals only a part of the mineral in the feces comes from that which is undigested in the feed. A significant amount is endogenous which means the mineral elements was absorbed, possibly used by the body and then secreted into the digestive tract and eliminated via the feces. A digestion trial, where mineral element intake in the feed and elimination in the feces is measured, may not give a true value of how much was available to the animal.

With certain mineral elements, the percentage absorbed decreases when higher levels are fed and increases with lower level intake. This makes it difficult to determine the true availability of a mineral element that can be used as a dependable value. Methods of processing, the temperature involved, solvents used, and other factors can affect the availability of mineral elements. Certain mineral elements have a very low availability. Iron oxide is virtually unavailable and probably no more than 2–4% of the iron is absorbed by the animal. The magnesium in immature, highly succulent forage has very little availability, sometimes as low as 5% for cattle, and is a key factor in grass tetany development. It is not known why the availability decreases in the immature forage under certain conditions. Phytin phosphorus in grain is not utilized as efficiently by the horse as by ruminants.

At one time it was thought that natural sources of minerals were more available to animals than inorganic mineral supplements. Present knowledge, however, indicates this is not always the case. The mineral elements in many mineral supplements are more available than the natural sources. A chemical analysis of a mineral element in a feed or mineral supplement may not be the same as the availability of the mineral element for an animal. To be of maximum value in meeting nutritional requirements, the mineral element must be biologically available to the animal involved. Biological availability may be defined as that portion of the mineral which can be used by the animal to meet its body needs. Unfortunately, there is still a great deal to be learned about mineral availability in animals.

The stage of maturity of a forage or the moisture level of a feed may affect mineral availability. For example, studies at Florida and Illinois show that the availability of phosphorus from high-moisture corn is greater than that from dry corn for the pig. Usually, the more mature a forage becomes, the lower the availability of its minerals to the animal.

XI. ACID–BASE BALANCE AND MINERAL NEEDS

More attention needs to be paid to acid–base balance (2). Many mineral interrelationships are involved in helping to regulate acid–base equilibrium. Minerals occur in fluids and in soft tissues.

One of their main functions is in maintaining osmotic pressure and acid–base balance. For optimum health and normal operation of the body system, the pH of the blood and other body fluids must be held within a narrow range. In a healthy man, for example, the blood pH is about 7.4 and the extremes in pH in which life is possible are between 7.0 and 7.8.

The feeds consumed by animals often show a wide range of potential acidity or alkalinity depending on their mineral composition. Fortunately, the body has very effective buffering systems which maintain the blood pH within normal levels even when feeds react physiologically as either alkaline or acid. Urine excretion of excessive alkaline or acid elements is one means of accomplishing this. Certain additives to the diet are also helpful in acid—base equilibrium. The total role of essential mineral elements in acid—base balance needs more clarification before mineral requirements can be more accurately determined.

XII. CHEMICAL FORM OF MINERALS

The chemical and physical form of a mineral element also affects its availability for animals. An example of the chemical form and its effect on bioavailability is iron oxide. Its iron is virtually unavailable, whereas the iron in ferrous sulfate is highly available for animals. Mineral elements exist in many chemical forms such as sulfates, carbonates, chlorides, oxides and others. Therefore, it is important to know the biological availability of each one. The form chosen for use, however, will also depend on its cost, availability in the area, its stability and effect in the type of diet used, and other factors.

Feed control regulations require that the level and the form of the mineral elements used in a mineral mixture be listed on the tag. This allows the reader to know whether the level used is satisfactory and if the form used is highly available for the animal.

Mineral elements usually exist as mineral salts. The animal's requirement for mineral is based on the mineral element itself and not on the weight of the mineral salt. For example, copper sulfate contains about 25.5% copper, and copper carbonate contains about 53.0% of the mineral element copper. Therefore, it takes about 2 lb of copper sulfate to provide the same amount of the element copper as 1 lb of copper carbonate. If there is a difference in the bioavailability of these two sources of copper for the animal involved, then a correction factor would also be used so that the same amount of available copper would be supplied. This kind of information is needed for all essential mineral elements used in different diets.

It is also important to know what combinations of mineral salts to mix together to avoid them reacting with each other and causing an adverse effect on a mineral mixture. Moreover, mineral mixes need to travel in trucks, railroad cars, and by other means under hot, cold, humid, dry, and other weather conditions. Loose minerals and mineral blocks also need to be stored and fed under a wide range of weather conditions and under varying levels of rainfall or snow. All of this is taken into consideration as mineral supplements or blocks are formulated. Therefore, proper formulation of mineral supplements requires considerable expertise.

XIII. PHYSICAL FORM OF MINERALS

The physical form of the various mineral salts varies and may affect palatability and bioavailability. It is necessary to determine the appropriate particle size to produce adequate mixing without smaller particles settling out. Some minerals used are in the form of a powder. Experts mixing them should wear gloves, protective eye goggles, and respiratory equipment (masks) to prevent inhalation of dusts which might cause adverse effects.

Those formulating certain mixtures need to consider the possible dustiness of their final product. The use of blackstrap molasses in small amounts and added as a very fine mist spray is one means of minimizing dustiness. It also improves palatability. Other products are also used. The final mineral product should not be dusty and should have the proper degree of palatability so the animal will consume, as close as possible, the correct amount to meet its nutritional needs. This is a big challenge ahead for those who produce mineral supplements to be self-fed to horses.

XIV. MINERAL PROCESSING METHODS

Methods of processing, the source of the mineral, the temperature involved, solvents used, and many other factors can affect the availability of mineral elements from various mineral salts. For example, iron carbonate can vary considerably in availability. Companies using it as a source of iron run tests to make sure their product is highly available. Processing methods also affect the palatability of mineral salts.

XV. FUNCTIONS OF MINERALS AND EFFECTS OF DEFICIENCY

Minerals are necessary for life and perform many important functions in the body. They are necessary for proper bone and teeth formation and maintenance. Minerals also serve the body in many other ways. Almost every process of the animal body depends for proper functioning on one or more of the mineral elements. Minerals are very important for growth, reproduction, and lactation.

A deficiency of minerals in the diet may cause any of the following symptoms: reduced appetite, poor weight gains, soft or brittle bones, stiffness or malformed joints, goiter, unthrifty looking animals, failure to come in heat regularly or to reproduce, poor milk production, weak or dead young, poor use of feed, and eventually death. Other deficiency signs will be specifically discussed for the different mineral elements later in this chapter.

XVI. THE MINERAL CONTENT OF THE ANIMAL BODY

Approximately 80% of the phosphorus and 99% of the calcium in the body are present in the bones and teeth. Calcium and phosphorus make up about 50% of the ash of the body. For the most part, these two mineral elements occur combined with each other in the body. An inadequate supply of either one in the diet will limit the utilization of the other.

Table 6.1 shows data on the mineral composition of the whole body of the horse (3). The information indicates that the calcium to phosphorus ratio in the whole body is about 2 to 1 on a fat-free basis. The fat-free dry matter of the horse is about 25% of the empty body weight, which, in turn, is about 92% of the full body weight (3).

TABLE 6.1

Mineral Composition of the Whole Body of Young Horses[a]

Age (months)	Full body weight (lb)	Empty body weight (lb)	Dry Matter[b]	Fat[b]	Fat-free matter[b]	Calcium[c]	Phosphorus[c]	Potassium[c]	Sodium[c]	Magnesium[c]	Iron[c]	Zinc[c]	Copper[d]	Manganese[d]	Ash[c]	Calcium to phosphorus ratio
4	288	268	31.1	5.3	25.8	74.4	36.2	8.4	6.3	1.6	0.22	0.13	19	4	213	2.06
12	845	799	41.9	16.4	25.5	69.3	33.4	7.5	6.8	1.7	0.47	0.14	18	6	203	2.08
24	926	845	40.2	14.6	25.6	65.7	33.1	7.8	7.4	1.5	0.59	0.14	13	5	204	1.97

[a]Data from Schryver et al. (3).
[b]Percentage of empty body weight.
[c]mg/g of fat-free tissue.
[d]mg/kg of fat-free tissue.

XVII. ESSENTIAL MINERAL ELEMENTS

Fourteen essential mineral elements have been shown to perform important functions in the body, and need to be considered as horse diets are developed. They are calcium, phosphorus, sodium, chlorine, copper, cobalt, iodine, iron, manganese, sulfur, potassium, magnesium, zinc, and selenium.

It is not known how many of these essential mineral elements need to be added to horse diets. Calcium, phosphorus, sodium, chlorine, and the trace minerals (copper, iron, iodine, manganese, zinc, and selenium) should be considered as possible additions. Cobalt is usually added with the other trace minerals, even though it may not be needed if the diet contains enough vitamin B12. Under certain conditions potassium and magnesium supplementation may be needed. It is doubtful whether sulfur supplementation is needed if the diet contains enough protein and, hence, adequate levels of the sulfur-containing amino acids methionine and cystine.

XVIII. MACRO- AND MICROMINERALS

The mineral elements are usually referred to as macro- and microminerals. The macrominerals are needed at much higher levels in the diet, while the microminerals are needed in trace amounts in the diet and are most often referred to as trace minerals. Those in each category are given in the following tabulation.

Macrominerals	Microminerals
Calcium (Ca)	Iron (Fe)
Phosphorus (P)	Copper (Cu)
Sodium (Na)	Iodine (I)
Chlorine (Cl)	Cobalt (Co)
Magnesium (Mg)	Manganese (Mn)
Potassium (K)	Zinc (Zn)
Sulfur (S)	Selenium (Se)

The symbols in parentheses are those for each of the mineral elements. Minerals are found in feeds in organic and inorganic combinations. For example, the mineral element sodium (Na) is not added to feeds as the element sodium. It is most frequently added as salt, which is a combination of sodium and chlorine. Sodium is also found in other combinations in feeds. The primary need for sulfur in the feed is as a constituent of the amino acids cystine and methionine. If they are not adequate in the diet, sulfur can be added as sodium sulfate or some other source. Table 6.2 gives some of the trace mineral salts used in animal feeding.

TABLE 6.2

Percentage of Mineral Element in Typical Mineral Salts

Mineral salt	Percentage mineral element	Mineral salt	Percentage mineral element
Cobalt		Zinc	
Cobaltous carbonate	49.5	Zinc carbonate	52.1
Cobaltous sulfate	24.8	Zinc sulfate	22.7
Cobaltous oxide	73.4	Zinc oxide	80.3
Copper		Iron	
Cupric carbonate	53.0	Iron oxide	69.9
Cupric oxide	80.0	Ferrous sulfate (7 H_2O)	20.1
Cupric sulfate	25.5	Ferrous sulfate (No H_2O)	36.7
		Ferrous carbonate	41.7
Manganese		Iodine	
Manganous carbonate	47.8	Potassium iodide	76.4
Manganous sulfate	32.5	Calcium iodate	60.0
Manganous oxide	77.4		
Selenium			
Sodium selenite	45.0		

It is important to note the different levels of the mineral elements in each of the mineral salts used. For example, cobaltous carbonate has twice (49.5%) as much of the cobalt mineral element as cobaltous sulfate (24.8%). Therefore, twice as much cobaltous sulfate would need to be added to supply the same level of cobalt as cobaltous carbonate. The requirements for minerals are based on the mineral element itself and not on the weight of the mineral salt. Therefore, the level to add to diets depends on the mineral salt used and how much of the mineral element it contains.

Many people are not aware that mineral salts vary in their mineral content. Many think that 1 lb of cobaltous sulfate, for example, supplies 1 lb of cobalt. But, since cobaltous sulfate contains only 24.8% cobalt, it takes a little over 4 lb of cobaltous sulfate to supply 1 lb of cobalt itself (elemental cobalt).

Converting Parts per Million to Other Terms

Many mistakes are made in converting parts per million (ppm) to other terms. Unless one understands how to do this properly, it is best not to get involved in it. This is one reason why horse people should not try mixing their own trace minerals unless they are especially well qualified to do so. Table 6.3 gives information in converting parts per million (ppm) into other terms or vice versa. It can be used as a guide in converting from one term to another.

TABLE 6.3

Table of Equivalents

ppm	Percent	g/kilo	g/lb	g/100 lb	g/ton	oz/100 lb	oz/ton	lb/100 lb	lb/ton
1.0	0.0001	0.001	0.00045	0.0453	0.907	0.0016	0.032	0.0001	0.0002
10,000	1.0	10.0	4.53	453.6	9072.0	16.0	320	1.0	20.0
1,000	0.1	1.0	0.45	45.3	907.0	1.6	32.0	0.1	2.0

XIX. CALCIUM AND PHOSPHORUS

These two minerals will be discussed together since there is a close relationship between them. Calcium and phosphorus make up nearly half the minerals in mare's milk and about half the minerals in the body of the horse. Thus, it is very important that these two minerals be adequately supplied in horse diets. Bone ash contains 35% calcium and 14–17% phosphorus (4). It also contains about 0.75% magnesium and small amounts of sodium, potassium, chlorine, fluorine, and traces of other mineral elements. Thus, proper bone formation and sound feet and legs depend to a high degree on proper calcium and phosphorus nutrition, although many other nutrients are also involved.

A. Role of Calcium

It is estimated that about 98% of the calcium is in the bones and teeth of the horse. But, the other 2% of the calcium in the body is involved in very critical functions. Calcium is needed for normal blood clotting. Calcium, along with sodium and potassium, is required for normal beating of the heart and is involved in the maintenance of acid–base balance within the very narrow range necessary for normal functions. Calcium is also involved in nerve and muscle tissue. It is needed for muscle and heart contraction, normal nerve function, the activation of certain enzymes, and the secretion of several hormones.

The role of calcium in body tissues is so critical that the level of calcium in the blood is very closely regulated by a complex set of body mechanisms. This regulation involves a number of hormones which include the parathyroid hormone, estrogens, calcitonin, and others. If the blood level of calcium starts to decrease, calcium is quickly mobilized from the bone to bring the level in blood back to normal. Therefore, calcium blood levels are not a reliable indicator of the calcium status of the horse diet. Calcium homeostatic mechanisms maintain blood serum calcium within a narrow range (5).

Effect of Calcium Deficiency

A lack of calcium in the developing foal can lead to rickets, which is characterized by poor mineralization of the osteoid tissue and the probability of enlarged joints and crooked long bones (5). In the mature horse, inadequate dietary calcium can result in weakening of the bones and an insidious lameness (6). In a severe deficiency, the tension of the muscles pulls the weak bones out of shape and the weight of the body causes the legs to bend and fracture.

An Ohio study (7) reported a negative linear relationship between dietary intake and perceived severity of metabolic bone disease in young horses. On farms with yearlings having the lowest incidence of metabolic bone disease the diets contained 1.2% calcium, whereas yearling with the most severe metabolic bone disease were on farms where the diets contained 0.2% calcium.

Horses fed low levels of calcium but excessive amounts of phosphorus develop a condition called nutritional secondary hyperparathyroidism (also called big head, miller's disease, bran disease, osteo fibrosa, and fibrous osteodystrophy). The excess phosphorus ties up calcium and thus decreases calcium absorption which causes the blood calcium level to drop. This drop causes the parathyroid gland to release a hormone to remove calcium from the bones to increase the blood calcium level. In the horse, this causes a great deal of calcium to be removed from the facial bones; fibrous connective tissue then invades this area causing enlarged facial bones.

A lactating animal responds to a deficiency of calcium or phosphorus by decreasing its level of milk production. But, it does not decrease the level of calcium and phosphorus in the milk. Even in an extreme deficiency the composition of the milk remains within normal limits. Therefore, a phosphorus- or calcium-deficient horse will decrease milk production in line with the degree of the deficiency.

The age of the horse determines the terms used for inadequate calcium and phosphorus nutrition. For the young, the term *rickets* is used to indicate that calcification of the growing bone is not taking place normally. In more mature horses the condition is often called *osteomalacia*.

B. Role of Phosphorus

Approximately 85% of the phosphorus in the body occurs in the bones and teeth. Phosphorus is an important part of many enzyme systems. It is an essential component of various organic compounds which are involved in almost every aspect of feed metabolism and utilization in the body. This includes a role in the utilization of energy, fat, carbohydrate, protein, and other nutrients. Phosphorus is also involved in muscle, nervous tissue, and skeletal growth and in normal blood chemistry. Therefore, phosphorus affects all aspects of the horse's life.

The horse does not have the same mechanism to mobilize phosphorus from bone as it does for calcium. However, when calcium is mobilized from bone some phosphorus is withdrawn along with it. Blood phosphorus levels appear to be more indicative of phosphorus deficiency than blood calcium levels are of calcium deficiency. But, some scientists disagree on how accurately blood phosphorus levels can indicate the status of phosphorus in the diet. Blood phosphorus levels, therefore, must be used with some degree of reservation.

EFFECT OF PHOSPHORUS DEFICIENCY

A lack of phosphorus will, like inadequate calcium and vitamin D, produce rachitic changes in growing horses and osteomalacic changes in mature horses

(5). A deficiency of phosphorus results in a loss of appetite. A more severe deficiency results in a depraved appetite which results in a craving for and eating or chewing of bones, wood, hair, rocks, clothing, and other materials. Some horses may chew up an entire mineral box. The animals also become weak, emaciated, lose weight, and eventually die. In certain range grazing areas, deaths occur from diseases that the weakened horses get from eating decayed bones from diseased animals.

Other symptoms of a phosphorus deficiency include retarded growth rate, reduced efficiency of feed utilization, lowered milk production, failure to exhibit estrus or heat, low conception rate, leg abnormalities, and reduced bone ash and bone strength.

A Canadian study (8) with 42 weanling horses showed that marginal phosphorus intakes did not alter productivity, feed intake, blood mineral concentration, or musculoskeletal abnormalities during the final 20 weeks of a 30-week trial. Mild to moderate physitis and flexure limb deformities occurred in 88% of the weanlings principally between week 6 and 8 of the study. The limb deformities had largely been resolved by week 12 of the study.

C. Calcium to Phosphorus Ratio and Vitamin D

To obtain proper calcium and phosphorus utilization, a number of conditions must be met. First, an adequate level of both calcium and phosphorus must be fed to the horse. Next, a suitable ratio between them must exist. Third, a sufficient amount of vitamin D must be available. The ratio of calcium to phosphorus will vary depending on the weight and age of the horse and the level of calcium and phosphorus in the diet. These two mineral elements are more efficiently utilized when they are present in a certain ratio to each other. If the calcium intake is too high and the phosphorus level is just adequate, then the excess calcium could cause a deficiency of phosphorus to occur. This is because the excess calcium will tie up a certain amount of phosphorus in the intestinal tract and prevent it from being absorbed.

Vitamin D is needed for maximum calcium and phosphorus absorption and utilization. Studies have shown that animals receiving no vitamin D excrete an excessive amount of calcium and phosphorus in the feces. Vitamin D is of more help as the ratio of calcium to phosphorus becomes imbalanced. However, if the calcium to phosphorus ratio becomes too abnormal, then vitamin D alone will not be of much help in correcting the situation.

Dr. E. A. Ott of the University of Florida proposed the calcium to phosphorus ratios shown in Table 6.4.

The ratio of calcium to phosphorus varies depending on the age and weight of the horse as well as the level of calcium and phosphorus in the diet. As the horse

TABLE 6.4

Calcium to Phosphorus Ratios for Complete Horse Diets[a]

Life cycle	Calcium to phosphorus ratio		
	Minimum	Maximum	Optimum
Nursing foal	1:1	1.5:1	1.2:1
Weanling	1:1	2.0:1	1.5:1
Yearling	1:1	3.0:1	2.0:1
Long yearling	1:1	3.0:1	2.0:1
Mature horse	1:1	5.0:1	2.0:1

[a]Proposed by E. A. Ott, University of Florida.

increases in age, it can tolerate a wider calcium to phosphorus ratio as is shown in Table 6.4.

In the past, the calcium to phosphorus ratio was overemphasized. Many recommended a ratio of 1:1 and indicated it should never exceed this ratio. The horse can tolerate a wider ratio (9,10) than was thought possible a few years ago as is shown in Table 6.4. A Minnesota study (10) showed that ratios of calcium to phosphorus as high as 6:1 in diets for growing horses may not be detrimental if phosphorus intake is adequate. Ratios less than 1:1 (when phosphorus intake exceeds calcium intake) may be detrimental to calcium absorption. Even if the calcium requirements are met, excessive calcium intake may cause skeletal abnormalities (11).

D. Avoiding Excess Calcium

Excessive levels of calcium can increase the need for other minerals. These include phosphorus, magnesium, zinc, manganese, copper, iodine, iron, and possibly others. Therefore, it is important to avoid feeding excessive levels of calcium in the diet. High levels of calcium also decrease the digestibility of calcium (12).

The horse has a regulatory mechanism that tries to keep the calcium in the blood at an adequate level. When blood calcium decreases to a certain level, the parathyroid gland releases a hormone, called *parathormone*, which causes the bone to release calcium to the blood to increase its level. The thyroid gland releases a hormone, called *calcitonin,* when the blood calcium level is too high. This hormone decreases the amount of calcium that is removed from the bone. A deficiency of calcium in the diet will eventually cause too much withdrawal of calcium from the bone resulting in weak, porous bones which then cause leg abnormalities because of the weight of the horse and pull of the muscles on a fragile bone.

E. Avoiding Excess Phosphorus

Feeding diets high in phosphorus and low or marginal in calcium for a prolonged period results in severely depleted skeleton and nutritional secondary hyperparathyroidism (13,14), a condition commonly called "big head" disease. This condition was quite common when horses were fed high levels of wheat bran, which is high in phosphorus (1.15%) and low in calcium (0.14%). A high phosphorus diet depresses the intestinal absorption of calcium, the concentration of calcium in the blood plasma, kidney calcium excretion, and calcium retention in the bone (13). Therefore, one should avoid feeding less calcium than phosphorus if at all possible. It is best to keep calcium and phosphorus levels as close to the requirement as possible. Excess phosphorus results in higher urinary phosphorus excretion (15).

F. Bone Not Static

An adult bone contains about 25% ash (mineral), 20% protein, 10% fat, and 45% water. Mineral salts are deposited in an organic matrix in bone, which consists of a mixture of proteins of which the main one is ossein (called bone protein by many). The exact mechanism of bone formation is still not well understood. There is little variation in the composition of bone ash. There is an approximate 2:1 calcium to phosphorus level in bone. The water content of bone decreases with age. The fat content varies depending on the condition of the animal, since bone marrow serves as a fat depot.

The protein matrix in bone is calcified when the proper levels of calcium, phosphorus, magnesium, and other minerals are present. However, bone is not static, and is continually being re-formed. Mobilization and restorage of calcium and phosphorus in bone occurs throughout life. There is a continuous interchange of calcium and phosphorus between the bone, the blood supply, and other parts of the body. If more leaves the bone than is being deposited, the bone eventually becomes weak, porous, and may be deformed or broken by the weight of the animal, the pull of the muscles, and the stress placed on the bone by running, racing, or walking. When more calcium and phosphorus is deposited than is leaving the bone, the bone will rebuild its strength and hardness. The main point to remember is that once a strong bone is formed, it does not remain that way forever. One must continually supply the horse with the calcium, phosphorus, and other nutrients necessary to maintain the bone in a strong condition. Fortunately, bone serves as a bank for borrowing calcium and phosphorus during emergencies when the need is greater than the diet provides. During heavy lactation, the bone is drawn upon to meet part of the need for minerals secreted in the milk. The stores in the bone may also be drawn upon during pregnancy. This depletion of reserves causes no apparent harm if it is not too heavy and if it is restored with an adequate diet as soon as the heavy demand is over. Therefore, the body's regulatory mechanism, which allows minerals deposited in the bone to be drawn upon during periods of emergency needs, is a good one. But, this borrowing from bone stores needs to be repaid as soon as possible in order to maintain strong bones. Therefore the horse must be continually supplied with the needed minerals throughout its life cycle. There is no period when one can relax proper mineral feeding.

A German study (16) showed that the skeleton of the horse showed a lower magnesium level, a higher calcium content, and a wider ratio of calcium to phosphorus than other domestic animals. The calcium content increased with age but phosphorus level increased only a small amount which resulted in an increase in the calcium to phosphorus ratio. The magnesium, potassium, and chlorine content decreased with increasing age while the zinc level did not show any relationship to age. The copper level increased until 8 years of age and decreased thereafter. Dry matter and crude ash content increased from birth to 60 months of age while the crude protein decreased.

G. Calcium and Phosphorus Requirement

The calcium and phosphorus requirements of the horse have recently received considerable attention. Unfortunately, there is still much to learn before more dependability can be placed on present estimates of their need by the horse. Many factors affect the calcium and phosphorus requirements, including breeding, growth rate, and productivity; stress of training and performance; level, ratio, and availability of the calcium and phosphorus in the diet; level of other nutrients in the diet; stage of the life cycle involved; the quality and maturity of the hay used; the kind of concentrates fed; environmental conditions; the level of minerals in the drinking water, the amount lost in the sweat; and many other factors. A Maryland study (17) showed that thoroughbred foals fed 130% of NRC protein recommendations exhibited increased urinary excretion of calcium and phosphorus. They suggested that supplementation with buffering substances such as sodium bicarbonate could ameliorate this effect. A Cornell study (18) showed that feeding horse foals a high protein level of 20% did not affect calcium absorption or calcium retention.

Table 6.5 gives the NRC publication recommendations on the requirements of the horse for calcium and phosphorus (5). These values do not contain a safety margin to take care of the many factors that can affect calcium and phosphorus needs.

Table 6.6 gives the 1989 NRC-recommended calcium and phosphorus re-

TABLE 6.5

Nutrient Concentrations in Total Diets for Horses and Ponies (Dry Matter Basis) (5).

	Digestible energy[a]		Diet proportions		Crude protein (%)	Lysine (%)	Calcium (%)	Phosphorus (%)	Magnesium (%)	Potassium (%)	Vitamin A	
	(Mcal/kg)	(Mcal/lb)	Conc. (%)	Hay (%)							(IU/kg)	(IU/lb)
Mature horses												
Maintenance	2.00	0.90	0	100	8.0	0.28	0.24	0.17	0.09	0.30	1830	830
Stallions	2.40	1.10	30	70	9.6	0.34	0.29	0.21	0.11	0.36	2640	1200
Pregnant mares												
9 months	2.25	1.00	20	80	10.0	0.35	0.43	0.32	0.10	0.35	3710	1680
10 months	2.25	1.00	20	80	10.0	0.35	0.43	0.32	0.10	0.36	3650	1660
11 months	2.40	1.10	30	70	10.6	0.37	0.45	0.34	0.11	0.38	3650	1660
Lactating mares												
Foaling to 3 months	2.60	1.20	50	50	13.2	0.46	0.52	0.34	0.10	0.42	2750	1250
3 months to weaning	2.45	1.15	35	65	11.0	0.37	0.36	0.22	0.09	0.33	3020	1370
Working horses												
Light work[b]	2.45	1.15	35	65	9.8	0.35	0.30	0.22	0.11	0.37	2690	1220
Moderate work[c]	2.65	1.20	50	50	10.4	0.37	0.31	0.23	0.12	0.39	2420	1100
Intense work[d]	2.85	1.30	65	35	11.4	0.40	0.35	0.25	0.13	0.43	1950	890
Growing horses												
Weanling, 4 months	2.90	1.40	70	30	14.5	0.60	0.68	0.38	0.08	0.30	1580	720
Weanling, 6 months												
Moderate growth	2.90	1.40	70	30	14.5	0.61	0.56	0.31	0.08	0.30	1870	850
Rapid growth	2.90	1.40	70	30	14.5	0.61	0.61	0.34	0.08	0.30	1630	740
Yearling, 12 months												
Moderate growth	2.80	1.30	60	40	12.6	0.53	0.43	0.24	0.08	0.30	2160	980
Rapid growth	2.80	1.30	60	40	12.6	0.53	0.45	0.25	0.08	0.30	1920	870
Long yearling, 18 months												
Not in training	2.50	1.15	45	55	11.3	0.48	0.34	0.19	0.08	0.30	2270	1030
In training	2.65	1.20	50	50	12.0	0.50	0.36	0.20	0.09	0.30	1800	820
Two year old, 24 months												
Not in training	2.45	1.15	35	65	10.4	0.42	0.31	0.17	0.09	0.30	2640	1200
In training	2.65	1.20	50	50	11.3	0.45	0.34	0.20	0.10	0.32	2040	930

[a] Values assume a concentrate feed containing 3.3 Mcal/kg and hay containing 2.00 Mcal/kg of dry matter.

[b] Examples are horses used in Western and English pleasure, bridle path hack, equitation, etc.

[c] Examples are horses used in ranch work, roping, cutting, barrel racing, jumping, etc.

[d] Examples are race training, polo, etc.

TABLE 6.6

Nutrient Concentrations in Total Diets for Horses and Ponies (90% Dry Matter Basis) (5).

	Digestible energy[a]		Diet proportions		Crude protein (%)	Lysine (%)	Calcium (%)	Phosphorus (%)	Magnesium (%)	Potassium (%)	Vitamin A	
	(Mcal/kg)	(Mcal/lb)	Conc. (%)	Hay (%)							(IU/kg)	(IU/lb)
Mature horses												
Maintenance	1.80	0.80	0	100	7.2	0.25	0.21	0.15	0.08	0.27	1650	750
Stallions	2.15	1.00	30	70	8.6	0.30	0.26	0.19	0.10	0.33	2370	1080
Pregnant mares												
9 months	2.00	0.90	20	80	8.9	0.31	0.39	0.29	0.10	0.32	3330	1510
10 months	2.00	0.90	20	80	9.0	0.32	0.39	0.30	0.10	0.33	3280	1490
11 months	2.15	1.00	30	70	9.5	0.33	0.41	0.31	0.10	0.35	3280	1490
Lactating mares												
Foaling to 3 months	2.35	1.10	50	50	12.0	0.41	0.47	0.30	0.09	0.38	2480	1130
3 months to weaning	2.20	1.05	35	65	10.0	0.34	0.33	0.20	0.08	0.30	2720	1240
Working horses												
Light work[b]	2.20	1.05	35	65	8.8	0.32	0.27	0.19	0.10	0.34	2420	1100
Moderate work[c]	2.40	1.10	50	50	9.4	0.35	0.28	0.22	0.11	0.36	2140	970
Intense work[d]	2.55	1.20	65	35	10.3	0.36	0.31	0.23	0.12	0.39	1760	800
Growing horses												
Weanling, 4 months	2.60	1.25	70	30	13.1	0.54	0.62	0.34	0.07	0.27	1420	650
Weanling, 6 months												
Moderate growth	2.60	1.25	70	30	13.0	0.55	0.50	0.28	0.07	0.27	1680	760
Rapid growth	2.60	1.25	70	30	13.1	0.55	0.55	0.30	0.07	0.27	1470	670
Yearling, 12 months												
Moderate growth	2.50	1.15	60	40	11.3	0.48	0.39	0.21	0.07	0.27	1950	890
Rapid growth	2.50	1.15	60	40	11.3	0.48	0.40	0.22	0.07	0.27	1730	790
Long yearling, 18 months												
Not in training	2.30	1.05	45	55	10.1	0.43	0.31	0.17	0.07	0.27	2050	930
In training	2.40	1.10	50	50	10.8	0.45	0.32	0.18	0.08	0.27	1620	740
Two year old, 24 months												
Not in training	2.20	1.00	35	65	9.4	0.38	0.28	0.15	0.08	0.27	2380	1080
In training	2.40	1.10	50	50	10.1	0.41	0.31	0.17	0.09	0.29	1840	840

[a]Values assume a concentrate feed containing 3.3 Mcal/kg and hay containing 2.00 Mcal/kg of dry matter.

[b]Examples are horses used in Western and English pleasure, bridle path hack, equitation, etc.

[c]Examples are horses used in ranch work, roping, cutting, barrel racing, jumping, etc.

[d]Examples are race training, polo, etc.

quirements on a 90% dry matter basis. Horse owners and trainers need to do some observing and experimenting on their own to determine if the level of calcium and phosphorus used is adequate. If they find their horses still have feet and leg problems, they may wish to increase their calcium and phosphorus levels. If they do this, it is recommended they continue with the new level for a year or so in order to allow enough time to determine the beneficial effect to the foals, the horses in training, and those that are racing or being used in other high-level performance activities. All changes should be made gradually and levels of nutrients should not be drastically changed at any one time.

A Cornell study (19) showed that horses and ponies are not able to balance their calcium and phosphorus needs properly from free-choice supplements. Their study and others indicated that it is best to feed diets that contain the proper levels of calcium and phosphorus and that a free-choice mineral mixture be used only for insurance (20). It may be that designing the complete mineral mixture to meet the salt needs of the horse might come close to meeting the calcium and phosphorus needs of the horse (see Section XXB in this chapter). Designing a proper complete mineral mixture for self-feeding to horses is a complex matter and a challenge for the future.

H. Calcium and Phosphorus in Feeds

Table 6.7 gives the calcium and phosphorus levels in some commonly used feeds in horse diets (5,21). These levels are average figures which can vary depending on the many factors that affect the calcium and phosphorus levels of feeds. The mineral content of the grains will usually vary less than that in the hays. But, even grains can vary considerably. Hays vary considerably in mineral content because of the stage of maturity when it is harvested, level of exposure to the sun after it is cut, whether or not rain fell on the cut hay, level of fertilization used, soil type and its fertility, climate and weather, genetics of the plant, etc. Therefore, the levels shown in Table 6.7 can serve as a guide to the average values of calcium and phosphorus in the more commonly used feeds for horses.

I. Calcium and Phosphorus Mineral Supplements

Tables 6.8 (22) and 6.9 list sources of calcium and phosphorus mineral supplements and the level of these two mineral elements in them. The composition of these mineral supplements can vary somewhat depending on the purity of the raw material and the method used in processing.

It should be noted that monosodium phosphate, disodium phosphate, and sodium tripolyphosphate contain high levels of sodium. Almost all of the phosphate sources contain some fluoride. Fluoride levels will be discussed later in this chapter.

TABLE 6.7

Calcium and Phosphorus in Some Widely Used Feeds (As Feed Basis) (5).

Feeding stuff	Calcium (%)	Phosphorus (%)
Protein concentrates		
Wheat bran, 15% protein	0.13	1.13
Cottonseed meal, 41% protein	0.18	1.22
Skim milk, dried, 34% protein	1.28	1.02
Linseed meal, solvent, 36% protein	0.39	0.80
Soybean meal, solvent, 44% protein	0.40	0.71
Corn gluten feed, 26% protein[a]	0.40	0.80
Fish meal, menhaden, 60% protein[a]	5.01	2.87
Meat and bone meal, 50% protein[a]	10.76	5.33
Feather meal, 86% protein[a]	0.33	0.55
Meat meal, 55% protein[a]	8.27	4.10
Peanut meal, solvent, 52.9% protein	0.32	0.66
Canola, solvent, 37.1% protein	0.63	1.18
Sunflower meal, solvent, 48.9% protein	0.45	1.02
Grains		
Barley	0.05	0.34
Oats	0.08	0.34
Corn, yellow	0.05	0.27
Sorghum	0.04	0.32
Wheat, hard winter or spring	0.04	0.38
Brewer's dried grains, 25% protein	0.30	0.50
Hays		
Alfalfa, early bloom, hay, sun-cured	1.28	0.19
Alfalfa, mid bloom, hay, sun-cured	1.24	0.22
Alfalfa meal, 17% protein, dehydrated	1.38	0.23
Bahiagrass, hay, sun-cured	0.45	0.20
Bermudagrass hay, sun-cured	0.30	0.19
Bluegrass hay, Kentucky, sun-cured	0.24	0.25
Brome hay, late bloom, sun-cured	0.25	0.25
Canarygrass, Reed, hay	0.32	0.21
Clover, Alsike, hay, sun-cured	1.14	0.22
Clover, Ladino, hay, sun-cured	1.20	0.30
Clover, Red, hay, sun-cured	1.22	0.22
Fescue, Meadow, hay, sun-cured	0.40	0.29
Lespediza, hay, sun-cured	1.07	0.17
Orchardgrass, hay, sun-cured	0.24	0.30
Pangolagrass, hay, sun-cured	0.53	0.19
Prairie hay, sun-cured	0.32	0.12
Timothy hay, early bloom	0.45	0.25
Trefoil, Birdsfoot, hay, sun-cured	1.54	0.21

[a]Sunde et al. (21).

TABLE 6.8

Typical Analyses of Phosphate Compounds[a]

Compound	Phosphorus content (%)	Calcium content (%)	Sodium content (%)	Nitrogen content (%)	Fluoride content (%)
Defluorinated phosphates manufactured from defluorinated phosphoric acid					
Monocalcium phosphate	21.0	16.0	—	—	0.16
Dicalcium phosphate	18.5	21.0	—	—	0.14
Defluorinated phosphate	18.0	32.0	—	—	0.16
Defluorinated wet-process phosphoric acid	23.7	0.2	—	—	0.18
Defluorinated phosphates manufactured from furnace phosphoric acid					
Monocalcium phosphate	23.0	22.0	—	—	0.03
Dicalcium phosphate	18.5	26.0	—	—	0.05
Tricalcium phosphate	19.5	38.9	—	—	0.05
Monosodium phosphathe (anhydrous)	25.5	—	19.0	—	0.03
Disodium phosphate	21.5	—	32.0	—	0.03
Sodium tripolyphosphate	25.0	—	30.0	—	0.03
Feed-grade phosphoric acid	23.7	—	—	—	0.03
High-fluoride phosphates					
Soft rock phosphate	9.0	17.0	—	—	1.2
Ground rock phosphate	13.0	35.0	—	—	3.7
Ground low-fluorine rock phosphate	14.0	36.0	—	—	0.45
Triple superphosphate	21.0	16.0	—	—	2.0
Wet-process phosphoric acid (undefluorinated)	23.7	0.2	—	—	2.5

[a]Data from Shupe et al. (22).

In general, the soluble phosphates such as sodium phosphate, phosphoric acid, and monocalcium phosphate are approximately equal having the highest biological availability, followed closely by dicalcium phosphate. These are followed by defluorinated phosphate and steamed bone meal, then low-fluorine rock phosphate, and finally soft phosphate. There are exceptions to this general statement as one studies the experimental information available. In general terms, however, the statement is an approximate average of the research information available to date. While this information has been obtained primarily with other

TABLE 6.9

Calcium Analyses of Some Mineral Sources

Ingredient	Calcium (%)
Limestone	38.0
Oyster shell	38.0
Calcite, high grade	34.0
Dolomitic limestone	22.0
Gypsum	22.0
Wood ashes	21.0

animals, it is the best guide to use with the horse until better information is obtained.

J. Phytin Phosphorus Availability

Phytin consists of the vitamin inositol combined with phosphorus and other minerals. Approximately one-half or more of the phosphorus in cereal grains and their by-products, including wheat bran, is in the form of phytin. The availability of phytin phosphorus is influenced by the level of calcium, the calcium to phosphorus ratio, vitamin D, zinc, alimentary tract pH, and other factors.

A Cornell study showed that ponies can use some of the phosphorus in wheat bran, but it was only one-half as available as the phosphorus in the supplements monosodium phosphate, dicalcium phosphate, or bone meal (23). The wheat bran phytin phosphorus availability was about 30% as compared to about 58% for the three inorganic phosphorus supplements used in this trial. Feeding vitamin D3 at a level of 83,000 IU daily (about 60 times the requirement) to the ponies increased phytin phosphorus utilization by about one-third. It is not known how vitamin D increased phytin phosphorus utilization. Such a low level of phytin phosphorus availability in wheat bran needs to be considered in balancing diets for horses. About two-thirds of the phosphorus in wheat bran is in the form of phytin phosphorus. Wheat bran is used extensively in horse diets and contains a high level of phosphorus (1.15%). The availability of the phytin phosphorus in the grains used for the horse is not known. Some of it may be partially available since a Cornell study (23) showed there is some phytase in the equine lower gut. Table 6.10 gives some data on the availability of calcium and phosphorus in some feeds used for horses (9). The low phytin phosphorus availability must be considered as one balances horse diets. One cannot depend on chemical analysis for total phosphorus alone since it gives the total phosphorus in the diet and does not differentiate phytin phosphorus and its availability for the

TABLE 6.10

Availability of Calcium and Phosphorus in Some Common Horse Feeds and Supplements[a]

Primary dietary source[b]	Calcium (%)	Phosphorus (%)
Corn	—	38
Timothy hay	70	42
Alfalfa hay	77	38
Linseed meal	68	45
Milk products	77	57
Wheat bran	—	34
Limestone	67	—
Dicalcium phosphate	73	44
Bone meal	71	46
Monosodium phosphate	—	47

[a]Data from Schryver et al. (9).

[b]In determining availability, the ingredients listed were the primary dietary source of either calcium or phosphorus in the experimental diets. The calcium and phosphorus content of the diets were close to the requirements of the experimental animals.

horse. About 20–50% of the phytin phosphorus is available to the pig (24). Until more research information is obtained, it might be well to consider phytin phosphorus in the grains and their by-products to be available for the horse in about the same range. There are some indications, however, that the horse may be more efficient than the pig in utilizing phytin phosphorus (25). It should also be pointed out that soybean meal, cottonseed meal, sesame meal, and other plant protein supplements also contain phytin phosphorus.

K. Oxalates and Calcium Availability

Calcium availability in feeds varies a great deal. Oxalates and phytates decrease the utilization of calcium (26). Two studies (27,28) demonstrated that the addition of oxalate to horse diets greatly impaired calcium utilization. Certain grasses grown in tropical and subtropical regions (para grass, buffel grass, kikuyu, green panic grass, pangola, setaria, and others) have high levels of oxalates and have caused calcium deficiencies in horses (29–31). Certain molds have high oxalate levels (26). Alfalfa contains oxalates which have been shown to reduce calcium availability in cattle (32) and sheep (33). The Kansas study (32) suggested that 20–33% of the calcium in alfalfa is in the form of oxalate and therefore apparently unavailable to ruminants. They concluded that calcium in alfalfa is only 50–75% as available to cattle as the calcium from inorganic sources. Dr. H. F. Hintz and co-workers at Cornell, however, did not find that the oxalates decreased the calcium availability for horses in the alfalfa samples they studied (34). A previous Cornell study (27) showed that 1% oxalic acid in equine diets reduced calcium absorption by approximately 66%.

One study (30) showed that oxalic acid also decreased phosphorus and magnesium utilization in the horse but a Cornell study (34) did not find this to occur in ponies fed alfalfa.

Grasses are low in calcium whereas alfalfa is high in it. Therefore, the ratio of calcium to oxalate may be important in determining the oxalate effect on calcium utilization. It may explain the difference between the Cornell study, where the horses were fed alfalfa, and the other studies, where the horses were fed grasses. When the calcium to oxalate ratio is 0.5:1.0 or lower, the pasture is considered inadequate (30) in calcium availability.

Total dietary oxalate concentrations of 2.6–4.3% produced a negative calcium balance, during which fecal calcium doubled and urinary calcium decreased in comparison to control horses (28). In another study (30) the same scientists observed similar negative calcium balances in horses fed various tropical grass hays with more than 0.5% total oxalate. They concluded that nutritional secondary hyperparathyroidism may occur when the calcium to oxalate ratio on a weight-to-weight basis is less than 0.5. Cornell workers reported no difference in absorption of calcium from alfalfa containing 0.5 and 0.87% oxalic acid, in which the calcium to oxalate ratios were 3 and 1.7, respectively (34).

L. Effect of Exercise on Calcium Needs

A Cornell study showed that exercise did not appear to change the proportion of calcium intake that was absorbed or retained (35). Urinary calcium decreased markedly and retention of calcium by the body increased during the exercise periods. The increase was due largely to decreased urinary excretion of calcium. The excretion of calcium in the feces was unaffected by exercise. The rate of calcium deposition in bone in horses exercised 10 miles per day increased to 15–20% above the level in resting horses. The rate of calcium deposition was not increased, however, when the horses were exercised only 6 miles per day. The results of this study indicate that bone tissue is more active in terms of calcium turnover during exercise than during rest. However, Cornell scientists state that the experiments were of short duration and left unanswered the question of the effect of exercise on the calcium requirement of the horse. This is the kind of study that is needed to determine the effect of exercise, degree of exercise, frequency of exercise, age of the horse when exercise is initiated, effect of the rider (and his weight), and the effect of temperature, humidity, and many other factors on calcium as well as other mineral and nutrient requirements of the horse. Bone density, breaking strength of the bone, and other criteria of bone integrity and strength are also needed. Until this and other information is obtained, there will continue to be bone problems with the horse. Since strong bone is so important to the value and usefulness of a horse, there is no doubt that information on proper bone formation and maintenance is of the highest priority.

The 1989 NRC report (5) states that any increase in the calcium requirement associated with exercise (work) appears to be readily met by an obligatory increase in calcium intake, as dry matter consumption increases to meet energy requirements.

M. Calcium and Phosphorus Absorption

Cornell studies indicate that the upper part of the small intestine is the major site of calcium absorption (36). The lower portion of the small intestine may also be a site of significant calcium absorption. Relatively little calcium is absorbed from the large intestine. Little or no calcium is absorbed from the cecum.

Phosphorus is absorbed from both the small and large intestine. However, a large amount of phosphate is secreted into the cecum and ventral colon, probably to buffer the volatile fatty acids that are synthesized in the large intestine. Much of the secreted phosphate is reabsorbed from the dorsal and small colon, which appear to be the major effective sites of phosphorus absorption (9, 37).

The true efficiency of calcium absorption is believed to decline with age and to range from as high as 70% in young horses to 50% in mature horses (5). For purposes of estimating calcium requirements in the 1989 NRC report (5), a calcium absorption efficiency of 50% was used for all ages of horse. In the 1989 NRC report (5) a true phosphorus absorption efficiency of 35% was used for all idle horses, gestating mares, and working horses, because they consume primarily plant sources of phosphorus. A value of 45% was used for lactating mares and growing horses because their diets are typically supplemented with inorganic phosphorus.

Certain fatty acids in the feed may form insoluble calcium soaps which decrease calcium absorption. Phosphorus absorption is likely to be higher in foals fed milk than in older horses (5). Table 6.10 shows the availability of calcium and phosphorus in some feeds.

N. Miscellaneous

A Cornell study (38) showed that it is difficult to change the level of calcium in hair by dietary means. Moreover, color of hair, age of animal, site of sample, season of the year, and other factors can influence the calcium level in hair. Therefore, hair analysis does not provide an accurate indicator of calcium status in the horse.

A study in Australia (39) based on urine samples from 229 Thoroughbred racehorses at Melbourne race tracks concluded that 40% of the horses were being fed inadequate levels of calcium.

A Florida study (40) showed that an inadequate protein intake in the weanling horse will limit growth and bone development even when calcium and phosphorus intakes are equal to or above NRC recommendations.

A Maryland study (41) indicated that their observations implied the NRC recommendations are very close to optimal for maximum cannon bone growth and development in well-managed young equines.

A Cornell study (42) showed that calcium concentrations in fecal and urinary samples could be useful in determining the calcium status in the horse but variation among samples could cause any one sample to be misleading. Plasma hydroxyproline could also be helpful but variation among samples causes difficulties in interpretation of the results.

XX. SALT

Salt contains both sodium and chlorine and is also called sodium chloride. The body contains about 0.2% sodium. A German study (43) indicated that about 50% of body sodium is located in the bone but only a small portion is withdrawn during a sodium deficiency. A large amount of sodium is found in fluids outside the body cells. Sodium occurs in considerable amounts in muscles and plays a vital role in muscle contraction. Chlorides play a very important role in the acid–base balance in the blood. Mare's milk contains from 161 to 364 ppm of sodium and 300 to 640 ppm of chlorine. Therefore, an adequate intake of salt is needed for optimum milk production. Salt serves as both a condiment and a nutrient. As a condiment, it stimulate salivary secretion. Saliva contains enzymes important in the digestion of feeds. When the salt intake is low, the body adjusts to conserve its supply. Urine output of sodium and chlorine nearly stops. A high salt intake triggers greater excretion of sodium and chlorine by the kidneys, and water needs are increased to flush out the excess salt. In this way, the body adjusts to a wide range of salt intake so long as ample water is available.

A. Effect of Salt Deficiency

A sodium deficiency reduces the utilization of digested protein and energy. A lack of salt will eventually decrease milk production (44). Without salt, feed is less efficiently utilized. A lack of sodium will ultimately decrease growth rate, skin turgor, and water intake, and eventually a cessation of eating will occur (45). Horses deficient in salt will lick mangers, fences, dirt, rocks, and other objects. They also develop depraved appetites and rough hair coats. Horses with a severe salt deficiency, brought about by considerable sweating, become fatigued and exhausted. Muscle contractions and chewing become uncoordinated, an unsteady gait occurs, and serum potassium increases with a sodium deficiency (45). Horses that sweat profusely react similarly to men in hot climates who perspire excessively in strenuous activity and, consequently, need to take extra salt tablets to avoid fatigue, exhaustion, and even collapse. Oregon State studies (46) showed that turkey poults had significantly greater body weight, bone ash, bone breaking strength, tibia weight, and lower bone abnormality scores as compared to poults fed lower sodium or sodium chloride levels. The study indicated that salt was concerned with bone integrity. Therefore, salt should always be available for horses to supplement any extra needs they may have.

B. Appetite for Salt

Horses have a specific appetite for salt if the diet is deficient in sodium (43, 47). Sodium-depleted steers develop the ability to detect sodium salts by smell. Their central nervous system function is altered to remember earlier locations of sodium salts (48). This may account for the behavioral activities of wild ruminant herbivores in locating and returning to salt sources. The 1984 NRC report on nutrient requirements of beef cattle stated "that minerals lacking in the diet can be provided by self-feeding a salt mineral mixture which is consumed in amounts to satisfy the animals appetite for salt." A dairy study at Cornell (49) showed an indication of a specific appetite for chloride because cows fed a low chloride diet ate nearly three times as much salt from a free-choice block as cows fed a chloride-supplemented diet.

These findings would suggest that self-feeding a salt–mineral mixture which is formulated to the animal's salt needs may be a good way to supply minerals to animals, especially when they are handled under grazing conditions. It must be stressed, however, that not all voluntary consumption of salt is related to the salt requirement of the horse (50). Habit and taste preference could also be involved. Developing mineral mixtures which can be self-fed to horses is a complex matter.

C. Requirements for Salt

The need for salt varies considerably depending on a horse's level of work, riding, or other activity. The horse sweats profusely, and its sweat contains about 0.7% salt. The more a horse exercises the more salt is lost via sweat and the more salt is required. This is one reason salt should be self-fed to horses, even though it is added to the diet.

Studies at Michigan (51) with 12 work horses during an 8-month period (May to December) showed a tremendous variation in average daily salt consumption for each animal (in ounces): 2.54, 3.26, 1.38, 1.69, 1.48, 1.53, 2.59, 2.72, 1.69, 2.09, 0.69, and 0.27. The horse consuming the most salt (3.26 ounces) ate

12 times as much as the lowest consumer (0.27 ounces). The Michigan researchers also measured salt consumption of all 12 horses during the harvest season in the hot months of the year from May to August. Salt consumption averaged 0.64 ounces per horse in May and rose each month to a high of 3.18 ounces per horse in August. Such variations in monthly salt intake and between each horse show why it is difficult to meet salt needs by feeding a specific amount per horse or a fixed percentage in the diet. The study also verifies why salt should be self-fed.

The Mississippi Station (44) kept records on 20 mules used for heavy farm work for a 4-year period. There was a marked difference in salt requirements between individual animals. There was also a marked variation in the amount of salt consumed during different seasons. Their findings are similar to those obtained in Michigan (51) with work horses.

A Cornell study (50) with mature unexercised horses showed that voluntary salt intake among horses was quite variable ranging from 19 to 143 g of salt per day and was inversely related to total salt intake (salt in feeds plus voluntary intake). Average daily salt consumption was 53 g. Metabolic studies using diets containing 1, 3 or 5% sodium chloride showed that urinary excretion was the major excretory pathway for sodium and chloride. Fecal excretion, intestinal absorption, and retention of sodium was not affected by the level of salt fed in this study. Urinary calcium excretion was unaffected by salt intake but calcium and phosphorus absorption and retention were enhanced when ponies were fed diets containing 3 or 5% sodium chloride. Magnesium and copper metabolism were unaffected by salt intake.

The 1978 NRC publication (52) states the following: "Prolonged exercise and elevated temperatures will increase the needs for sodium and chloride, since sweat contains considerable quantities of salt. The requirements have not been determined, but if salt is fed at a rate of 0.5–1.0% of the diet or is ingested free-choice, a deficiency is not likely to occur. Generally, chlorine requirements will be met if sodium needs are supplied by sodium chloride." The 1989 NCR report states that salt is often added to concentrates at rates of 0.5–1.0% or fed free-choice as plain, iodized, or trace-mineralized salt (5). The 1978 NRC publication recommends 0.35% sodium in the total diet of the horse (52). This is equivalent to 0.90% salt (sodium chloride). The 1989 NRC report recommends 0.3% sodium in the diet for working horses and 0.1% sodium for all other classes of horse (5).

Salt requirements increase with perspiration losses. Horses at moderate work can lose 50–60 g of salt in the sweat and 35 g in the urine (53). This is approximately 0.2 lb of salt. Another report (54) indicated that sodium losses are estimated to range from 8.25 to 82.5 g daily. Therefore, under warm or hot climatic conditions, horses need to have an extra daily supply of salt to prevent heat stress.

Table 6.11 gives the 1989 NRC-suggested requirements for sodium (5).

The chloride requirements of horses have not been specifically established (5). Chloride requirements are presumed to be adequate when the sodium requirements are met with sodium chloride (5). A chloride deficiency in horses has not been described, but if it occurred it should result in blood alkalosis because of a compensatory increase in bicarbonate during the chloride deficit (5, 55).

D. Salt Feeding Recommendations

It is recommended that 0.5–1.0% salt be added to horse concentrate feeds. If the concentrate feed is used as a small part of the diet it should contain 1% salt. If it is used as the major part of the diet (such as a complete feed), then 0.5% salt should be added. In all cases, salt should also be self-fed to horses so they can consume additional salt if the level in the diet is not adequate.

E. Effect of Excess Salt

The 1978 NRC publication (52) states "Excessive salt intake may result in high water intake, excessive urine excretion, digestive disturbances; or death from cramps. There is little likelihood that horses will consume toxic amounts of salt unless a salt-starved animal is suddenly exposed to an unlimited amount of salt or high levels of salt are fed without adequate water."

TABLE 6.11

1989 NRC-Suggested Requirements of Minerals for Horses and Ponies (On a Dry Matter Basis)

	Adequate Concentrations in Total Diets				
	Maintenance	Pregnant and lactating mares	Growing horses	Working horses	Maximum tolerance levels
Sodium (%)	0.10	0.10	0.10	0.30	3[a]
Sulfur (%)	0.15	0.15	0.15	0.15	1.25
Iron (mg/kg)	40	50	50	40	1,000
Manganese (mg/kg)	40	40	40	40	1,000
Copper (mg/kg)	10	10	10	10	800
Zinc (mg/kg)	40	40	40	40	500
Selenium (mg/kg)	0.1	0.1	0.1	0.1	2.0
Iodine (mg/kg)	0.1	0.1	0.1	0.1	5.0
Cobalt (mg/kg)	0.1	0.1	0.1	0.1	10

[a]As sodium chloride.

Horse owners can easily prevent this occurrence by not giving salt-starved horses free access to salt until they start leaving some behind in the mineral box and making sure ample water is always available to drink.

F. Salt as Good Carrier for Trace Minerals

Since horses continually need salt, it serves as a dependable and safe carrier for the trace minerals—copper, iron, cobalt, zinc, manganese, iodine, and selenium. Trace minerals are needed in only minute amounts in horse diets. Very few farmers have the facilities and expertise for weighing and mixing the trace minerals in their diets. The horse owner is wise to buy trace mineralized salt, or a trace mineral mixture, rather than try to mix the trace minerals on the farm.

G. Salt Needs Vary

Salt needs are not always the same. They will vary because of different feeding and production situations. Sometimes horses need more than anticipated and sometimes less. Some of the factors that influence salt needs are presented below:

1. The kind of concentrates, pasture, or hay being fed. Considerable differences in salt consumption can occur with different feeds.

2. The level of salt or other minerals in the water being used. These levels can vary considerably between areas.

3. The animal's life cycle stages. Salt needs are different during various stages of the cycle.

4. Genetic differences in animals. The breeding of the animal may affect salt requirements.

5. The growth rate, level of production, reproduction rate, and level of milk production. High producing or rapidly growing animals need more salt. Heavy milk producers need more salt since milk contains a considerable amount of salt.

6. The temperature and/or humidity in the area. Animals in warm areas lose large amounts of salt through sweating. In warm areas, forage seeds and seed products are also low in sodium.

7. The level of potassium in the diet. An excess of potassium can aggravate a deficiency of sodium just as too much sodium can heighten the effects of a potassium deficiency. This can occur when high roughage diets are fed. For example, certain pastures have up to 18 times more potassium than sodium. It has also been shown that forages grown on sandy soils heavily fertilized with potassium have lower forage sodium levels. Recent European studies verify early work showing there is an interrelationship between potassium and sodium (56).

The studies indicate that appetite for salt is related to potassium level and metabolism.

8. The availability of sodium and chlorine in feeds. Unfortunately, there is little experimental information available on this subject.

9. The sodium content of feeds. Recent analyses of feeds for sodium indicate that some of the old sodium values reported in feed tables are too high. Feeds produced today have lower sodium values. This could account for some of the differences in response obtained in different trials on salt needs of animals.

These factors, and others, indicate that salt needs vary between localities and with different feeding and management conditions. Therefore, a trial to ascertain the salt requirements of a horse under one set of conditions should not be used to generalize for all conditions and all localities.

H. Salt in Manure

There is much comment over the environmental effect of salt or salts in manure. The reference should really be made to total mineral salts in the manure and not just sodium chloride (salt).

Many universities have conducted studies which show that adding animal manure to the soil at the proper level has no harmful effect on the soil or crops. Typical are the Michigan studies (57) from 1963 through 1971 which showed that the optimum rate for applying manure to a sandy loam soil was 10 tons per acre yearly. Higher rates posed the hazard of nitrate contamination of ground water and caused a buildup of available nitrogen and potassium. The loam soil handled 10–15 tons of manure per acre annually, without nutrient accumulation, especially when the corn was harvested as silage.

Manure is a valuable resource. Now that more concern is being expressed about environmental problems, research is underway on the benefits of adding manure back to the soil when it is practical to do so. Many other kinds of studies are also being done to find other uses for manure.

Eventually, manure, which heretofore has been regarded primarily as a waste, will become a valuable resource and various means will be found to utilize it. In the meantime, it is recommended that horse producers continue to use the salt level needed in their diets. It is not feasible to have horses deficient in salt because the manure may have an increased salt level. Rather, the horses should be provided their salt needs and, if in a few isolated situations the level in the manure is high, then the manure should be handled in a manner to take care of this problem. As more studies are conducted, methods of making sure that problems are not encountered with salt or other nutrients in the manure will be found.

I. Salt to Control Feed Intake

A 1983 study by H. F. Hintz and co-workers at Cornell University involved studying the effect of salt on limiting the intake of a grain mixture (57a). They used yearling ponies which were fed an amount of grain adequate for 3 days but given to them at one feeding. When the grain mixture contained 16% salt, the ponies ate one-third of the total each day. The intake was distributed equally over the 3-day period. But, when only 8 or 12% salt was added to the grain mixture, the ponies ate the 3-day diet in 2 days or less. This indicated the salt level was not high enough to limit their feed intake over a 3-day period. Their studies showed that the addition of 16% salt to the grain mixture caused a 5-fold increase in water consumption and urine excretion. The additional salt did not cause an increase in blood pressure during the 30-day trial.

The Cornell study indicates that salt may be used to control supplement intake for horses under pasture or range conditions where it is difficult to feed a supplement daily. For the present, however, it is not recommended that high-salt feed supplements be used with racing or high-level performance horses.

The level of salt to use may vary with (1) the size of the horse, (2) the level of feed intake one wants the horse to consume daily, (3) the condition of the horse, (4) the amount and palatability of the other feeds available to the horse, and (5) other factors. Therefore, one might want to modify the 16% level of salt used in the Cornell study to fit the conditions involved. It is of extreme importance, however, to make sure that horses have plenty of water to drink at all times.

XXI. IODINE

A. Introduction

Iodine is important in the animal's diet because it is a constituent of the thyroid hormones produced by the thyroid gland. The iodine which goes to the thyroid is trapped and via complex reactions combines with the amino acid tyrosine forming the thyroid hormones which consist primarily of thyroxine and triiodothyronine. Since thyroxine is the major thyroid hormone produced, for the purposes of this chapter it will be the one referred to. Thyroxine contains approximately two-thirds iodine. It controls the rate of energy metabolism and the level of oxidation in all cells. About 75% of the iodine in the animal's body is the thyroid gland. Older animals rarely show any symptoms of a lack of iodine, but goiters are usually found in the young at birth as a result of a deficiency of iodine in the diet of the mother during the gestation period (Fig. 6.1).

Iodine deficiencies occur in certain areas of the world where soil and water are deficient in iodine. The main deficient areas in the United States are in the Great

Fig. 6.1 Iodine-deficient foal. Foals deficient in iodine are born weak and usually die because of their inability to stand and suckle the mare. (Courtesy of the late J. W. Kalkus, Washington State Agricultural Experimental Station.)

Lakes region and westward to the Pacific Coast. It is possible, however, that other borderline iodine-deficient areas exist in the U.S.

B. Effect of Deficiency

Goiter is an enlargement of the thyroid gland which is located along each side of the trachea in the upper neck area. The enlargement occurs when a lack of iodine causes the thyroid to overwork in an effort to produce more thyroxine or in response to an increased demand by the body for additional thyroxine. When an iodine deficiency occurs and less thyroxine is produced, the anterior pituitary gland releases a thyroid-stimulating hormone (TSH) which causes the thyroid to overwork in an attempt to produce more thyroxine. But, an enlargement of the thyroid gland may be caused by factors other than an iodine deficiency. Feeding excess iodine to the dam, for example, caused an enlarged thyroid in the foal at Cornell University (61). However, this situation seldom occurs. Goitrogens in feeds are capable of producing an enlarged thyroid by interfering with thyroxine synthesis. A high calcium level in the feed or water may cause goiter, especially if the iodine level in the diet is borderline in meeting body needs.

To illustrate the importance of the thyroid, the gland has been removed experimentally shortly after birth in all animal species and its removal causes decreased mental, physical, and sexual development.

The iodine-deficient foal may be stillborn or show extreme weakness at birth which results in difficulty in standing and suckling. Foals will also show labored breathing and rapid pulse rate. Occasionally, a stillborn foal may be hairless. Foals born alive with a well-developed goiter will usually die or the few that live will remain weaklings. No treatment has been found especially effective. Iodine will prevent goiter but may not have much, if any, effect once the goiter is developed. There is some indication that the incidence of "navel ill" in foals may be decreased by feeding iodine to the brood mares, but, this needs more verification.

European studies show a decline in libido (sex drive) and semen quality in stallions lacking in iodine but this also needs more verification. Iodine deficiency in mares may cause abnormal estrus cycles (5, 52, 58). Most mares will not show a goiter even though their foals do. This is also the situation with many other animal species.

C. Iodine Requirement

The recommended iodine requirement for the horse is 0.1 ppm in the total diet (5). This is usually supplied by iodized salt or mineral mixtures containing iodine. If the horse diet contains goitrogenic substances, many horse owners will increase the level of iodine in the diet to 0.2 ppm or slightly higher, depending on the level of goitrogenic activity. But, one must not go too high, since excessive levels of iodine may be detrimental. A level of 3.5–4.8 ppm of iodine in the diet has been shown to be harmful to a small percentage of foals (52, 59–61). Affected foals have enlarged thyroids, may be weak, and have skeletal deformities. But, even this level is 35–48 times the iodine requirement. The 1989 NRC estimates that the maximum amount of iodine that is tolerable in the diet is 5 ppm (5). So, there is a good safety factor between the required level of iodine and that which might cause harmful effects. Most other animals do not show harmful effects from excess iodine until a level of 200–600 ppm of iodine is fed in the total diet.

Table 6.11 gives the iodine requirement of the horse as recommended by the 1989 NRC Committee (5). Table 15.10 gives the suggested iodine level to use in a trace mineralized salt for horses.

D. Sources of Iodine

Iodine is supplied in a number of ways. Potassium iodide is used but requires the addition of stabilizers such as sodium thiosulfate and calcium hydroxide to keep it from being destroyed. Many prefer iodate compounds such as potassium and calcium iodate because they have a higher level of iodine stability and a stabilizer is not needed. Pentacalcium orthoperiodate (PCOP), sodium iodide, EDDI, and others are also used as sources of iodine.

XXII. IRON

A. Introduction

Iron is necessary for hemoglobin formation. Hemoglobin carries oxygen from the air in the lungs to all parts of the body and brings back carbon dioxide on the return trip. Hemoglobin contains about 0.33% iron. All red blood cells contain hemoglobin, and a lack of iron causes a lack of red blood cells which causes anemia. About 60% of the total iron in the body is in the blood hemoglobin. The remainder is in other compounds as well as in some important enzyme systems. Thus, iron is very important and affects every organ and tissue in the body.

Bone marrow provides an important reserve for iron and is one of the last body reserves to be depleted. So, bone marrow serves as an important diagnostic measure of body iron status. The red blood cells which contain hemoglobin are formed in the bone marrow in a process called hematopoiesis. The red blood cells and their hemoglobin are continually being destroyed and replaced. So, iron is constantly needed throughout life and especially during rapid growth when the body is increasing in size and a corresponding increase in total red blood cells and hemoglobin is needed to serve the body properly. When red blood cells are normally destroyed, their iron can be reused to form new ones. But, in certain diseases the iron cannot be reused and additional iron is needed. If the red blood cells are not renewed as rapidly as they are destroyed, anemia results. A measure of the adequacy of iron in the diet can be determined by measuring the level of red blood cells or hemoglobin in the blood. But other factors can also affect this level.

B. Effect of Deficiency

The primary signs of an iron deficiency are microcytic and hypochromic anemia (5). In severe cases of anemia the breathing of the horse becomes heavy and labored. The horse breathes harder in an attempt to make the available hemoglobin carry as much oxygen as possible to the various tissues of the body. An anemic horse, therefore, has difficulty performing any routine task such as walking, working, riding, or performance. It will tire quickly and is unable to do much. The horse becomes weak, inactive, and lacks a healthy pink color. Its blood will look watery if the horse dies and a necropsy is performed. An anemic horse is more susceptible to stress factors and diseases. This causes losses from secondary effects such as reduced growth rate, diarrhea, and pneumonia.

An iron-deficiency anemia may occur when horses are heavily parasitized. The parasites cause bleeding and loss of blood which in turn causes loss of iron. Increased iron requirements should be carefully watched for in areas where parasitism is heavy. This is especially the case where the climate is hot and humid. Parasite infestation is greater on pastures which are too heavily grazed and the horses are forced to consume forages too close to the ground. The parasite problem is still a very serious one since most horses are usually treated for parasites once a month in areas such as Florida and every 2 months in drier areas such as California.

Mare's milk is low in iron (62). Unfortunately, the iron level of the milk cannot be increased by feeding iron to the lactating mare. Therefore, there is the possibility of a low or borderline iron intake in the foal if it depends too heavily on mare's milk for its total feed intake. Fortunately foals will quickly consume other feed and increase their iron intake via that means. Many horse producers supply the foals with a creep feed which usually provides a good source of iron.

C. Requirement

The 1989 NRC recommended 40 ppm of iron in the diet for mature horses and 50 ppm for growing foals and pregnant and lactating mares (5) (Table 6.11). Table 15.10 gives the suggested iron level to use in a trace mineralized salt for horses. An excessive level of iron should be avoided to prevent harmful effects since excess iron is especially toxic to young foals (63). A Florida study (64) showed that adding 1 g of iron (as ferric citrate) per kg of diet did not appear to affect growth or bone development of young ponies, nor did it affect various blood constituents.

D. General Information

Iron is only one of many nutrients needed for hemoglobin formation. Copper, other minerals, vitamins, and protein are also needed. High phytate and phosphorus levels have been shown to bind iron and make it unavailable. The kind of fiber and other carbohydrates in the feeds used can also affect iron utilization. Very high levels of iron can also decrease zinc utilization as well as phosphorus absorption although this would be a rare occurrence. Therefore, many nutrient interrelationships are involved in proper iron utilization. Moreover, the form of iron used is important. For example, iron oxide is virtually unavailable to the horse. Iron sulfate has good iron availability and is used extensively. Some samples of iron carbonate have low iron availability and those who use it run tests on it beforehand to make sure the iron has high availability.

Smith *et al.* (65) showed that serum iron concentration alone cannot be used to evaluate the iron status of the horse. Dietary iron absorption in nonruminants fed adequate iron is likely to be 15% or less (5). Iron utilization increases in iron-deficient diets and diminishes with higher than normal intakes of cobalt, copper, manganese, zinc, and cadmium (66).

XXIII. COPPER

A. Introduction

A trace of copper is necessary to serve as a catalyst before the body can utilize iron for hemoglobin formation. Copper is also needed in many animals for growth, reproduction, bone development, tissue respiration, and skin pigmentation. Whether the same is true for the horse is not yet known, but it may be similar. Copper is a constituent of several enzyme systems in the body and therefore, very important for the horse.

Dr. W. J. Miller of the University of Georgia reported that determining copper requirements is a very complex matter since molybdenum, zinc, sulfate, calcium, iron, cobalt, manganese, selenium, and other minerals and nutrients can affect copper metabolism and needs.

B. Effect of Deficiency

A Texas A & M study by Dr. Charles H. Bridges and co-workers (67) indicated that osteochondritis dissecans in Thoroughbred foals is clinically similar to the lameness that has been associated with a simple copper deficiency. Their data implicated copper as a significant factor in the pathogenesis of osteochondrosis in suckling foals. They also stated that excess zinc may have played a secondary role by reducing the absorption of copper in the foals. A New Zealand study (68) also indicated that a lack of copper may be associated with osteochondrosis and osteodysgenesis. Studies (69, 70) with horses reared in pastures near a zinc smelter showed that they developed lameness and swollen joints. Both studies reported osteochondrosis and osteoporosis in the foals. Some have suggested that the lesions may have been the result of high zinc levels lowering the copper levels in the foals. These studies implied that excessive levels of zinc may increase copper requirement, as it does in the pig.

Another Texas A & M study (71) showed that copper-deficient foals developed intermittent, but nondebilitating diarrhea. All the foals developed stilted gaits and ultimately walked on the front of their hooves.

An Ohio survey (7) of horse farms in Ohio and Kentucky revealed a high incidence of osteochondrosis, epiphysitis, and other skeletal disorders in growing horses. They observed indications of low copper in the forages and feeds, suggesting that the skeletal disorders may be related to deficiencies of copper, zinc, and possibly other minerals. But, in many of their survey sites there were adequate amounts of copper and zinc in the feeds and forages but there was still a high incidence of skeletal disorders in the foals. The Ohio study, and others, indicates that skeletal disorders have a number of causes and are a complex matter. But, it also indicates that one should pay attention to the copper and zinc levels of the diet since they could be involved in certain situations. In another study by Knight *et al.* (72) mares were fed 13 and 32 ppm of copper in the diet and their foals were fed a creep feed containing 15 and 55 ppm of copper. All foals were healthy, grew normally, and showed no signs of lameness or ataxia. Histological lesions were reported for each group, however.

A Texas A & M study (73) showed that a diet containing up to 580 ppm of zinc had no apparent effect on copper absorption. A study at Cornell (74) showed that a level of 1200 ppm of zinc in the diet of weanling pony colts did not show signs of zinc toxicity or copper deficiency nor signs of osteochondrosis. It is apparent, therefore, that the horse can tolerate a fairly high level of zinc. It also appears that copper and zinc are interrelated with many other minerals and nutrients. These interrelationships may account for different results obtained under different conditions. It is apparent too that osteochondrosis and other bone disorders are complex and that nutrition, genetic predisposition, disease, management, and other factors may be involved. So, a research finding under one set of conditions may not apply under other conditions.

It appears that the horse is tolerant to a high level of copper. A Minnesota study (75) showed that feeding copper at a level of 791 ppm for 6 months resulted in copper accumulation in the liver of mares and foals but there was no evidence that the horses were adversely affected. The 1989 NRC report estimated that 800 ppm of copper in the diet is the maximal tolerable level (5).

C. Requirement

The 1989 NRC recommends 10 ppm of copper in the diet as the requirement for the horse (5). French and German scientists also recommend 10 ppm of copper in the diet for all ages of horse, regardless of degree of work or stage of reproduction (5). A Cornell study (76) estimated the copper requirement for maintenance of mature ponies to be 3.5 ppm in the diet. A University of Florida study (77) involved feeding three levels of copper, zinc, iron, and manganese to yearling horses. The horses receiving the highest level of the four trace minerals showed the greatest increase in bone mineral from the start to completion of the 140-day trial. This indicated that the level of one or more of the four trace minerals required may be higher than the 1978 NRC recommendations (52). Another Florida study (78) showed that copper supplementation enhances the level of bone mineralization in growing horses.

A Colorado State University study (79) showed that increasing the zinc level from 24 to 53 ppm and the copper level from 4 to 12 ppm in the diet did not change the milk content of zinc and copper. There was a trend, however, for the foals of mares fed the higher level of zinc and copper to have a greater gain in weight.

A number of studies indicate that forage levels of zinc may be less than 30 ppm in many areas, which is lower than the 1989 NRC requirement of 40 ppm of zinc in the diet of the horse (5). Dr. H. F. Hintz of Cornell University (80) reported that in New York the analysis of more than 5000 samples of forages by the New York dairy forage testing laboratory showed an average of less than 7 ppm of copper. Sometimes the values were lower than 5 ppm. Other studies indicate that the copper level in forages in the U.S. is frequently lower than the 10 ppm that the 1989 NRC recommended (5). Dr. Hintz suggested that the grain mixture for horses should contain 20–25 ppm of copper. But, he also cautioned that this level might not be adequate if only limited amounts of grain are fed or if the forage or environment contains high levels of factors that decrease copper absorption or increase the copper requirement. When horses are fed primarily on forages with little or no concentrate feeding, then the copper, zinc, and other minerals need to be supplied by either trace mineralized salt or a complete horse mineral supplement.

Table 6.11 gives the 1989 NRC recommendations on copper requirements of the horse (5). Table 15.10 gives the suggested level of copper to use in a trace mineralized salt.

D. General Information

Horses are tolerant of fairly high levels of molybdenum; they have not shown any signs of harmful effects on high-molybdenum pastures which severely affect cattle. One study (81), however, showed that a level of 5–25 ppm of molybdenum in forages caused disturbances in copper utilization in horses. Extra copper counteracts the high level of molybdenum in cattle. Sulfur also helps counteract high levels of molybdenum but it has not been studied for the horse. A Cornell University study (82) with ponies showed that the addition of molybdenum to the diet decreased copper absorption and retention as a consequence of increased excretion of dietary copper in the feces and increased excretion of absorbed copper in the bile.

Studies in Ireland (83) showed that fetal livers contained 371 ppm of copper, foal livers 219 ppm of copper, and adult horse livers 31 ppm of copper (all on a dry matter basis). The study indicated that the copper level in the liver decreases with age. A Kentucky study (84) indicated an apparent relationship between low blood serum copper levels and hemorrhaging in aged parturient mares, suggesting either reduced absorption of copper as horses get older or reduced ability to mobilize copper stores.

A study at Michigan State University (85) indicates that copper has a sparing effect on the toxicity of selenium in the horse.

Colostrum is substantially higher in copper than milk and there appears to be a decline in the copper level of milk throughout lactation.

XXIV. ZINC

A. Introduction

Zinc is concerned with many enzyme systems involved in protein, carbohydrate, and lipid metabolism. It is also involved with many hormones and their production and activity in the body. Zinc is necessary for a properly functioning immune system which is needed to decrease the body's susceptibility to infections. Various kinds of stress and disease cause a lower than normal zinc level in the blood plasma. A lack of zinc causes serious consequences in all livestock including the horse.

B. Effects of Deficiency

The symptoms of a zinc deficiency in the horse are very similar to those in the pig, cattle, sheep, and goat. Foals fed a zinc-deficient diet stopped growing between the 6th and 7th week (86). At this time lesions began to appear in the hoof area. These lesions were initially characterized by hair loss and flaking of the dried outer layer of the skin. The lesions extended progressively upward in the legs and by the 90th day were seen on the abdomen (belly) and thorax (chest). In some cases, hair along with portions of the dried outer layer of the skin could be removed in small sheets. In advanced states of the deficiency, areas of the skin devoid of hair on the legs became covered with a rough, crusty layer of serous exudate and desquamated epithelium (this would resemble what one might call a mangelike skin condition). Small, ordinarily insignificant abrasions on the legs healed poorly. Infections of the lower legs occurred occasionally and required treatment with antibiotics (applied on the surface of the leg as well as by injection). Face lesions around the muzzle and nose developed in the horses between the 70th and 80th day of the deficiency. The foals also showed a hair loss, reduced tissue and blood zinc levels, and reduced blood alkaline phosphatase activity. The foals were slaughtered at the end of 90 days and various tissues in the body were studied. Therefore, no information was obtained on the effect of zinc treatment of the deficient animals. This has been done in the pig, however. Zinc-deficient pigs with their body completely covered with lesions caused by the deficiency (which resemble a mangy pig) recovered completely, with the skin and hair returning to normal within 35 days after zinc was added to the diet. In 1 or 2 days after zinc supplementation, one could see changes occurring. Chances are good that the same might happen with the horse, provided that the zinc deficiency is not long standing. Zinc is needed for the development and maintenance of hair and skin. Since horse owners like to have horses with beautiful looking skin and hair, it is essential that zinc levels in the diet be adequate.

C. Requirement

The 1989 NRC publication on nutrient requirements of the horse recommends a level of 40 ppm of zinc in the diet (5). Research in Kentucky (86) and Cornell (3) indicated that a level of 40 and 41 ppm of zinc in the diet, respectively, was adequate under their study conditions. The addition of 5 ppm of zinc to a diet containing 35 ppm of zinc prevented a decrease in blood serum zinc in pregnant mares and a decrease in milk zinc during late lactation (58). These three investigations indicate that 40 ppm of zinc in the diet is near the requirement. A later study with ponies (87) suggested that their maintenance requirement was less than 3–5 ppm of zinc in the diet when it contained about 3 Mcal of digestible energy per kg of dry matter.

Table 6.11 gives the 1989 NRC recommendations on the zinc requirements for the horse. Table 15.10 gives a suggested level of zinc to use in a trace mineralized salt for horses.

Phytin in feeds produced from plants also affects zinc needs. About 70% of the phosphorus in plants is present as phytin phosphorus. About 50% or more of the phosphorus in cereal grains, and their by-products, is in the form of phytin phosphorus. Soybean meal, cottonseed meal, and other plant protein supplements are also high in phytin phosphorus. Unfortunately, zinc combines with phytin and forms zinc phytate which is insoluble in the digestive tract and remains unabsorbed. This was well demonstrated in swine studies which showed that the pig needed 50 ppm of zinc when fed soybean meal protein as compared to 18 ppm of zinc when fed the milk protein casein which contains no phytin phosphorus. Thus, the phytin in grains and their by-products and in soybean meal, cottonseed meal, and other plant protein supplements will increase zinc needs. How much the phytin phosphorus increases the need for zinc in the horse is not known, but it should be taken into consideration as one evaluates zinc requirements for the horse.

Recent studies in Florida, Idaho, and some foreign countries show that zinc deficiencies occur in beef cattle grazing forages containing 20–30 ppm of zinc. Other studies indicate that borderline zinc deficiencies may be more common than previously realized. This raises the question about zinc deficiencies occurring in horses fed primarily on pastures or hay from forages containing less than 40 ppm of zinc which is the 1989 NRC zinc requirement for horses (5).

The grains are low in zinc levels which vary from 10 to 30 ppm depending on whose analytical figures one uses. Even most plant protein supplements vary from 30 to 70 ppm of zinc. Thus, a combination of grains and the protein supplements (which are a minor part of the concentrate mixture) will require zinc supplementation. Most commonly used hays, including alfalfa hay as well as dehydrated alfalfa meal, contain less than 30 ppm of zinc. A few hays contain above 30 ppm of zinc. It appears, therefore, that diets commonly fed to horses need zinc supplementation in order to meet NRC zinc requirements.

D. Zinc Toxicity

The danger of excess zinc is low since the use of 700 ppm of zinc in the diet was not detrimental to mares or their foals (88); excessive levels should, however, be avoided. The 1989 NRC has set 500 ppm of zinc in the diet as the maximum tolerable level. Foals fed 90 g of zinc per day (equivalent to 2% of the diet or a level of 20,000 ppm) developed enlarged epiphyses followed by stiffness, lameness, and increased tissue levels (89).

An Australian study (90) showed that zinc levels above 1100 ppm induced a copper deficiency under field conditions. Foals grazing near a zinc smelter for several months showed signs of zinc toxicity. The pasture being grazed contained 500 ppm of zinc. In another study (70) the foals from pregnant mares grazed near a zinc smelter had generalized osteochondrosis in many joints of the skeleton.

E. General Information

The highest concentration of zinc occurs in the choroid and iris of the eye and in the prostate gland (5). Intermediate concentrations occur in the epidermal tissues, such as skin and hair. Table 6.12 shows data on the zinc concentration in certain body tissues (86). Traces of zinc also occur in the blood, muscles, bones, and in various organs. The colostrum of the mare is rich in zinc and contains 2–3 times the level found in milk (62).

Studies with the pig showed that zinc requirements appear to be higher for the male than for the female. A zinc deficiency was easier to produce on a diet fed in a dry form as compared to a wet mash. The autoclaving, or heating, of a diet increased the availability of zinc in the pig. Excess calcium increases the need for zinc in the pig. If high enough, calcium may double or even triple the zinc requirement. A high copper level depletes zinc stores in the liver and increases zinc needs in the pig. Pigs deficient in iron are more susceptible in zinc toxicity. Studies with other animals indicate that genetics may be involved in zinc requirements since some animals are more seriously affected than others by a deficiency of zinc. There is also considerable variation in the availability of zinc in various feeds depending on stage of maturity, processing methods used, and many other

TABLE 6.12

Zinc Levels in Tissue[a]

	Liver	Kidney	Pancreas	Lung	Brain	Heart	Spleen	Skeletal muscle	Aorta
Control	121.1	104.8	190.3	60.4	50.6	112.6	114.6	154.0	82.8
Control + 40 ppm zinc	198.8	122.1	293.3	76.0	61.6	118.3	119.8	128.8	93.9

[a]All levels in ppm on a dry matter basis. Data from Harrington et al. (86).

factors. It is apparent, therefore, that many factors affect zinc requirements. All of this must be considered as one decides on the level of zinc supplementation to use.

XXV. MANGANESE

A. Introduction

Manganese has some very important functions in the body. It is involved in a number of enzyme systems concerned with protein, fat, and carbohydrate utilization. Manganese is concerned with proper bone development and formation. It is also involved with growth, reproduction, and lactation. So, manganese plays a major role in proper nutrition and animal production.

B. Effect of Deficiency

The effects of a deficiency of manganese in the horse are not known. With animals other than the horse, bone abnormalities such as lameness, enlarged hock, crooked and shortened legs, overknuckling, ataxia, skeletal deformities, slipped tendon, and others have been obtained with a lack of manganese. The bone becomes lower in mineral content, density, and breaking strength and so breaks occur more easily. In pigs, it sometimes takes two generations on a manganese-deficient diet for the leg abnormalities to show up. The bone abnormalities are more apt to occur if the diet is too high in calcium and phosphorus which reduces the amount of manganese that is absorbed. This may be due to the absorption of the manganese by the precipitated calcium phosphate in the intestinal tract before it is eliminated in the feces.

The exact role of manganese in bone formation is still not known. There is evidence, however, that it is concerned with enzyme activity. It has been shown that manganese is needed for the synthesis of chondroitin sulfate which is a major constituent of bone cartilage. Manganese is also needed for the formation of mucopolysaccharides and glycoproteins. A lack of these two products leads to decreased cartilage formation which might be involved in the skeletal abnormalities obtained with many animals. The organic matrix of the bone is composed largely of mucopolysaccharides. Therefore, while all the complex chemistry may not yet be understood, it is definite that proper bone formation requires an adequate level of manganese.

Research studies with other animals have shown that manganese is also concerned with reproduction. In the female, there appear to be three levels or stages of a manganese deficiency. In the least severe deficiency stage, the animal gives birth to live young, some or all of which show incoordination or loss of equilibrium. In the second, or more severe deficiency stage, the young are resorbed, born dead, or die shortly after birth. In the third, or most severe deficiency situation, the estrus cycle is absent or irregular, and the animals do not mate. In the male, sterility and an absence of sexual desire occurs. It is associated with a degeneration of testicle germinal epithelium and a lack of sperm production. Whether these deficiency symptoms occur with the horse is not known. They are given, however, so that horse owners are aware of what might occur with a deficiency of manganese.

C. Requirement

The manganese requirement for the horse is not known but the 1989 NRC (5) stated that, on the basis of data obtained with other species, 40 ppm in the diet should be adequate.

Table 6.11 gives the 1989 NRC requirement for manganese in the horse. Table 15.10 gives a suggested level to use in a trace mineralized salt for horses.

D. General Information

Some scientists feel that the absorption of manganese from the diet is poor in all animals. A University of Pennsylvania report (91) showed that in southern New Jersey the manganese level in the soil, grass, and hay is very low.

Congenitally enlarged joints, twisted lets, and shortened forelimb bones have been associated with "smelter smoke syndrome" in Oklahoma (92). It is thought that the extensive liming required to offset the acidic effects of smelter effluent on the soil markedly reduces the availability of manganese.

XXVI. COBALT

A. Introduction

Vitamin B12 contains about 4% cobalt in its molecule. The only known role played by cobalt is in the synthesis of vitamin B12, which takes place in the digestive tract and which is aided by intestinal microorganisms (93). This means that a deficiency of cobalt is a deficiency of vitamin B12. Horses have remained in good health on pastures so low in cobalt that ruminants confined to them died (5, 52, 94). The horse's requirement for cobalt, therefore, is very low.

B. Effect of Deficiency

No information is available for the horse on the symptoms caused by a deficiency of cobalt. In cattle and sheep, which are grazing animals, a lack of cobalt results in a loss of appetite and weight, weakness, anemia, emaciation, listlessness, and finally death. The anemia is a normocytic, normochromic anemia. The only sure diagnosis of a cobalt deficiency is to give the animal cobalt and determine if it responds. Response is usually rapid with appetite returning within a week. Remission of the anemia, however, will take a longer period of time. All of the cobalt-deficiency symptoms can also be cured with vitamin B12 which is evidence that a cobalt deficiency is actually a vitamin B12 deficiency. Giving the cobalt-deficient animals either cobalt or vitamin B12 may not cure all the symptoms if the deficiency is a long-standing one and irreparable body tissue damage has occurred.

The appearance of a cobalt-deficient animal is similar to that of a starved one. The starvation that occurs from a cobalt deficiency results, at least partially, from the inability of the ruminant to metabolize proprionate, since they lack vitamin B12, which is needed for its utilization by the body. Proprionate is a volatile fatty acid, which is a product of rumen fermentation in ruminants and cecum fermentation in the horse. It is an important source of energy for both the ruminant and the horse. Other volatile fatty acids may require vitamin B12 for their utilization since acetate as well as proprionate (both volatile fatty acids) need vitamin B12 for their utilization in sheep.

It is known that the horse produces volatile fatty acids, especially from forages, in the cecum and that these volatile fatty acids may supply about one-fourth of the energy used by the horse. Therefore, it is reasonable to assume that the horse needs vitamin B12 for the utilization of proprionate and possibly other volatile fatty acids. Since grains, forages, and other plant feeds do not contain vitamin B12, the horse needs to get it from: (1) microorganism synthesis from cobalt in the digestive tract; (2) some animal protein supplement; (3) diet supplementation; or (4) some combination of these three sources. There is no doubt that the horse needs the vitamin for many body functions. The big questions are whether supplementation is needed and what role does cobalt play in meeting the vitamin B12 needs of the horse.

C. Requirement

There is experimental evidence to show that the horse synthesizes vitamin B12 in the digestive tract from cobalt in the diet and that some absorption of the synthesized vitamin B12 occurs. It would appear, therefore, that the horse has a requirement for cobalt even though that level may be quite low. Many people use 0.1 ppm cobalt in the diet or a trace mineralized salt which contains cobalt as a means of ensuring an adequate intake of cobalt for vitamin B12 synthesis by the horse. The still unanswered question is how much of the vitamin B12 needs of the horse is supplied by synthesis in the digestive tract. This is still not known, so many owners with high-level performance horses supplement their diets with vitamin B12 to ensure that this vitamin is not deficient in their animals.

The 1989 NRC publication on nutrient requirements for the horse states that an intake of 0.1 ppm of cobalt in the diet should be adequate for the horse (5). It is doubtful if the horse needs cobalt if the diet contains an adequate level of vitamin B12. To be on the safe side, however, it is best to have the diet adequate in cobalt as well.

Table 6.11 gives the 1989 NRC recommendations on cobalt needs of the horse (5). Table 15.10 gives a suggested level of cobalt to use in a trace mineralized salt.

A condition called polycythemia, caused by excess cobalt, occurs in some animals. Presumably it might also occur in the horse. Polycythemia results in an abnormally high level of red blood cells. This condition should be avoided since it is harmful. A very wide margin exists between the level of cobalt needed to meet nutritional requirements and that which is toxic. Ruminants require from 0.05 to 0.1 ppm of cobalt in the total diet. Harmful effects from excess cobalt may occur at levels of 10–200 ppm. The 1989 NRC (5) has recommended 10 ppm of cobalt in the diet as the maximum tolerable level to use for horses. This is about 100 times the requirement.

D. General Information

Cobalt is normally not stored in significant quantities in the body. What is stored does not easily pass back out into the rumen or into the intestinal tract of the horse where it can be used for vitamin B12 synthesis. As far as is known, there is no synthesis of vitamin B12 from cobalt in the body tissues. It is only synthesized in the digestive tract by microorganisms located there. Therefore, ruminants need a continuous supply of cobalt in the daily diet in order to ensure a continuous supply of vitamin B12 from microorganism synthesis. The same may be true for the horse.

A requirement for cobalt is found only in certain bacteria and algae and they need it to produce vitamin B12-like compounds. Vitamin B12 is not produced by higher plants and animals. Therefore, it appears that only certain microorganisms are vulnerable to a cobalt deficiency.

In animals, the majority of cobalt in the body is excreted via the urine. Small amounts of cobalt are lost by way of the feces, sweat, and hair. The highest concentration of cobalt in the body occurs in the muscle, bone, and kidney, but these levels are very low. Supplementary cobalt is ineffective in raising the vitamin B12 concentration in cow's milk if the diet already contains sufficient cobalt. However, cobalt supplementation of cobalt-deficient or cobalt-marginal diets significantly increases the vitamin B12 content of both the milk and colostrum. Information is not available on what the situation is with the horse.

In dairy cattle, only about 3% of the cobalt in the diet is converted to vitamin B12 which occurs primarily in the rumen, and only 1–3% of the vitamin B12 produced is absorbed by the animal. These are very low levels of vitamin B12 synthesis and absorption. The situation in the horse is not known.

XXVII. SELENIUM

A. Introduction

Selenium is part of the enzyme glutathione peroxidase which aids in detoxification of lipo- and hydrogen peroxides which are toxic to cell membranes. The loss of unsaturated fatty acids by peroxides causes extensive damage to cell membranes and affects their permeability and function, and causes cell wall fragility. So, glutathione peroxidase protects the cells in the body from oxidative stress. Most feeds contain compounds that can form peroxides. Unsaturated fatty acids are a good example. Rancidity in feeds causes formation of peroxides that destroy nutrients. Vitamin E, for example, is easily destroyed by rancidity. Selenium spares vitamin E by its antioxidant role as a constituent of glutathione peroxidase.

There is an interrelationship between selenium and vitamin E. Both are needed by animals and both have metabolic roles in the body in addition to an antioxidant effect. In some instances, vitamin E will substitute in varying degrees for selenium, or vice versa. However, there are deficiency symptoms that respond only to selenium or to vitamin E. Although selenium cannot replace vitamin E in nutrition, it reduces the amount of vitamin E required and the onset of vitamin E deficiency symptoms. Selenium may also have other functions such as in sulfur amino acid synthesis and in increasing immune response in animals (94a).

The sulfur amino acids, methionine and cystine, protect against several diseases associated with low intakes of selenium and vitamin E. It is thought that this is due to their antioxidant activity. So, the sulfur amino acids can spare vitamin E and selenium via their antioxidant role.

Plants do not need selenium and can grow normally and produce optimum yields even though they contain an insufficient amount of selenium to meet animal requirements. Therefore, an excellent looking pasture can be deceiving and result in selenium deficiencies in horses grazing it.

Selenium deficiencies are occurring more frequently throughout the United States and in all areas of the world. Selenium-deficient areas have been found in 44 U.S. states and throughout most of Canada (Fig. 6.2). Moreover, feeds produced in selenium-deficient states are shipped to other states and to many foreign countries. Therefore, selenium deficiencies are likely to occur almost anywhere. Countries that import U.S. grain and soybeans are likely to encounter selenium deficiencies, even though their soils may be adequate in selenium. This will especially be the case if their diets consist largely of imported feeds.

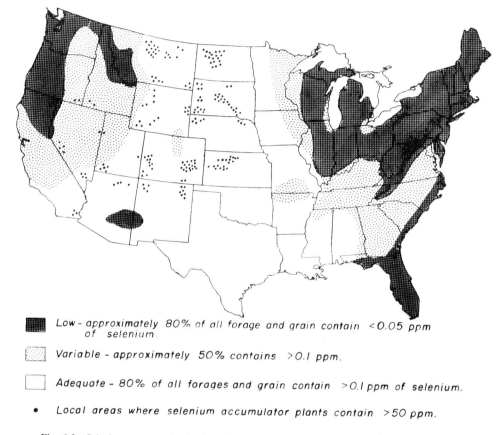

Low - *approximately 80% of all forage and grain contain <0.05 ppm of selenium.*

Variable - *approximately 50% contains >0.1 ppm.*

Adequate - *80% of all forages and grain contain >0.1 ppm of selenium.*

• Local areas where selenium accumulator plants contain >50 ppm.

Fig. 6.2   Selenium concentration in plants in various parts of the United States. The Pacific Northwest, the Northeast, and Florida are characterized by selenium deficiency, while areas of toxic selenium concentrations are found in the Great Plains States. (Courtesy of J. Kubota, U.S. Plant, Soil and Nutrition Laboratory, Ithaca, N.Y.)

B. Effect of Deficiency

Deficiency symptoms associated with a lack of selenium such as white muscle disease, muscular dystrophy, yellow fat disease, and others have been observed in horses. One report (95) indicated that the myopathy in horses results in weakness, impaired locomotion, difficulty in suckling and swallowing, respiratory distress, and impaired cardiac function. Therefore, even though research information for horses is considerably less than for other animals, enough data are available to warrant selenium supplementation under most conditions.

A number of reports show that a lack of selenium causes white muscle disease or muscular dystrophy in foals (96–99). The age of onset may vary from birth or as late as 5–7 months of age (96). The affected foals show muscular stiffness, are disinclined to move, have a stiff gait and difficulty in walking, and may be unable to stand up or follow the mare to nurse. In severe cases the body temperature is elevated and the heart and respiratory rate are accelerated. The foals appear dejected, lie on their sides, and may die within a day, or death may be delayed for a week. Some of the foals may excrete myoglobin, which gives the urine a brownish color. The dead foal may show some hair loss, and, on necropsy, show some small hemorrhages, muscular dystrophy, and yellowish-brown fat. When blood analyses are made, they show a decreased level of blood serum selenium and a higher level of blood serum GOT (glutamic oxalacetic transaminase). A few reports indicate that, when foals show these symptoms, giving the mares selenium prevents further outbreaks. If this is not possible, one report indicates that prevention of muscular dystrophy in foals can be accomplished by giving the foals 1 mg of selenium per day by oral or parenteral supplementation.

Tying-up usually occurs after a few days of rest from heavy work, training, racing, or performance schedule. It causes a lameness and rigidity of the muscles in the loin area. One report (100) stated that 2–5% or more of all horses at the race track may get tying-up disease. It may appear in horses of all ages. Some scientists feel that tying-up is due to excessive accumulation of lactic acid in the muscle. They feel that the oxidation, or breakdown, of the glycogen stored in the muscle can increase the production of large amounts of lactic acid in a relatively short time. A number of studies (100–103) indicate that tying-up may be related to a lack of selenium and/or vitamin E. A single injection of 25 mg of selenium and 250 mg of vitamin E will cure tying-up in most horses, but a second dose may be needed (100). Some horses are cured quickly, whereas others may respond partially or take a longer time to respond. A small percentage of horses may not respond at all. This could be due to irreparable damage having occurred or to the dose involved or when the injection was given during the tying-up syndrome. It is apparent, however, that it is worth trying selenium and vitamin E injections in horses suffering from tying-up. Whether proper levels of selenium and vitamin E in the diet would prevent the tying-up disease is not known. There are a number of reports indicating that the tying-up syndrome does not correlate with vitamin E or selenium status (104–106). This indicates the complexity of this syndrome.

Selenium deficiencies also affect reproduction in horses (107, 108). Such horses have shown signs of heart failure and skeletal muscular dystrophy. In some cases the blood serum selenium levels are low and the serum vitamin E levels appear to be normal. The older horses show anorexia (loss of appetite), emaciation, generalized muscular weakness, tachycardia (rapid heart beat of 80–100 beats per minute), and diarrhea. Enlargement of the liver, kidney, and heart is seen on necropsy. The horses also show an elevation of blood serum GOT.

C. Requirements

The 1989 NRC publication indicates that the selenium requirement of the horse is estimated to be 0.1 ppm in the diet (5). A Washington State University study (109) with 4-year-old ponies fed a dry cull pea and pea straw diet supplemented with vitamins and minerals showed that a level of 0.078 ppm selenium in the diet prevented selenium deficiency signs over a period of 217 days. It appears, on the basis of present knowledge, that a level of 0.1 ppm of selenium in horse diet is adequate (110).

Table 6.11 gives the 1989 NRC requirement for selenium for the horse (5). Table 15.10 gives the selenium level to use in a trace mineralized salt for the horse.

Selenium in excess can be harmful, so one needs to be careful in its use. Selenium can be added at levels of at least 50 times the recommended level (or 5 ppm) before toxic effects appear (52). Therefore, there is a wide safety margin between the level needed and that which may cause harmful effects. The 1989 NRC reported that 2 ppm is the maximal tolerable level of selenium for the horse (5).

D. General Information

There are some areas of seleniferous soils in South Dakota, Wyoming, Montana, North Dakota, Nebraska, Kansas, Colorado, Utah, Arizona, and New Mexico that produce selenium-accumulator plants which are toxic to livestock (Fig. 6.2). The occurrence of the toxic selenium-accumulator plants is most widespread in Wyoming and South Dakota. However, these accumulator plants probably add very little to the selenium content of animal feeds because they normally grow in dry nonagricultural areas according to the 1983 NRC publication on "Selenium in Nutrition." The publication also states that the impact of the selenium-accumulator plants on the livestock industry in the seleniferous areas is small because of the widespread adoption of practical techniques for controlling the problem.

Generally, higher levels of protein, sulfur, and arsenilic acids will partially protect against a toxicity of selenium. Selenium is eliminated rapidly from the body when animals are fed a low-selenium diet. One should avoid using high-level selenium diets, however. The amount of selenium used should be close to the requirement level. In the horse, a toxicity of selenium results in a loss of condition, erosion of the long bones, elongation and cracking of hooves, and a partial or total loss of hair in the mane and tail (111). Acute selenium toxicity (blind staggers) is characterized by apparent blindness, head pressing, perspiration, abdominal pain, colic, diarrhea, increased heart and respiration rates, and lethargy (5, 112). Chronic selenium toxicity (alkali disease) is characterized by alopecia, especially about the mane and tail, as well as cracking of the hooves around the coronary band (5, 112).

The selenium available to plants varies a great deal among soils from different locations. Moreover, the total selenium level in soils does not accurately predict how much of it is available to plants. Almost all of the selenium excreted by animals in the feces and urine is in an insoluble form that is unavailable to plants.

The major selenium compounds in seeds or forages consumed by livestock appear to be selenocystine, selenocysteine, selenomethionine, and selenium-methyl-selenomethionine. Selenium supplements which are added to animal diets are mainly sodium selenite and sodium selenate.

There are considerable differences in the availability of selenium in various feeds. Data in Table 6.13 were recalculated by Dr. G. F. Combs, Jr., of Cornell University from earlier data obtained by Dr. M. L. Scott and others at Cornell.

The data shown in Table 6.13 indicate that feeds of plant origin are much higher in selenium availability than the animal protein source feeds. The selenium in alfalfa meal and brewer's grains has the highest availability. Sodium selenite and sodium selenate have the highest availability of the inorganic selenium compounds.

The data shown in Table 6.13 show that analysis of feeds for total selenium

TABLE 6.13

Biological Availability of Selenium in Feed Sources by Chick Bioassay[a]

Feed source	Percent availability[b]	Plant sources	Percent availability[b]
Plant			
Soybean meal	70	Cottonseed meal	91
Distillers grains and solubles	75	Corn	97
Brewer's grains	87	Brewer's grains	119
Wheat	90	Alfalfa meal	208

Feed source	Percent availability	Inorganic compounds	Percent availability[b]
Animal			
Condensed fish solubles	6	Elemental Se	3
Meat and bone meal	15	$Na_2 Se$	44
Menhaden meal	16	$Na_2 SeO_4$	74
Herring meal	23	$Na_2 SeO_3$	100[b]
Poultry by-product meal	25		
Tuna meal	31		

[a]Data from Drs. M. L. Scott and G. F. Combs of Cornell University.

[b]The availability figures shown in this table are based on giving Na_2SeO_3 a value of 100 and using it as the reference standard.

can be misleading. It means that many diets thought to be adequate in selenium may be deficient when the availability of selenium is considered. This accounts for more selenium deficiencies being detected. Moreover, the use of new highly refined and sensitive chemical methods make it possible to determine very low levels of selenium in feeds and body tissues. This has made it possible to determine the selenium-deficient feeds and areas where they occur and thus more adequately evaluate the selenium status in an area.

XXVIII. POTASSIUM

A. Introduction

Potassium, along with sodium and chloride, occur largely in the fluids and soft tissues of the body. All three minerals function in osmotic pressure and acid–base balance which is very important in the passage of diet nutrients into body cells as well as in body water balance. The body has a continual need for potassium, sodium, and chloride since it has very little storage capacity for them and an excessive intake is rapidly excreted. For optimum health and normal operation of the body system, the pH of the blood and other body fluids must be maintained within a narrow range. Fortunately, the body has a very effective buffering system which maintains the blood pH within normal levels even when feeds react as either alkaline or acid. Urine excretion of excessive alkaline or acid products is one means of accomplishing this. But, there are limits to how much the acid–base balance can vary before it starts affecting the performance of the animal. Unfortunately, there is still too little known about acid–base balance and methods to keep it in the best equilibrium for optimum health and performance of the horse.

In addition, potassium has many other important functions in the body. It activates and functions as a cofactor in a number of enzyme systems which include energy transfer and utilization, oxygen and carbon dioxide transport in blood, protein synthesis, and carbohydrate metabolism. It is also involved in nerve and muscle function. Muscle tissue has the highest concentration (56%) of potassium in the body.

B. Effect of Deficiency

A summary of potassium deficiency symptoms which occur in all animals are as follows: decreased growth rate, decreased feed intake, muscular weakness, stiffness and paralysis, heart abnormalities such as cardiac arrhythmias, intra-cellular acidosis, degeneration of vital body organs, and nervous disorder. Only a limited amount of potassium research has been conducted with the horse. A study at the University of Kentucky (113) showed that young foals deficient in potassium had a reduced feed intake, decreased rate of growth, unthrifty appearance, and hypokalemia (decreased blood serum level of potassium).

A study with ponies at Cornell University (114) showed that the small intestine appears to be the greatest site of potassium absorption but a significant amount is also absorbed from the large intestine. The kidney was the primary pathway of potassium excretion.

C. Requirement

A Cornell University study (115) showed that the maintenance requirement of the horse for potassium was 48 mg per kg of body weight. This is equivalent to about 0.4% of potassium in the diet. This requirement does not include sweat and dermal (skin) losses plus increased needs for performance. They stated that the total potassium requirement of the horse is probably less than 0.6% of the diet. Until more data become available, they indicated that the use of 0.6% potassium in the diet appears reasonable. But, they also stated that the need for potassium for top level performing horses may be even higher because of the extra potassium they lose in the sweat and the fact there are reports that their blood potassium levels can be quite low. In a study at the University of Kentucky (116) it was recommended that foals (1–3-months old) fed a purified diet should have 1.0% potassium in the diet in order to have normal levels of various blood constituents.

Table 6.5 gives the 1989 NRC recommendations on the potassium requirement of the horse (5). The potassium requirement level varies from 0.3 to 0.43% in the diet. The highest requirement (0.43%) occurs during intense work and with lactating mares (0.42%) from foaling to 3 months of lactation.

D. General Information

Many factors can increase potassium needs. Vomiting or diarrhea or loss of electrolytes during dysentery, scours or hauling to various locations may increase potassium needs. Stress conditions such as weather, hard work or performance, diseases, injury, sweating, high water intake and other factors can cause potassium depletion and increased needs.

The effect of hot weather in increasing potassium needs of high producing dairy cows has been shown recently by University of Florida and Texas A & M scientists. The University of Florida studies showed that high producing dairy cows also needed higher levels of potassium and sodium than previously recommended in order to milk to their full genetic potential. This is reasonable to expect since potassium is the most abundant mineral element in milk (1.5%) and sodium is also contained in milk at a high level. These studies with dairy cows should be considered as one evaluates the potassium needs of the horse. High-level performance horses are certainly exposed to hot weather and heat stress during performance and training. Moreover, they sweat considerably more than other classes of livestock, and significant amounts of potassium are lost in sweat. The high milk producing mare should also need adequate potassium intake in order to milk to her full genetic potential. This discussion indicates the need to obtain more information on the potassium needs of the horse.

At the University of Florida, a potassium deficiency with high producing dairy cows resulted in dramatic reductions in feed and water intake, milk yield, and blood potassium concentrations within 3–5 days after feeding a potassium-deficient diet. A near-complete loss of appetite and vigor, a craving for unnatural feeds (pica), and death of three cows occurred. The rapid deficiency effects show the lack of potassium storage in the body. Giving the deficient cows potassium reversed the condition within 12–24 hours. If the stress of lactation is combined with heat stress, then the potassium requirement appears to be greater. High humidity would further increase sweating at high temperatures. These studies indicate the importance of a continual adequate supply of potassium in horse diets. A periodic decrease in potassium intake because of diet inadequacy would adversely affect optimum health and performance in the horse.

The blood serum concentration of potassium does not always reflect the proper potassium balance. Much body potassium can be lost before the serum level decreases. There are some reasons for this. The main one is that potassium may shift from inside the body cells to the fluid outside the cells and thus keep up the level of potassium in the blood serum. But, the animal's body system could be short of potassium.

Research at the University of Nebraska showed that dry winter ranges result in forages very low in potassium (117). Some of the forages contained as little as 0.09–0.10% potassium. We need to know if this is the case in other areas with dry range forages.

A recent cooperative study between the Tennessee and Texas Agricultural Experiment Stations showed that increasing the potassium concentration from 1 to 1.5% in the diet of steers that were stressed by shipping resulted in an increased weight gain of 11 lb per head in the first 28 days in a feedlot in Texas. Potassium is concerned with water balance in the body and considerable body water loss occurs during shipping cattle. This study should be followed by others to verify the role of potassium in helping to minimize shipping stress and losses with horses.

While the research data cited above are with cattle grazing dry range forages low in potassium, this finding should alert horse owners and research scientists to determine whether or not horses might be affected similarly if grazing such forages or if consuming hay made from the forages.

Cornell scientists found that an increased intake of potassium did not influence the digestibility or retention of calcium, phosphorus, or magnesium (118). A high level of calcium (3.4% in the diet) did not influence potassium retention or digestibility. They studied three levels of magnesium to determine its effect on potassium digestibility. The levels of magnesium were 0.16, 0.31, and 0.81% of the diet. The highest level of magnesium (0.81%) increased potassium absorption. The reason for the increased absorption is not known. The Cornell scientists stated that it appears unlikely that high levels of calcium or magnesium would induce a potassium deficiency in the horse. There have been no reports of potassium toxicity in the horse (25) but excessive levels should be avoided.

There seems to be an antagonism between sodium and potassium. Excess sodium increases potassium needs by causing more of it to be excreted by the kidneys. Excess potassium likewise has the same effect on sodium excretion. This theory is believed to be true, but some scientists question whether it always occurs. Pasture grass sometimes contains 18 times as much potassium as sodium. If this ratio of potassium to sodium causes excess sodium excretion, it emphasizes the need to make sure that the horse gets enough sodium.

Potassium loss accompanies persistent diarrhea. Young animals with diarrhea develop acidosis and a potassium deficit more rapidly than mature animals. Their metabolic rate is greater and their kidneys do not conserve potassium as well.

XXIX. MAGNESIUM

A. Introduction

It is estimated that about 60–70% of the magnesium in the body is present in the bone. Bone contains about 0.75% magnesium. Magnesium is closely associated with calcium and phosphorus and serves as a constituent of bone. Very little is known, however, as to the actual function of magnesium in bone. It must have some function since a lack of magnesium in the diet causes weak pasterns in the pig and leg weakness in some other animals.

About 30–35% of the magnesium in the body occurs in the fluids and soft tissues where magnesium has a vital role in certain enzyme systems. It activates enzymes which transfer phosphate to various compounds and thus influences most, if not all, metabolic processes in the body. So, magnesium is important in almost all functions in the body.

B. Effect of Deficiency

Symptoms of a deficiency of magnesium include nervousness, muscular tremors, ataxia followed by collapse, with hyperpnea, (hard breathing), sweating, convulsive paddling of legs and, in some cases, death (119, 120). A Kentucky study with foals showed degeneration in the lung, spleen, skeletal muscle, heart muscle, and certain areas of the heart aorta (121, 122). Skeletal muscle degeneration was consistently found in all the foals fed the magnesium-deficient diet for 71 days or longer, although it was not extensive. In this study, a sharp decrease in blood serum magnesium levels was detected within 24–48 hours after the foals were given the magnesium-deficient diet. The University of Kentucky study (121, 122) showed that the amount of magnesium which can be mobilized from bone depends on the age of the horse and the rate of bone growth. Magnesium availability from bone decreases with age and may be as low as 2% in adult animals. There are also differences in the sensitivity of various bones to depletion of their magnesium level. The Kentucky study with horses showed that the rate of reduction of bone magnesium varied not only between different bones but between different areas of the same bone. A number of studies with different animals show that the young animal can mobilize about a third of the magnesium from bone during periods of inadequate magnesium intake. But, mature animals in response to a magnesium deficiency frequently show little or no depletion of bone magnesium. Present information, therefore, would indicate that bone is not a reliable source of magnesium if the animal diet lacks this mineral during certain periods.

C. Requirement

Cornell studies showed that the horse requires about 13 mg/kg bodyweight daily of magnesium for maintenance (123) and the pony requires 12.8 mg/kg bodyweight for the same purpose. The 1989 NRC (5) recommends that 0.09 to 0.13% magnesium in the diet (dry matter basis) for all classes of horses (Table 6.5). On a 90% dry matter basis, the 1989 NRC report (5) recommends from 0.08 to 0.12% magnesium for all classes of horses (Table 6.6).

D. General Information

Grass tetany is increasing in frequency throughout the United States. Just a few years ago, tetany was thought to be a problem only in a few areas. Under certain conditions, the magnesium in forages decreases in availability to a very low level for ruminants. Sometimes only 5% of the magnesium in the forage is available. Whether this is so for the horse is not known. However, since horses consume about 50% of their total feed as forage, it is important to be on the lookout for this possibility occurring. The 1973 NRC publication (53) on nutrient requirements of the horse stated: "Outbreaks of tetany that respond to magnesium therapy have been reported from humid grassland areas. The addition of 5% magnesium oxide to the salt mixture has been a helpful protective measure." The

1989 NRC states that pastures that are conducive to magnesium deficiency, tetany, and death in ruminants do not affect horses similarly.

Studies by Cornell University (123) showed that the main site of magnesium absorption in the horse is the small intestine. The lower half of the small intestine appears to be more effective in absorbing magnesium than the upper half. Small amounts of magnesium may also be absorbed from the large intestine. About 5% of the magnesium absorption occurred in the cecum, dorsal colon, and small colon.

Diets high in phosphorus decrease magnesium absorption. Horses fed a diet with 20% milk products absorbed more magnesium than horses fed linseed meal. This would indicate that milk products may enhance magnesium absorption or that the magnesium of milk products is more available than that of linseed meal. The addition of lysine to linseed meal did not improve magnesium absorption.

Magnesium is excreted in both the feces and the urine, but the major part is excreted in the feces. An excessive intake of calcium will increase the excretion of magnesium in the urine. Therefore, excess calcium in the diet or water will increase magnesium requirement.

Excess potassium may also increase magnesium needs. This is due to a decreased absorption or increased excretion of magnesium or both. It appears, therefore, that excess intake of calcium and phosphorus (and possibly potassium) can increase the requirements for magnesium. It has also been shown that a moderate excess of magnesium, in a mineral supplement, feed, or water, will not markedly disturb calcium retention, although it may tend to slightly increase the requirements for calcium and phosphorus in the diet. Thus, it appears that a balance between calcium, phosphorus, and magnesium is necessary.

Magnesium oxide has about 53% magnesium, whereas magnesium carbonate has 28.8% magnesium; magnesium chloride has 12.0% magnesium, and magnesium sulfate has 9.9% magnesium. So, magnesium oxide has the highest percentage of magnesium. The form of magnesium to use would depend on the way it is to be used, its availability for the horse, and the cost per unit of magnesium.

There are no controlled studies on the effect of excess magnesium on the horse. Although the maximum tolerable level of magnesium in the diet for the horse was estimated to be 0.3% (from data on other species) by NRC (124) some alfalfa hays with magnesium levels of 0.5% have been fed to horses without apparent ill effects in Cornell studies (125).

XXX. MOLYBDENUM

A. Introduction

Molybdenum is an essential nutrient since it is a constituent of the enzyme xanthine oxidase as well as several other enzymes. The level of molybdenum required is very low, which makes it difficult to demonstrate a need for it in horse diets. Most of the concern with molybdenum is when it occurs at excessive levels in the soil and feed.

Molybdenum occurs in soils, plant, and animal tissues. The molybdenum level of forages used for grazing can vary a great deal and is affected by soil molybdenum level, soil pH, and season of the year, as well as by contamination by industrial and mine wastes.

The highest and the most variable level of molybdenum in feeds fed to animals occurs in legumes and their seeds. Usually, the molybdenum level of mixed pastures varies according to the proportion of legumes in the forage. The level of molybdenum in many forages increases steadily during the growing season. It is known that alkaline soils increase molybdenum availability but decrease copper availability to the plant. Therefore, alkaline soils increase the severity of the problem in areas of excess molybdenum. Soil wetness in poorly drained soils may also increase the molybdenum level in the forages produced there. Forages usually contain levels of 3–5 ppm of molybdenum which is considered about normal. In molybdenum areas of excess, however, forages may contain 20–100 ppm of molybdenum and sometimes as high as 200 ppm. With ruminants, a copper deficiency may occur when forages have a copper level below 5 ppm and the molybdenum levels goes above the 3–5 ppm level. In areas of excess molybdenum one can usually counteract the excess molybdenum by increasing the copper level in the diet by 2–3 times the required level with ruminants. But, sometimes higher copper levels are needed, depending on the many interrelationships of molybdenum–copper–sulfates–other nutrients–type of soil and the kind of diet being fed. It is a complex matter with no exact answers yet.

B. Effects of Deficiency

There are species differences in the response to excess molybdenum. Cattle are the least tolerant, followed by sheep, whereas horses and pigs are the most tolerant farm animals. The high tolerance of horses is apparent since they have not shown symptoms of molybdenum toxicity on pastures that severely affected cattle. But, foals have been reported to develop rickets when their dams were maintained on high-molybdenum pastures (81). This could be due to the high molybdenum level causing some disturbance in phosphorus metabolism, which has occurred in cattle and sheep in some areas, or it could be due to a lack of copper or to a lack of both copper and phosphorus.

Symptoms obtained with cattle due to molybdenum toxicity include the following: weight loss; scouring; disturbance in phosphorus metabolism giving rise to lameness; joint abnormalities; osteoporosis; high serum phosphate levels; difficulties in conceiving; lack of libido in males; and damage to the testes and spermatogenesis. Connective tissue changes and some spontaneous bone fractures have also been observed in sheep on high-molybdenum pastures.

Since cattle and sheep suffer bone abnormalities with high molybdenum intake levels, it is not surprising that foals have been reported to suffer from rickets when their dams were on high-molybdenum pastures. It appears that copper has a role in proper bone metabolism. Dr. E. A. Ott, at the University of Florida, showed that copper enhances bone mineralization in growing horses (78). Therefore, even though the males did not exhibit apparent harmful effects from grazing high-molybdenum pastures (which were very harmful to cattle) their foals showed the effects.

It is known that the molybdenum content in the milk of cows, ewes, and goats reflects the level of molybdenum in the diet. It is not known if the mare's milk would also reflect the diet molybdenum level. But, if it did, it may account for the foals showing rickets when consuming both the mare's milk and forage both of which were high in molybdenum. It has also been reported that molybdenum accumulates in the equine liver and that it may cause rickets in foals grazing pastures containing 5–25 ppm of molybdenum (81). It may be logical to assume, therefore, that even though mares may not show apparent harmful effects when grazing high-molybdenum forages that the foals may be adversely affected. This possibility should be explored in high-molybdenum areas.

C. Requirement

There is no known requirement for molybdenum in horse diets.

It appears that the adequacy of copper in the diet of mares and foals grazing high-molybdenum pastures needs study. The 1989 NRC recommends a level of 10 ppm of copper in the diet of horses (5). This level might be increased somewhat to counteract the high-molybdenum forage. Copper should not be increased more than 2–3 times the normal requirement for horses. Since horses appear to be more tolerant to excess molybdenum than cattle, doubling the copper requirement might be adequate to counteract reasonable levels of excess molybdenum in the forage. Studies at Cornell (82) showed that the addition of molybdenum to the diet of ponies decreased copper absorption and retention and caused increased dietary copper in the feces and increased excretion of absorbed copper in the bile. Therefore, copper needs were increased by molybdenum in the diet. Urinary excretion of molybdenum by ponies effectively removed most of the absorbed molybdenum (81). Doubling, or even tripling, the copper level in the diet should not be harmful since the horse has been fed 791 ppm of copper in the diet with no apparent adverse effect (75). But, one should not feed any more copper than necessary to counteract the excess molybdenum level. It is always prudent to avoid feeding nutrients much above required levels.

D. General Information

Under normal conditions, the liver contains about 4 ppm of molybdenum. This level can increase to 25–30 ppm with excess molybdenum intake. The level in the liver returns to normal when excess molybdenum intake ceases. If the excess molybdenum intake is prolonged, a depletion of tissue copper levels occurs and a deficiency of copper results. If the sulfate intake is high, it helps counteract the effect of high molybdenum levels in the forage. The degree of molybdenum absorption and retention in the body tissues is decreased by the sulfates. This means that if the copper intake of the diet is about normal, the sulfates can counteract a slight excess of molybdenum. The high levels of molybdenum in the liver decline rapidly if the molybdenum intake decreases and the copper and sulfate in the diet is increased. There are limits, however, to the extent that sulfur sources decrease the adverse effects of increased levels of molybdenum intake. In some animals, at a certain level, the sulfide sulfur formed within the intestinal tract or in body tissues may actually exacerbate the effects of high molybdenum. This further complicates the molybdenum–copper–sulfate interrelationship. But, with few exceptions, the sulfate ion, whether it originates in the diet or from sulfur amino acid (cystine and methionine) breakdown in the body, limits molybdenum uptake, increases its excretion, and therefore increases the animal's tolerance to a high molybdenum level in the diet.

Understanding molybdenum and its effect on animals is still a very complex matter. There is the interrelationship between molybdenum, copper, and sulfur. But, dietary protein, zinc, iron, vitamin E, vitamin C, lead, and tungstate have all been reported to affect the level of molybdenum in tissues.

It should be noted that not all manifestations of molybdenum toxicity are due to a lack of copper. In certain situations, and with different species of animals, symptoms occur which are not typical of a copper deficiency. This may be due to the many interrelationships of other nutrients besides the molybdenum–copper–sulfate relationship.

XXXI. SULFUR

No information is available on the sulfur needs of the horse. Sulfur is an important constituent in the sulfur-containing amino acids methionine and cystine. It is also present in insulin and in the vitamins, thiamin and biotin, as well as in heparin and chondroitin sulfate. If the diet is adequate in protein, it will usually provide at least 0.15% sulfur. According to the 1989 NRC, the 0.15% sulfur level in the diet appears to be adequate (5). Table 6.11 gives the 1989 NRC recommendation on sulfur requirement for the horse (5). One report (126) noted harmful effects from excess sulfur accidentally fed to 5–12-year-old horses at a level of 200–400 g of flowers of sulfur which contains over 99% sulfur. Two of the 12 horses died following convulsions.

XXXII. FLUORINE

A. Introduction

Many scientists classify fluorine as an essential mineral element since it is beneficial to teeth and decreases tooth decay. The big problem encountered with livestock is not a deficiency of fluorine but rather an excess.

B. Effect of Excess Fluorine

Utah State University scientists (127) reported that horses with moderate to marked fluorosis appeared unthrifty even when they had an ample supply of good-quality feed (Fig. 6.3). The hair coat was rough and dry in appearance, and the horse's winter coat was slow to shed in the spring. The skin became taut and

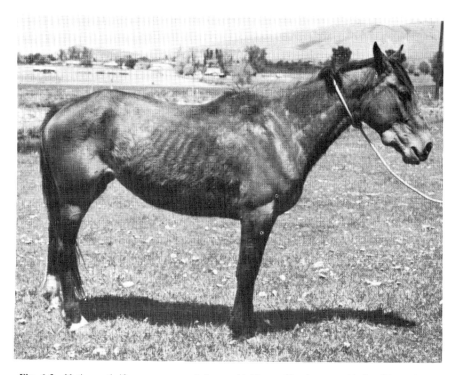

Fig. 6.3 Notice unthrifty appearance of 5-year-old Thoroughbred mare with fluoride toxicosis. Enlargement on mandible is due to abscessation. (Courtesy of James L. Shupe, Utah State Agricultural Experimental Station.)

was less pliable than normal. When tooth abrasion and wear became excessive, feed utilization became poor and "slobbering" of poorly masticated feed was common. As the deficiency increased and horses developed marked clinical fluorosis, they became lame and were unable to walk, run, or jump normally (Fig. 6.4). They took shortened steps and, then, only reluctantly. The horses' lameness frequently intensified with use, exercise, or work. Even when the affected horses were not worked, there was apparent pain, since these animals often stood with their feet in unnatural positions with one forefoot placed in front of the other. Moreover, the horses shifted the position of their feet frequently as if

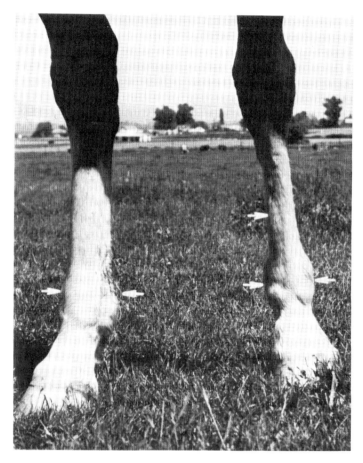

Fig. 6.4 Fluorosis in the hind legs of a horse. Notice fluoride-induced enlargements (arrows) around fetlock joints and on the medial aspects of metatarsals. (Courtesy of James L. Shupe, Utah State Agricultural Experimental Station.)

they were trying to relieve the pain. Many of the horses examined had severe dental fluorosis with excessive wear of the teeth and abrasion (Fig. 6.5). The teeth were irregular and uneven. These irregularities resulted in poor chewing of the feed as well as occasional biting and injury to the surface of the cheek and gum. As the teeth became more affected, some of the teeth broke down and allowed feed material to be forced through the hole in the teeth and into the pulp cavity. This material formed abscesses. These horses then developed a lumpy jaw appearance (Fig. 6.6).

Bones are affected by excessive fluorine intake. The jaw bones often become thickened. Bones in the nose often thicken and cause a "Roman nose" appearance. In severe chronic fluorine cases, all bones are affected to some degree. The bones appear chalky white with a rough, irregular surface and are thicker than normal bones.

Fluorine has a remarkable affinity for deposition in the bone. The amount of fluorine in the bone can increase to a certain level without causing any apparent

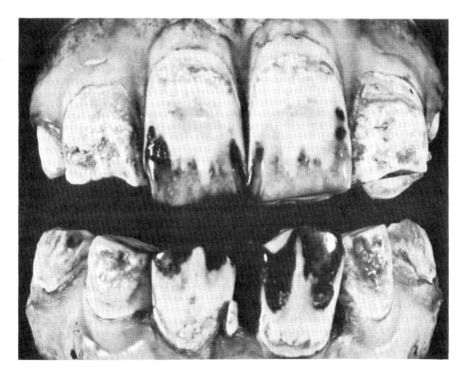

Fig. 6.5 Notice abnormal discoloration, abrasion, and pitting of enamel of permanent incisor teeth from the 5-year-old mare with fluoride toxicosis shown in Fig. 6.4. (Courtesy of James L. Shupe, Utah State Agricultural Experimental Station.)

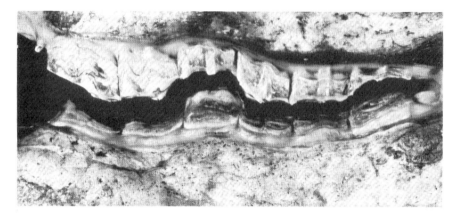

Fig. 6.6 Notice irregular and excessive abrasion in permanent cheek teeth from 5-year-old mare with fluoride toxicosis shown in Fig. 6.4. (Courtesy of James L. Shupe, Utah State Agricultural Experimental Station.)

changes in structure and function. However, once a certain level of fluorine deposition occurs, toxic effects take place and are more apparent.

Fluorosis in horses can be diagnosed clinically by well-qualified persons. If excess fluorine intake occurs during the period of tooth formation, the teeth will show a mottled enamel. This condition is characterized by chalky-white patches on the surface of the teeth. The teeth may become stained and show a discoloration which varies from yellow to black. This mottling is primarily a defect of the permanent teeth, which occurs during their formation. However, if the permanent teeth are already in, this mottling does not occur. This can be seen when older horses, with their permanent teeth, are brought into an area or situation where excess fluorine intake occurs.

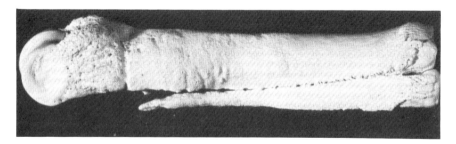

Fig. 6.7 Medial aspect of right 3rd and 2nd metacarpal bones of a 2.5-year-old Thoroughbred horse. Note extensive fluoride-induced periosteal hyperostosis. (Courtesy of James L. Shupe, Utah State Agricultural Experimental Station.)

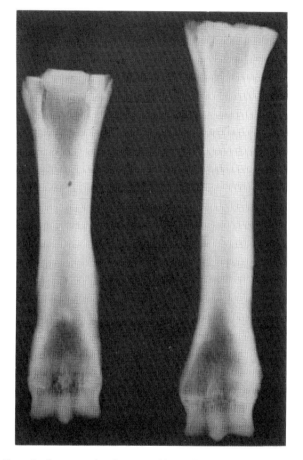

Fig. 6.8   Radiograph of metacarpal and metatarsal bones from a 2.5-year-old Thoroughbred horse with fluoride toxicosis. Note thickened cortex with irregular periosteal surface and abnormal trabecular pattern. (Courtesy of James L.Shupe, Utah State Agricultural Experimental Station.)

The studies by Utah State University (127) show that lesions of fluorosis develop in horses in a similar way to cattle, sheep, and other animals. The teeth and bones are the major sites of specific lesions. The fluorine gradually accumulates in the bones and teeth, although no harm is immediately apparent. Approximately 99% of the excess fluoride retained in the body is stored in the bone. This deposition of fluorine in the bone is nature's way of protecting the soft body tissues against excessive fluorine deposition. However, once the bones become saturated with fluorine, the unabsorbed fluorine is free to produce its toxic effects on the various organs and soft tissues in the body. Growth, reproduction, lactation, and the well-being of the animal are interfered with.

C. Safe Level of Fluorides

The safe level of fluoride intake is shown in Table 6.14. These are the levels recommended by the 1974 NRC report on "Effect of Fluorides in Animals" (22). The values in Table 6.14 are presented as ppm in diet dry matter. In the first column, the values reflect the use of a soluble fluoride such as NaF. In the second column, it is assumed the fluoride in the diet is present as some form of defluorinated rock phosphate. When this is the case, the fluoride tolerance level is increased by 50%. This is because the fluoride in NaF is more readily absorbed and sometimes is twice as toxic as the fluoride in rock phosphate. Since NaF is not always twice as toxic, a figure of 50% increased toxicity is used.

The fluoride tolerance levels shown in Table 6.14 assume that fluoride is being continually ingested by the animal. It is not known what the tolerance level

TABLE 6.14

Dietary Fluoride Tolerances for Domestic Animals[a]

Animal	Performance tolerance level[b]		Pathology tolerance level[c]	
	ppm of F as soluble fluoride such as NaF	ppm of F as some form of defluorinated rock phosphate	ppm of F as soluble fluoride such as NaF	ppm of F as some form of defluorinated rock phosphate
Beef or dairy heifers	40	60	30	45
Mature beef or dairy cattle[d]	50	75	40	60
Finishing cattle	100	150	NA[e]	NA
Feeder lambs	150	225	ID[e]	ID
Breeding ewes	60	90	ID	ID
Horses	60	90	40	60
Finishing pigs	150	225	NA	ID
Breeding sows	150	225	100	150
Growing or broiler chickens	300	450	ID	ID
Laying or breeding hens	400	600	ID	ID
Turkeys[f]	400	600	ID	ID
Growing dogs	100	150	50	75

[a]Data from Shupe et al. (22).
[b]Levels that, on the basis of published data, can be fed without clinical interference with normal performance.
[c]At this level of fluoride intake pathological changes occur. The effects of these changes on performance are not fully known.
[d]Cattle are first exposed to this level at 3 years of age or older.
[e]NA, Not applicable; ID, insufficient data.
[f]This level is safe for growing turkey females. Very limited data suggest the tolerance for growing male turkeys may be lower.

would be if an intermittent intake of fluoride or varying levels were consumed periodically.

The majority of feed grade phosphates that originate from rock phosphate deposits have fluoride levels ranging from 2 to 5% and average about 3.5%. In the United States, a phosphate that is to be classified as defluorinated phosphate must contain no more than 1 part of fluorine to 100 parts of phosphorus (see Table 6.8).

The information shown in Table 6.8 can be used as a guide in determining the calcium and phosphorus levels of various phosphorus sources for horse feeding. Moreover, the fluorine levels can be used as a guide to determine the safety of using certain phosphate sources in horse feeding. A level of up to 60 ppm of fluorine in the total diet of the horse has been established as safe by the NRC Committee on Fluorine for Animals (22). Although evidence is not available, the committee also stated that "it is felt that Thoroughbreds, Quarter horses and other breeds being trained and developed for racing at an early age, during the period of rapid bone growth, may have a lower tolerance for fluoride in the diet." The 1989 NRC states that horses can tolerate 50 ppm of fluorine in the diet for extended periods without detrimental effects (5). This level has been fed to horses for extended periods by Utah State University scientists (127) without any detrimental effects. Horses evidently are more tolerant of excess fluorine than cattle.

XXXIII. SELF-FEEDING MINERALS

Many horse people do not self-feed minerals. They feel that using a feed that has all the minerals needed by the horse means there is no need to self-feed extra minerals. This is the best way to meet mineral requirements. But, not all horses eat the same amount of feed, have the same mineral requirements, and are fed the same level and quality of pasture and/or hay. Horses vary in their mineral requirements depending on their heredity or breeding, their rate of growth, level of exercise or work, reproduction rate, level of milk production, level of total feed intake, quality of the hay or pasture consumed, level of minerals in the water, and how much they drink of it, plus many other factors. Therefore, there is considerable variation in the minerals needed by individual horses. This means a diet may not supply all the minerals needed by every horse under all conditions. The concentrate mixture (or feed that has the minerals added) can come close to meeting the mineral needs of many horses if it is properly developed for a particular situation. There is considerable variation in the level of concentrate mixture fed to horses. Some are fed very little or no concentrates at various periods of the year. Moreover, the concentrate mixtures are fed with a wide variety of pastures and/or hay in various stages of maturity. Some horses will be on grass, others on grass–legume pastures, and some on just an exercise area (since the pasture contains very little or no vegetation). Moreover, there is a large variation in the amount and quality of hay fed. Therefore, the mineral content and intake from pasture and hay varies a great deal. Since pasture and/or hay usually make up at least one-half of the total feed intake of horses, it is apparent their affect on the level of minerals needed in the concentrate mixture is substantial.

Salt needs, for example, will depend to some extent on how much a horse is worked or exercised. A horse worked moderately can lose 50–60 g of salt in the sweat and 35 g in the urine. Sweat contains about 0.7% salt. Unless this lost salt is replaced, the horse will soon show signs of fatigue or overheating just as the human does. This is why it is so important to self-feed salt to horses so they can consume extra salt if they need it. In a Michigan study (51), it was found that the horse consuming the most salt ate 12 times as much salt as the lowest consumer.

See Sections XIXG1 and XXB for problems involved in developing complete mineral mixtures to be self-fed to horses. It is a complex matter since the mineral mixture must be formulated to meet approximate needs and must not be under- or overconsumed by the horse.

There is some objection to the use of a mineral box because of the possibility of injury if a horse runs into it, although experience has shown that this is a rare occurrence. This possibility can be eliminated by placing the mineral box in a corner of the pasture, pen, or stall.

It is preferable to use a plastic rather than a wooden mineral box to preclude the possibility of head injury. The added benefits in better mineral nutrition and superior performance and reproduction would more than pay for any cost or inconvenience involved with self-feeding minerals to horses.

XXXIV. WHY HORSES EAT DIRT OR CHEW WOOD

It is known that animals that are short of minerals will eat dirt and chew wood and other objects. Therefore, the first step is to prevent this from occurring and to make sure the diet is adequate in minerals. However, the addition of the known essential mineral elements to the diet will not always solve the problem.

For example, pigs on well-balanced diets will chew each other's tails. Some producers have cut their pigs' tails off at birth. When this is done, some of the pigs start chewing each other's ears and other parts of the body. No one definite remedy will eliminate this problem. Giving the pig more space, placing a tire or some other object in the pen for it to play with, and other management improvements will minimize the problem.

Some horses chew each other's tails when they are fed only a completely pelleted diet. This may be due to a lack of enough pasture or hay. It could be due to a lack of some minerals, or it could be due to some other factors.

All classes of livestock have some animals that chew wood. Some of it may be due to boredom. Even under the best of nutrition and management conditions, chewing wood and eating dirt occurs with some animals (see Fig. 6.9). Why it happens is not known. Maybe there are some minerals lacking that we do not think are needed by animals. Maybe some animals have higher nutritional requirements than others, while others may have just developed bad habits. We know that soil has some factor (or factors) that benefits the pig fed all the essential mineral elements. Therefore, we need to keep an open mind on the possible need for minerals, or other factors, we think do not need to be added to livestock diets at the present time. Reports that cadmium, chromium, molybdenum, vanadium, tin, and other new trace mineral elements that have metabolic roles in the body have appeared. They may be involved in the problem of wood chewing and eating dirt under some conditions. At least we should eliminate that possibility. There is no definite answer to wood chewing and eating dirt, but making sure horses have an adequate mineral intake and a well-balanced diet is very important. One also needs to look at housing and management. Housing is

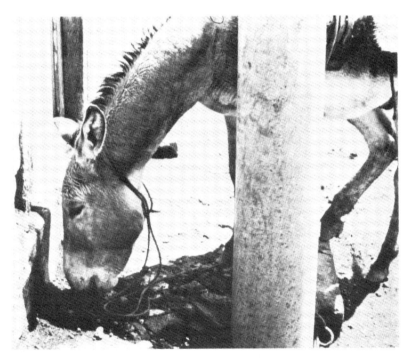

Fig. 6.9 Note the burro eating dirt. Sometimes this is due to a lack of minerals in the ration. (Courtesy of Lee McDowell and Joe H. Conrad, University of Florida and INIAP, Pichilingue, Ecuador.)

usually based on our preferences, without taking into consideration a horse's likes and dislikes. Maybe part of the problem is that the horse is bored or dissatisfied with its housing and management. Until this is taken into consideration, wood chewing, eating dirt, breeding problems, and handling problems may persist.

XXXV. HAIR ANALYSIS AS NUTRITIONAL INDICATOR

Horse owners and trainers are interested in whether hair analyses can be used as an indicator of the nutritional adequacy of the horse. In fact, many are routinely obtaining blood and hair analyses periodically throughout the year.

As one considers this matter there are a number of questions that need to be answered. The first is where to take the hair sample from the horse's body. This should be studied to determine if it is the same area for the various nutrients. It may be that certain parts of the body are better for different nutrients. Also, any difference in analysis of hair that is regrowing, after being clipped, compared to hair that has never been clipped should be studied. The best indicator might be hair that is regrowing. The age of the animal may also influence the level of minerals in hair.

Another unanswered question is how to treat the hair that is being clipped for analysis. Should it be brushed to take out the dirt and sweat that has accumulated on it? If so, how should this be done to get the best hair sample for analysis. One might want to wash the hair. If this is done, how should one do it to prevent the nutrients from being washed out of the hair? Which solvents should be used? These questions are important if one is to get a hair analysis that represents what the hair itself contains. One does not want to analyze the dirt and sweat in the hair since it can vary greatly, nor does one want to lose nutrients from the hair by washing it.

There may be differences in hair analyses during various seasons of the year. This could be due to changes in environmental temperature or to hormonal changes that occur during the year. The effect of the season on hair analysis has already been shown for certain nutrients (128). Therefore, one has to follow a regular system of analysis with at least one (and possibly more) analysis during each season of the year. This means one cannot take just one hair sample, analyze it, and expect to have a good evaluation of the nutritional status of the horse.

There will be differences in results obtained in hair analyses between animals. This has been shown to be the case particularly for calcium and phosphorus. Therefore, results obtained with other animals cannot be relied upon to be the same for horses, but should be used primarily as a guide. Hair color may affect certain minerals in hair. Cornell showed that colored hair from pinto ponies had higher levels of calcium than did white hair taken from the same ponies (129). But, iron, copper, and molybdenum concentrations were similar in both colors of hair.

Horse owners and trainers are interested in hair analyses because it would save having to obtain blood samples. They prefer not to use a needle in their horses unless they have to. If hair analyses are similar to blood analyses, they could sometimes be used as an indicator of the nutritional status. It seems logical to assume that the horses blood supply is the main source of the nutrients that are deposited in the hair. Blood analyses sometimes will not show a nutrient as being low in the diet until a severe deficiency exists. Therefore, if the blood supply of a nutrient is adequate, it may also be adequate in the hair, since the hair obtains these nutrients from the blood. This discussion does not imply that hair analyses should not be studied—they should. Hair analyses may be a good indicator for certain nutrients. This discussion should emphasize, however, that much needs to be learned about this type of analytical work. It should be done properly if it is to be of value to horse owners interested in using it as a guide on the nutritional status of their horses.

One study at Louisiana (130) indicated it was doubtful whether hair analyses could be used to determine the calcium and phosphorus status of the horse. Another study at Virginia Polytechnic Institute (128) involved studies on phosphorus supplementation of horses. Their analytical work on hair showed that phosphorus supplementation increased the phosphorus level in the hair of the weanlings but not in the mares.

A Cornell study (38) showed that excess molybdenum increased hair molybdenum concentration, but did not influence the level of copper and iron. They were not able to make any conclusions concerning the effect of excess molybdenum on hair calcium or phosphorus.

G. J. Fosmire, Pennsylvania State University (129), stated that "hair analysis is not only inaccurate but also misleading in the nutritional assessment of a given individual human, and the technique should not be used as a routine measure for assessing nutritional status."

It is evident that much more needs to be learned before hair analysis can be used as a reliable indicator of nutritional status in the horse.

XXXVI. CONCLUSIONS

As soils decline in fertility, new minerals may need to be added to horse diets. Looks can be deceiving as to the nutritional adequacy of a beautiful looking pasture. Plants do not need selenium, iodine, and cobalt, and sodium is essential only for certain plants. Moreover, certain plants grow normally and produce optimum yields even though the level of these four and other essential minerals are not adequate for the horse. Declining soil fertility, increased horse productivity, and intensified and confined conditions increase the need to supply adequate minerals in the diet.

Mineral interrelationships, availability, chemical and physical form of minerals, mineral processing methods, and acid–base balance affect the mineral needs of the horse. Mineral requirements are a very complex matter and much remains to be learned about the these requirements in the horse.

The 14 essential mineral elements are calcium, phosphorus, potassium, magnesium, sulfur, sodium, chlorine, copper, iron, iodine, manganese, zinc, cobalt, and selenium. Deficiency signs, requirements, their role in body metabolism, and general information are provided on each essential mineral element for the horse.

Bone is not static once it is formed. There is a continuous interchange of calcium and phosphorus between the bone, the blood supply, and other parts of the body. So one needs to provide the horse with its mineral needs continually throughout its lifetime. Bone consists not only of calcium and phosphorus but other minerals, protein, vitamins, and other nutrients. Therefore, sound bone requires a well-balanced diet.

Excess molybdenum in forages may be a problem with foals whose dams graze these pastures. Extra copper supplementation may counteract the excess molybdenum but more information is needed. Excess fluorine in the diet needs to be avoided since it is harmful to teeth, bones, and various organs and tissues of the body. Safe levels of fluorine and molybdenum are recommended.

Formulating complete mineral mixture for self-feeding to horses is a complex and challenging area. The mixtures must be designed to meet approximate mineral needs so that they are not under- or overconsumed by the horse.

Horses eat dirt and chew wood. Some factors causing it are known but others are not.

Very little is known regarding the use of hair analyses as an indicator of the adequacy of nutrients in the diet. More needs to be learned before hair analysis can be used as a reliable indicator of nutrient adequacy.

REFERENCES

1. Conrad, J. H., C. Y. Deyoe, L. E. Harris, P. W. Moe, R. L. Preston, and P. J. Van Soest. "U.S.-Canadian Tables of Feed Composition." NAS-NRC, Washington, D. C., 1982.
2. Topliff, D. R., M. A. Kennerly, D. W. Freeman, R. G. Teeter, and D. G. Wagner. *Proc. Equine Nutr. Physiol. Symp., 11th,* p. 1 (1989)
3. Schryver, H. F., H. F. Hintz, J. E. Lowe, R. L. Hintz, R. B. Barber, and J. T. Reid. *J. Nutr.* **104,** 126 (1974).
4. El Shorafa, W. M., J. P. Feaster, and E. A. Ott. *J. Anim. Sci.* **49,** 979 (1979).

5. Ott, E. A., J. P. Baker, H. F. Hintz, G. D. Potter, H. D. Stowe, and D. E. Ullrey, "Nutrient Requirements of Horses." NAS-NRC, Washington, D. C., 1989.
6. Krook, L., and J. E. Lowe. *Pathol. Vet.* **1,** Suppl. 1, 1 (1964).
7. Knight, D. A., A. A. Gabel, S. M. Reed, R. M. Embertson, W. J. Tyznik, and L. R. Bramlage. *Proc. 31st Annu. Conv. Am. Assoc. Equine Pract.* p. 445 (1985).
8. Cymbaluk, N. F., and G. I. Christison. *J. Anim. Sci.* **67,** 951 (1989).
9. Schryver, H. F., H. F. Hintz, and J. E. Lowe. *Cornell Vet.* **64,** 493 (1974).
10. Jordon, R. M., V. S. Myers, B. Yoho, and F. A. Spurrell. *J. Anim. Sci.* **40,** 78 (1975).
11. Schryver, H. F., H. F. Hintz, and P. H . Craig, *J. Nutr.* **101,** 1257 (1971).
12. Schryver, H. F., P. H. Craig, and H. F. Hintz. *J. Nutr.* **100,** 955 (1970).
13. Schryver, H. F., H. F. Hintz, and P. H. Craig. *J. Nutr.* **101,** 259 (1971).
14. Krook, L. *Cornell Vet.* **58,** 60 (1968).
15. Schryver, H. F., H. F. Hintz, and P. H. Craig. *J. Nutr.* **101,** 1257 (1971).
16. Voges, F., E. Kienzle, and H. Meyer. *Proc. Equine Nutr. Physiol. Symp., 11th,* p. 139 (1989).
17. Glade, M. J., D. Beller, J. Bergen, D. Berry, E. Blonder, J. Bradley, M. Cupelo, and J. Dallas. *Nutr. Rep. Int.* **31,** 649 (1985).
18. Schryver, H. F., D. W. Meakim, J. E. Lowe, J. Williams, L. V. Soderholm, and H. F. Hintz. *Equine Vet. J.* **19,** 280 (1987).
19. Schryver, H. F., S. Van Wie, P. Daniluk, and H. F. Hintz. *J. Equine Med. Surg.* **2,** 337 (1978).
20. Ott, E. A. *Proc. 30th Annu. Pfizer Res. Conf.* p. 101 (1982).
21. Sunde, M. L., J. R. Couch, L. S. Jensen, B. E. March, E. C. Naber, L. M. Potter, and P. E. Waibel. "Nutrient Requirements of Poultry." *N.A.S.–N.R.C.,* Washington, D.C. (1973).
22. Shupe, J. L., C. B. Ammerman, H. T. Peeler, L. Singer and J. W. Suttie. "Effects of Fluoride in Animals." *N.A.S.–N.R.C.,* Washington, D.C. (1974).
23. Hintz, H. F., A. J. Williams, J. Rogoff, and H. F. Schryver. *J. Anim. Sci.* **36,** 52 (1973).
24. Cunha, T. J., J. P. Bowland, J. H. Conrad, V. W. Hays, R. J. Meade, and H. S. Teague. "Nutrient Requirements of Swine." *N.A.S.–N.R.C.,* Washington, D.C. (1973).
25. Evans, J. W., A. Borton, H. F. Hintz, and L. D. Van Vleck. "The Horse." Freeman, San Francisco, California, 1977.
26. Hintz, H. F. *Feed Manage.* **36**(8), 18 (1985).
27. Swartzman, J. A., H. F. Hintz, and H. F. Schryver. *Am. J. Vet. Res.* **39,** 1621 (1978).
28. McKenzie, R. A., B. J. Blaney, and R. J. W. Gartner. *J. Agric. Sci.* **97,** 1621 (1981).
29. Walthall, J. C., and R. A. McKenzie. *Aust. Vet. J.* **52,** 11 (1978).
30. Blaney, B. J., R. J. W. Gartner, and R. A. McKenzie, *J. Agric. Sci.* **97,** 507 (1981).
31. Gartner, R. J. W., B. J. Blaney, and R. A. McKenzie. *J. Agric. Sci.* **97,** 581 (1981).
32. Ward, G. M., L. M. Harbers, and J. J. Blaha. *J. Dairy Sci.* **62,** 715 (1979).
33. Ward, G. M., L. H. Harbers, and A. J. Kahno. *Nutr. Rep. Int.* **26,** 1123 (1982).
34. Hintz, H. F., H. F. Schryver, J. Doty, C. Lakin, and R. A. Zimmerman. *J. Anim. Sci.* **58,** 939 (1984).
35. Schryver, H. F., H. F. Hintz, and J. E. Lowe. *Am. J. Vet. Res.* **39,** 245 (1978).
36. Schryver, H. F., P. H. Craig, H. F. Hintz, D. E. Hogue, and J. E. Lowe. *J. Nutr.* **100,** 1127 (1970).
37. Schryver, H. F., H. F. Hintz, P. H. Craig, D. E. Hogue, and J. E. Lowe. *J. Nutr.* **102,** 143 (1972).
38. Cape, L., and H. F. Hintz. *Am. J. Vet. Res.* **43,** 1132 (1982).
39. Caple, I. W., J. M. Bourke, and P. G. Ellis. *Aust. Vet. J.* **58,** 132 (1982).
40. Ott, E. A., and R. L. Asquith. *Fla. Agric. Exp. Stn., Res. Rep.* **A1-1982-2** (1982).
41. Glade, M. J., N. K. Luba, and H. F. Schryver. *J. Anim. Sci.* **63,** 1432 (1986).
42. Hintz, H. F., H. F. Schryver, S. Brown, and J. Williams. *Proc. Equine Nutr. Physiol. Symp., 11th,* p. 126 (1989).

43. Meyer, H., A. Linder, M. Schmidt, and E. Teleb. *Proc. Equine Nutr. Physiol. Symp., 8th*, p. 16 (1983).
44. Templeton, G. S., *Miss., Agric. Exp. Stn., Bull.* **270** (1949).
45. Meyer, H., M. Schmidt, A. Lindner, and M. Pferdekamp. *Z. Tierphysiol., Tierernaehr. Futtermittelkd.* **51,** 182 (1984).
46. Egwuatu, C. O., J. A. Harper, D. H. Helfer, and G. H. Arscott. *Poult. Sci.* **62,** 353 (1983).
47. Ralston, S. L. *J. Anim. Sci.* **59,** 1354 (1984).
48. Bell, F. R. *J. Anim. Sci.* **59,** 1369 (1984).
49. Coppock, C. E., R. A. Aguirre, L. E. Chase, G. B. Lake, E. A. Oltenccu, R. E. McDowell, M. J. Fettman, and M. E. Woods. *J. Dairy Sci.* **62,** 723 (1979).
50. Schryver, H. F., M. T. Parker, P. D. Daniluk, K. I. Pagan, J. Williams, L. V. Soderholm, and H. F. Hintz. *Cornell Vet.* **77,** 122 (1987).
51. Hudson, R. S. *Mich. Agric. Exp. Stn., Q. Bull.* **8,** 8 (1926).
52. Hintz, H. F., J. P. Baker, R. M. Jordon, E. A. Ott, G. D. Potter, and L. M. Slade. "Nutrient Requirements of Horses." *N.A.S.–N.R.C.*, Washington, D.C. (1978).
53. Tyznik, W. J., P. V. Fonnesbeck, H. F. Hintz, E. A. Ott, W. H. Pfander, and M. C. Stillions. "Nutrient Requirements of Horses." *N.A.S.–N.R.C.*, Washington, D.C. (1973).
54. Meyer, H. *In* "Equine Exercise Physiology" (J. R. Gillespie and W. E. Robinson, eds.), ICEEP Publ., Vol. 2, p. 650. University of California, Davis, California, 1987.
55. Tasker, J. B. *In* "Clinical Biochemistry of Domestic Animals" (J. J. Kaneko, ed.), 3rd ed., p. 425. Academic Press, New York, 1980.
56. Mitchell, A. R. *Br. Vet. J.* **128,** 76 (1972).
57. Anderson, E. D., *Farm Q., Fall Ed.* p. 44 (1972).
57a. Parker, M. I. M. S. Thesis. Cornell University, 1984.
58. Kruzkova, E. *Tr. Vses. Inst. Konevodstvo* **2,** 28 (1968); cited in *Nutr. Abstr. Rev.* **39,** 807 (1968).
59. Baker, H. J., and J. R. Lindsey. *J. Am. Vet. Med. Assoc.* **153,** 1618 (1968).
60. Drew, B., W. P. Barber, and D. G. Williams. *Vet. Res.* **97,** 93 (1975).
61. Driscoll, J., H. F. Hintz, and H. F. Schryver. *J. Am. Vet. Med. Assoc.* **173,** 858 (1978).
62. Ullrey, D. E., W. T. Ely, and R. L. Covert. *J. Anim. Sci.* **38,** 1276 (1974).
63. Mullaney, T., and C. Brown. *Equine Vet. J.* **20,** 119 (1988).
64. Lawrence, L. A., E. A. Ott, R. L. Asquith, and G. J. Miller. *Proc. Equine Nutr. Physiol. Symp., 10th*, p. 563 (1987).
65. Smith, J. E., J. E. Cypriano, R. DeBowes, and K. Moore. *J. Am. Vet. Med. Assoc.* **188,** 285 (1986).
66. Underwood, E. J. "Trace Elements in Human and Animal Nutrition." Academic Press, New York, 1977.
67. Bridges, C. H., J. E. Womack, E. D. Harris, and W. L. Scrutchfield. *J. Am. Vet. Med. Assoc.* **185,** 173 (1984).
68. Carberry, J. T. *N. Z. Vet. J.* **26,** 280 (1984).
69. Haskam, E. G., G. J. Graaf, and H. J. Over. *Tijdschr. Diergeneeskd.* **107,** 672 (1982).
70. Kowalczyk, D. F., D. E. Gienson, C. R. Shoop, and G. C. F. Ramber. *J. Environ. Res.* **40,** 285 (1986).
71. Bridges, C. H., and E. D. Harris. *J. Am. Vet. Med. Assoc.* **193,** 215 (1988).
72. Knight, D. A., S. E. Weisbrade, L. M. Schmall, and A. A. Gavel. *Proc. 33rd Annu. Conv. Am. Assoc. Equine Pract.* p. 191 (1988).
73. Young, J. K., G. D. Potter, L. W. Greene, S. P. Webb, J. W. Evans, and G. W. Webb. *Proc. Equine Nutr. Physiol. Symp., 10th*, p. 153 (1987).
74. Coger, L. S., H. F. Hintz, H. F. Schryver, and J. E. Lowe. *Proc. Equine Nutr. Physiol. Symp., 10th*, p. 173 (1987).

75. Smith, J. D., R. M. Jordon, and M. L. Nelson. *J. Anim. Sci.* **41,** 1645 (1975).
76. Cymbaluk, N. F., H. F. Schryver, and H. F. Hintz. *J. Nutr.* **111,** 87 (1981).
77. Ott, E. A., and R. L. Asquith. *Proc. Equine Nutr. Physiol. Symp., 10th,* p. 185 (1987).
78. Thomas, M. L., E. A. Ott, J. D. Pagan, P. W. Poulos, and C. B. Ammerman. *Proc. Equine Nutr. Physiol. Symp., 10th,* p. 165 (1987).
79. Baucus, K. L., S. L. Ralston, G. Rich, and E. L. Squires. *Proc. Equine Nutr. Physiol. Symp., 10th,* p. 179 (1987).
80. Hintz, H. F. *Feed Manage.* **36**(3), 50 (1985).
81. Walsh, T., and L. B. O'Moore. *Nature (London)* **171,** 1166 (1953).
82. Cymbaluk, N. F., H. F. Schryver, H. F. Hintz, D. F. Smith, and J. E. Lowe. *J. Nutr.* **111,** 96 (1981).
83. Egan, D. A., and M. P. Murrin. *Res. Vet. Sci.* **15,** 147 (1973).
84. Stowe, H. D. *J. Nutr.* **95,** 179 (1968).
85. Stowe, H. D. *Am. J. Vet. Res.* **41,** 1925 (1980).
86. Harrington, D. D., J. Walsh, and V. White, *Proc. Equine Nutr. Physiol. Symp., 3rd,* p. 51 (1973).
87. Schryver, H. F., H. F. Hintz, and J. E. Lowe. *J. Anim. Sci.* **51,** 896 (1980).
88. Graham, J., J. Sampson, and H. R. Hester. *J. Am. Vet. Med. Assoc.* **97,** 41 (1940).
89. Willoughby, R. A., A. E. MacDonald, and B. J. McSherry. *Am. J. Vet. Res.* 91, 382 (1972).
90. Eamens, G. J., J. F. Macadam, and E. A. Laing. *Aust. Vet. J.* **61,** 205 (1984).
91. Donoghue, S. *Proc. 26th Annu. Conv. Am. Assoc. Equine Pract.* p. 65 (1980).
92. Cowgill, V. M., S. J. States, and J. E. Marburger. *Environ. Pollut., Ser. A* **22,** 259 (1980).
93. Alexander, F., and M. E. Davies. *Br. Vet. J.* **125,** 169 (1969).
94. Filmer, J. F. *Aust. Vet. J.* **9,** 163 (1933).
94a. Knight, D. A., and W. J. Tyznik. *Proc. Equine Nutr. Physiol. Symp., 9th,* p. 136 (1985).
95. Dill, S. G., and W. C. Rebkun. *Compend. Contin. Educ. Pract. Vet.* **7,** S627.
96. Dodd, D. C., A. A. Blakely, R. S. Thornbury, and H. F. Dewes. *N. Z. Vet. J.* **8,** 45 (1960).
97. Hartley, W. J., and A. B. Grant. *Fed. Proc., Fed. Am. Soc. Exp. Biol.* **20,** 678 (1961).
98. Schougaard, H., A. Basse, G. C. Nielsen, and M. G. Simesen. *Nord. Veterinaermed.* **24,** 67 (1972).
99. Wilson, T. M., H. A. Morrison, N. C. Palmer, G. G. Finley, and A. A. Van Dreumel. *J. Am. Vet. Med. Assoc.* **169,** 213 (1976).
100. Hill, H. E. *Mod. Vet. Pract.* **43,** 66 (1962).
101. Hartly, W. J., and A. B. Grant. *Fed. Proc., Fed. Am. Soc. Exp. Biol.* **20,** 679 (1961).
102. Stewart, J. M. *N. Z. Vet. J.* **8,** 82 (1960).
103. Smith, H. *Horse Rider* **25**(6), 62 (1986).
104. Lindholm, A., and A. Ashein. *Acta Agric. Scand., Suppl.* **19** (1971).
105. Gallagher, D., and H. D. Stowe. *Am. J. Vet. Res.* **41,** 1333 (1980).
106. Blackmore, D. J., C. Campbell, D. Cant, J. E. Holden, and J. E. Kent. *Equine Vet. J.* **14,** 139 (1982).
107. Weinandy, D. *Mod. Vet. Pract.* **66,** 895 (1985).
108. Heimann, E. D., L. W. Garrett, W. Loch, and W. H. Pfander. *Proc. Equine Nutr. Physiol. Symp., 7th,* p. 62 (1981).
109. Nygard, K. R., R. J. Johnson, J. A. Froseth, and R. C. Piper. *J. Anim. Sci.* **42,** 1559 (1976).
110. Stowe, H. D. *J. Nutr.* **93,** 60 (1967).
111. Moxon, A. L. *Tech. Bull.—S. D., Agric. Exp. Stn.* p. 311 (1937).
112. Rosenfeld, I., and O. A. Beath. "Selenium: Geobotony, Biochemistry, Toxicity and Nutrition." Academic Press, New York, 1964.
113. Stowe, H. D. *J. Nutr.* **101,** 629 (1971).
114. Hintz, H. F., and H. F. Schryver. *J. Anim. Sci.* **42,** 637 (1976).

115. Hintz, H. F., and H. F. Schryver. *J. Anim. Sci.* **42,** 637 (1976).
116. Stowe, H. D. *J. Nutr.* 101, 629 (1971).
117. Karn, J. F., and D. C. Clanton. *J. Anim. Sci.* **45,** 1426 (1977).
118. Hintz, H. F., and H. F. Schryver. *J. Anim. Sci.* **37,** 927 (1973).
119. Harrington, D. D. *Am. J. Vet. Res.* **35,** 503 (1974).
120. Harrington, D. D., C. R. Marroquin, and V. White. *J. Anim. Sci.* **33**(1), 231 (1971).
121. Harrington, D. D., *Br. J. Nutr.* **34,** 45 (1975).
122. Harrington, D. D., C. R. Marroquin, and V. White. *J. Anim. Sci.* **35,** 1105 (1972).
123. Hintz, H. F., and H. F. Schryver. *J. Anim. Sci.* **35,** 755 (1972).
124. Ammerman, C. B., J. P. Fontenot, M. R. S. Fox, H. D. Hutchinson, P. Lepore, H. D. Stowe, D. J. Thompson, and D. E. Ullrey, "Mineral Tolerance of Domestic Animals." NAS-NRC, Washington, D.C., 1980.
125. Lloyd, K., H. F. Hintz, J. D. Wheat, and H. F. Schryver. *Cornell Vet.* **77,** 172 (1987).
126. Corke, M. J. *Vet. Rec.* **109,** 212 (1981).
127. Shupe, J. L., and A. E. Olsen. *J. Am. Vet. Med. Assoc.* **158,** 167 (1971).
128. Carle, C. J., J. P. Fontenot, and K. E. Webb, Jr. *J. Anim. Sci.* **40,** 180 (1975).
129. Fosmire, G. *J. Nutr. Notes* Dec., p. 10. (1985).
130. Wysocki, A. A., and R. H. Klett. *J. Anim. Sci.* **32,** 74 (1971).

7

Protein and Amino Acid Requirements

I. INTRODUCTION

Protein generally refers to crude protein, which is defined for feedstuffs as the nitrogen content $\times$ 6.25. This definition is based on the assumption that the nitrogen content is 16 g of nitrogen per 100 g of protein. In general this is accurate enough for feedstuffs, but there are certain proteins in which the nitrogen content varies considerably from 16%.

Proteins are made up of many amino acids combined with one another. Protein in the diet is not required as such, but it is needed as a source of essential (now usually referred to as indispensable) amino acids and nitrogen for nonessential (now usually referred to as dispensable) amino acid synthesis. The amino acids are put together in various combinations to form body proteins, and are often referred to as the building blocks of proteins. Every protein has a definite amino acid composition and no two are alike.

Amino acids contain nitrogen combined with carbon, hydrogen, oxygen, and sometimes sulfur, phosphorus, and iron. The nitrogen is in the form of an amino group (NH_2); it is from this that the name amino acid is derived. Amino acids are commercially synthesized, and a few are available in large quantities.

A lack of protein is frequently a limiting factor in the diet. This is because farm grains and their by-products are deficient in both quantity and quality of protein for horses. And since protein supplements are usually more expensive feeds, some people still tend to feed too little protein.

Animals continually use protein either to build new tissues, as in growth and reproduction, or to repair worn out tissues. Thus, horses require a regular intake of protein. If adequate protein is lacking in a diet, the horse suffers a reduction in growth or loss of weight. Ultimately, protein will be withdrawn from certain tissues to maintain the functions of the more vital tissues of the body as long as possible. Protein is needed to form milk, muscle, hide, hoof, hair, hormones, enzymes, blood cells, and other constituents in the body. Thus, protein affects almost every body function. It has been shown also that animals are more resistant to infections if they are fed an adequate protein diet. This is because the compounds in the bloodstream that help resist disease are proteins. So, adequate protein in the diet is one way of helping keep animals resistant to disease.

More emphasis is now placed on amino acid requirements rather than on protein needs since protein is not required *per se*. But, protein level is still used and is a good guide for meeting amino acid needs (1).

Protein is not stored for later use as some other nutrients are. Thus, a protein-deficient diet should not be fed for any length of time.

In actual feeding practice, a diet consists of a mixture of many different proteins. Each feed ingredient, such as a grain, a protein supplement, or a hay, contains many different kinds of protein. The proteins that the horse consumes are digested or broken down to amino acids in the digestive tract. There are differences in the digestibility of proteins. They can result from structural differences in the proteins as they occur naturally in the feed or from variation in processing treatment. The temperature used in processing protein supplements can affect the amino acid availability. For example, too low as well as too high a temperature during processing can adversely affect amino acid digestibility.

II. AMINO ACIDS

There are 22 different amino acids. Proteins are made up of many amino acids in different combinations. The various combinations and the different levels of amino acids are what characterize proteins and make one different from another. For example, one protein may have hundreds of amino acids, whereas another may have thousands of amino acids in its molecular makeup.

Some of these amino acids are labeled *indispensable* and others are labeled *dispensable*. The dispensable amino acids are so named because they can be synthesized in adequate amounts in the body and do not need to be supplied from an outside source. This means they are dispensable as constituents of a diet. However, they are essential to the animal after they are synthesized in the body. Physiologically, all amino acids found in animal tissues are essential otherwise tissue could not be formed. The indispensable amino acids are synthesized at too low a level, or not at all, by the body and, hence, a certain amount must be supplied preformed in the diet. This means they are an essential part of the diet and must be included at adequate levels to supply the animal's need.

Amino acids are made up of both D and L forms, which refer to the stereo configurations. Amino acids occurring in nature are in the L form. However, both D and L forms can be synthesized in the laboratory. Animals can change the D to the L form to a certain extent with some amino acids but not with others. For the most part, only the naturally occurring L form of the amino acid is utilized by the animal. There are some differences between animals in this regard, however. There is no information available for the horse to indicate to what extent, if any, it can utilize the D form of certain amino acids. This is important to know if synthetic amino acids are to be used to supplement horse diets. If the synthetic amino acid contains equal amounts of the D and L forms, then it is only one-half as effective as the L form. It takes twice as much of the DL-amino acid as of the L-amino acid to meet the animal's needs.

III. INDISPENSABLE AMINO ACIDS

It is not known how many indispensable amino acids the horse requires. It is known, however, that the amino acid lysine can benefit the young horse when added to the diet. University of Florida studies (2,3) indicated that maximum growth rate occurred when the lysine content of the diet was at least 1.9 g/Mcal of digestible energy. It appears that the lysine requirement is at least 0.6—0.7% in the total diet for weanlings and 0.4% for yearlings (4–6). The younger foal needs a higher lysine level (7). A Florida study (8) showed that 38–40 g of lysine daily is adequate to support maximum growth in yearly thoroughbred and Quarter Horse foals. The 1989 NRC (1) report states that the lysine requirement varies from 0.28 to 0.61% depending on the stage of the life cycle of the horse (Table 6.5). The highest requirement is 0.61% for the rapidly growing weanling horse. Therefore, lysine is an indispensable amino acid for the young horse, since it cannot synthesize enough to meet its body requirements. How many other amino acids are indispensable for the horse is not known. Table 7.1 gives a guide to the indispensable and the dispensable amino acids required by the pig. It can also be used as a guide for the horse.

Animals can synthesize certain amino acids from other amino acids or other nutrients in the diet. Present knowledge indicates that the young horse definitely requires only one indispensable amino acid; the rest of the 21 amino acids shown in Table 7.1 are dispensable for the horse. It may be, however, that the young foal or young horse under the heavy stress of training or racing may require other indispensable amino acids, since it may not be able to synthesize all it needs under high-stress conditions.

An adequate amount of dispensable amino acids in the diet lessens the need for certain indispensable amino acids. For example, cystine is synthesized from methionine. Thus, if the cystine content of the diet is low, there must be enough methionine in the diet to supply not only the body's need for methionine but also a sufficient quantity to be used for the synthesis of cystine as well. It is more efficient, therefore, to supply the dispensable amino acids in the diet than to create them by synthesis from indispensable amino acids.

Of special importance is that body protein synthesis requires that all of the indispensable amino acids involved be present simultaneously. Any indispensable amino acid that is present in inadequate amounts will limit the utilization of all the others. For example, if one indispensable amino acid is present at about 80% of its required level, it will limit the use of all the other indispensable amino

TABLE 7.1

Amino Acid Classification for the Pig

Indispensable amino acids	Dispensable amino acids
Lysine	Glycine
Tryptophan	Serine
Methionine	Alanine
Valine	Norleucine
Histidine	Aspartic acid
Phenylalanine	Glutamic acid
Leucine	Hydroxyglutamic acid
Isoleucine	Cystine[b]
Threonine	Citrulline
Arginine[a]	Proline
	Hydroxyproline
	Tyrosine[c]

[a]Partially synthesized, but not during early growth.
[b]Can replace at least 50% of the methionine requirement.
[c]Can replace at least 50% of the phenylalanine and tyrosine requirement.

acids to about 80% of their needed level in the diet. Therefore, it is important that diets contain the proper level of all the indispensable amino acids for maximum amino acid utilization and protein production. The mechanism of protein synthesis in the body is governed by the completeness and balance of the amino acids supplied from the diet and those currently available in the body from amino acid synthesis. Amino acids are not stored in the body as such. An incomplete mixture of amino acids is broken down and used as a source of energy, i.e., they are not held for any length of time in the tissues waiting for the arrival of the missing amino acid. This means that protein supplements should be mixed with the grain at the proper level so the horse receives a well-balanced concentrate diet each time it eats.

An Illinois study (9) reported that mares fed supplemental methionine showed a trend toward increased serum levels of methionine, lysine, histidine, arginine, tryptophan, leucine, isoleucine, and valine. This may indicate that methionine is an indispensable amino acid under certain conditions.

A Florida study (10) reported results indicating that threonine may be the second limiting amino acid for yearling horses receiving grass forage-based feeding programs. Inadequate threonine appears to reduce feed intake and weight gain.

A report by M. J. Glade (11) suggested that specific amino acid supplementation during conditioning may enhance aerobic work capacity. His work involved using leucine, isoleucine, valine, glutamine, and carnitine.

TABLE 7.2

Lysine Level in Some Commonly Used Feeds[a]

Feed	Percent dry matter	Percent crude protein	Percent lysine
Corn	88	9.1	0.25
Sorghum	90.1	11.5	0.26
Oats	89.2	11.8	0.39
Barley	88.6	11.7	0.40
Wheat	89.9	13.0	0.40
Fish meal, menhaden	91.7	62.2	4.74
Soybean meal	89.9	48.5	3.09
Dried skim milk	94.1	33.4	2.54
Canola meal	90.8	37.1	2.08
Sunflower meal	92.5	45.2	1.68
Cottonseed meal	91.0	41.3	1.68
Peanut meal	92.4	48.9	1.45
Linseed meal	90.2	34.6	1.16
Wheat bran	89.1	15.4	0.56
Citrus pulp	91.1	6.1	0.20
Dehydrated alfalfa meal	91.8	17.4	0.85
Dried brewer's yeast	93.1	43.4	3.23

[a]Adapted from 1989 NRC report (1).

Table 7.2 gives the lysine levels in some of the feeds commonly used for horses. Of the grains, corn and the sorghum have the lowest level of lysine. Of the plant protein supplements, soybean meal has the highest level. Fish meal, dried skim milk, and dried brewer's yeast are also excellent sources of lysine.

IV. QUALITY OF PROTEIN

Feeds that supply the proper proportion and amount of the various indispensable amino acids supply a "good-quality" protein whereas those feeds that furnish an inadequate amount of any of the indispensable amino acids are considered "poor-quality" protein sources. This is best illustrated in a Cornell University study (12) where linseed meal and a blend of milk products (containing dried whey, dried whey fermentation solubles, cheese rind, and dried buttermilk) were fed to colts that were $5\frac{1}{2}$ months old and weighed 420 lb. The milk products contained 25% protein and the linseed meal 35% protein.

The linseed meal and milk products supplied about 40% of the total protein in the diet. The results obtained during a 76-day trial are given in the following tabulation.

Protein supplement	Average daily gain (lb)	Average daily feed intake (lb)	Feed per lb of gain (lb)
Milk products	2.09	11.53	12.14
Linseed meal	1.32	10.07	16.79

The colts fed the milk products consumed 14% more feed, gained 56% more weight, and were 20% more efficient in the conversion of their feed to body weight. Since both protein supplements supplied the same amount of total protein, these results indicate that a lack of some amino acid was involved. In the next experiment the Cornell workers added 0.4% lysine to the linseed meal diet. The addition of lysine caused the horses fed linseed meal to do as well as those fed the milk products. This indicated that the big difference between linseed meal and the milk products was the fact that linseed meal is low in lysine. The linseed meal diet contained between 0.3 and 0.4% lysine, whereas the milk products diet had between 0.65 and 0.70% lysine. This trial also verified another study (13) in which it was found that 475–lb horses needed 0.60% lysine in the diet. Borton *et al.* (14) found that foals weaned early and fed dried skim milk grew faster than those fed soybean meal diets.

The diet that has the highest protein quality is the one that supplies all the indispensable amino acids needed in the proportions most nearly like those in which they exist in the body protein of the horse. This kind of a diet will meet the protein needs with a minimum intake of feed protein. Grains are lacking in lysine and, therefore, contain "poor-quality" protein, as does linseed meal, which is also low in lysine. Soybean meal, on the other hand, is a "good-quality" protein supplement, since it has a good level and balance of the indispensable amino acids (see Table 7.2).

The majority of the amino acids are absorbed in the small intestine. There is also some absorption (15,16) of amino acids of microbial origin from the cecum and large intestine (microbial synthesis). It is thought, however, that the amino acids synthesized by the organisms in the cecum and large intestine are not too efficiently utilized by the horse. If this is the case, then protein fed to a young horse needs to be of good quality. It means the young horse will be dependent, to a considerable extent, on the balance and level of the amino acids in the feed and cannot depend on the synthesis of amino acids in the cecum to take care of its amino acid needs.

V. BALANCE OF INDISPENSABLE AMINO ACIDS

A balance of indispensable amino acids in the diet is very important for horses in much the same way as for other nonruminants (17). Amino acids need to be fed at the right level, in the right proportion, and at the right time with the other indispensable amino acids for maximum response. The complexity of the amino acid supplementation problem accounts for the inconsistent results that have been obtained by many workers supplementing diets with amino acids. Sometimes they get beneficial effects and other times they do not. To obtain good results from adding amino acids to diets, one must be careful to avoid imbalances. All the limiting amino acids need to be added for optimum response. If two amino acids are lacking in the diet, the addition of the second limiting amino acid without adding the first limiting amino acid (the one most lacking) can cause a depression in the rate of gain and feed intake. When the first limiting amino acid is also added, however, the growth depression is overcome. Once the amino acid supplementation problem is well understood, it will be possible to use lower protein diets as well as have a better balance of amino acids in the diet.

High-quality protein supplements should be used in diets as much as possible until more is learned about the amino acid needs of the horse. This should especially be the case with the young, rapidly growing horse that is being developed at an early age for racing or high-level performance.

To determine definitely the need for an amino acid, various levels of the amino acid as well as diets of different protein content—and in some cases other possible limiting amino acids in various combinations and levels—must be fed to test all possibilities. In other words, determining the need for amino acids in diets is difficult and complex. Moreover, the requirement for amino acids varies not only with the type of diet used and the relative supply and availability of the amino acids, but also with the dietary supply of other essential substances which can, in case of need, be made from an amino acid.

In studying the amino acid needs of the horse one must be aware that if an amino acid is added to the diet and the horse does not respond, it is not a definite assurance that the diet is adequate in that amino acid. The horse might have responded if the correct amino acid level was fed and if the balance and level of other possible limiting amino acids was also met.

VI. AMINO ACID ANTAGONISM

Clarification of amino acid antagonisms requires further research but one must be aware of it as an area of possible concern. This antagonism might occur when one amino acid affects the requirement for another by interfering with its metabolism. This might occur among amino acids that are structurally related.

These interrelationships indicate the complexity of determining the requirement for indispensable amino acids and whether or not supplementation may be needed. Amino acid requirements will vary a great deal depending on the feeds used, their quality, their method of processing, amino acid and other nutrient interrelationships, and many other factors.

In a Minnesota study (18) with 92-day-old pony foals fed a low-protein (13%) diet plus 100 ppm niacin and 0.6% added lysine gained significantly faster than foals fed the low-protein diet or the low-protein diet with either niacin or lysine added alone. No explanation was give for these results where a possible synergism occurred.

VII. AMINO ACID INTERRELATIONSHIPS

The study of amino acid needs of the horse is in its infancy. It is known that lysine supplementation is helpful in the young horse. There is a possibility that methionine and threonine supplementation might also be needed in certain situations.

Some amino acid interrelationships in the pig are presented as a means of clarifying factors affecting amino acid requirements. These interrelationships need to be considered as one studies amino acid needs of the horse, especially with the young foal, even though there is no assurance that the swine data will entirely apply to the horse.

A. Tryptophan

The pig can use tryptophan and synthesize niacin from it, but it cannot convert niacin back to tryptophan. It is estimated that each 50 mg of tryptophan in excess of the tryptophan requirement will yield 1 mg of niacin. It appears that when studying niacin needs, one should use diets that are just adequate in tryptophan. When studying tryptophan needs one should use diets that are adequate in, or contain a small excess of, niacin. D-Tryptophan has a biological activity of 60–70% that of L-tryptophan for the growing pig.

B. Methionine

Methionine can be converted to cystine but cystine cannot be converted back to methionine. Cystine can satisfy at least 50% of the total need for sulfur amino acids. Methionine can meet the total need for sulfur amino acids (methionine + cystine) in the absence of cystine. It appears that methionine supplementation may be beneficial with improperly processed soybean meal but may be of no help with a properly processed meal.

Methionine can furnish methyl groups for choline synthesis. Choline is therefore effective in sparing methionine, because, with an adequate level of choline in the diet, methionine is not needed for choline synthesis. In a diet which is mildly deficient in both, adding either one will improve growth.

DL-Methionine can replace the L-form in meeting the needs for methionine, but the D-form is used less effectively than the L-form by the very young pig.

C. Phenylalanine

Phenylalanine can meet the total requirement for phenylalanine and tyrosine because it can be converted to tyrosine. Tyrosine cannot be converted to phenylanine. If the diet contains sufficient tyrosine, phenylalanine will not be used for tyrosine synthesis. Tyrosine can satisfy at least 50% of the total need for the aromatic amino acids tyrosine plus phenylalanine for growth.

D. Arginine

Swine can synthesize enough arginine for maintenance and pregnancy, but not enough for early growth in the young, in which situation diet arginine is needed.

E. Proline

Proline is not yet considered to be an indispensable amino acid, but the young pig (2.2–11 lb) appears to be unable to synthesize it rapidly enough for its requirements and so, in this situation, a dietary source is needed. If further studies verify this, proline may then possibly be considered as an indispensable amino acid for the pig.

VIII. EXCESS PROTEIN

Excess proteins are deaminated (the nitrogen removed as ammonia and urea). The remainder of the protein molecule serves as a source of energy or is stored as fat through complex mechanisms in the body. Thus, excess protein is not entirely wasted. However, excess protein should not be fed to supply energy and fat, since it is usually too expensive to do so. Rather, grains and their by-products and other high-energy feeds, which are cheaper sources of energy, should be used.

A Colorado study (19) with endurance-type horses showed that increased sweating and high pulse and respiration rates occurred when excess protein was fed to their horses. A survey of seven trainers at seven thoroughbred race tracks in New Jersey showed that the average time for their horses to finish a race increased by 1–3 seconds for every 1000 g of crude protein ingested in excess of the 1978 NRC-recommended protein requirements (20). While this was a very high level of protein intake, it indicates a need for further attention. Another study (21) found that feeding 130% of the 1978 NRC-recommended amount of protein increased urinary excretion of calcium and phosphorus as compared to foals fed 70 or 100% of the NRC-recommended protein requirement. A Cornell University study (22) reported no harmful effects from feeding 24% protein diets. A VPI study (23) found no detrimental effects from feeding an excessive level of protein (20%) in the diet to 2-year-old horses which were fitness-trained on a treadmill for 5 weeks. But, they also found no beneficial effects from the 20% protein diet as compared to a 10% protein diet. A Utah State University study (24) reported that the higher protein levels used in their trial were of no metabolic hindrance to competitive performance. A Kentucky study with 6-month-old colts fed 10, 13, 16, and 19% protein showed an increased rate of gain with each increase in protein for 112 days (25).

These studies would indicate no appreciable, if any, harmful effect from feeding a reasonable amount of protein above the NRC requirement levels. But one should be careful and not exceed these levels to any appreciable degree. It is best to err on the side of feeding a small excess of protein rather than run the risk of a lack of protein for the high-level performance horse. This is especially important because there is so much variability in the level, availability, and quality of protein in the hays used.

Dr. H. F. Hintz of Cornell University has stated that there is no scientific evidence that the level of protein is related to the incidence of bumps in horses (26).

IX. ENERGY–PROTEIN RATIO RELATIONSHIP

There is an optimum calorie to amino acid ratio for each stage of growth as well as for the entire life cycle of the horse. If the calorie to amino acid ratio is correct, it maximizes energy utilization in the horse. As the energy content of the diet increases, there is an increased need for amino acids, or protein, in the diet. A lack of protein in the diet, therefore, decreases energy utilization, increases the feed required per pound of gain, and decreases rate of gain or performance level. Practically no information is available on the calorie (or energy) to amino acid (or protein) ratios for the horse. The poultry industry is the most advanced in making use of more sophisticated energy values, calorie to amino acid ratios, and other means of maximizing energy efficiency and utilization. Eventually, horse nutrition will need to reach the same degree of expertise. In the meantime, the level of protein used should be slightly more liberal than the experimentally determined minimum protein requirements. This will ensure provision of an efficient protein to calorie ratio. It will also provide a needed safety margin since there are considerable differences in the quality of protein among the hays and other feeds fed to horses.

More attention to calorie–amino acid ratios will be needed as fat is added to high-level performance diets as a means of increasing diet energy density. Since more feed is consumed when the diet energy level is low and less when higher energy density diets are fed, the dietary protein levels need to be adjusted upward to the higher energy level.

X. EFFECT OF PROCESSING ON AMINO ACIDS

The method of processing has considerable effect on the amino acids and quality of protein supplements. An excellent review by Meade (27) presents the following facts:

1. Severe overheating of protein supplement feeds results in seriously depressed availability of all amino acids. Lysine appears to be more sensitive than some of the other amino acids to overheating.

2. Amino acids are highly available from properly processed soybean meal, cottonseed meal, and peanut meal. This is very important since soybean meal and cottonseed meal are widely used in horse diets.

3. The amino acid content of fish meals was not found to vary greatly, but availability did vary greatly, with serious depression of availability if meals were overheated or scorched in processing and shipping.

4. Amino acid composition of meat and bone meal varies greatly depending on the starting material. Availability of amino acids is depressed if the meals are overheated in processing.

5. There is a need for a rapid method for determining the availability of amino acids.

Methionine availability is lowered when the soybean meal is underheated. Raw soybean meal contains some factors that are destroyed by heat, such as trypsin inhibitor, which reduces its nutritional value.

XI. SYNTHETIC AMINO ACIDS

Until recent years, the study and use of synthetic amino acids in horse diets has been at a low level because of the limited amino acid supply available, the high price for some amino acids, and the limited research available on amino acids for the horse.

It now appears that new technological breakthroughs are occurring in amino acid synthesis via genetic engineering, recombinant DNA techniques, and other means. As these new developments occur, amino acids will be produced more efficiently in sufficient quantities and at prices that will encourage their study and use in horse diets. When this happens, lower protein and amino acid levels will be required in horse diets because of the optimum amino acid balance provided by proper amino acid supplementation.

As a rule of thumb, 3 lb of feed-grade lysine monohydrochloride (78.4% L-lysine) can replace 100 lb of 44% soybean meal per ton of diet for the pig (28). Costwise, it will be more economical to use when 3 lb of lysine + 97 lb of grain costs less than 100 lb of soybean meal (28). These figures can be used as an approximate guide for the horse until more adequate data become available.

XII. NONPROTEIN NITROGEN (NPN) AS PROTEIN SUBSTITUTE

Urea is a nonprotein nitrogen compound. It consists of carbon dioxide and ammonia and contains over 46% nitrogen. Thus, 1 lb of urea is equivalent to 2.87 lb of protein. Urea is the end product of protein or nitrogen metabolism in mammals (including man). Microorganisms in the rumen of ruminant animals can ingest urea, which is then used to synthesize protein. When the microorganisms travel from the rumen to the stomach they are digested by enzymes in the digestive tract. Thus, the microorganisms serve as a source of protein and are digested just like other protein sources. Urea is an important protein substitute for feeding beef, dairy, sheep, goats, and other ruminant animals.

For many years there has been considerable interest in whether the horse can utilize urea as a source of protein. The horse has a large cecum and many scientists have wondered whether urea conversion into protein by microorganisms in the cecum and the protein digestion and subsequent absorption of its amino acids in the large intestine is a possibility.

Cornell University scientists (29) observed equal nitrogen retention (or utilization) by mature ponies from urea, soybean meal, and linseed meal when they were added to a low-protein diet. They concluded that the equine can utilize urea to increase nitrogen retention if a low-protein diet is fed. They also stated that the efficiency of utilization of the absorbed nitrogen from urea is considerably less than that of nitrogen from a protein supplement.

A 1981 study by L. M. Slade and R. G. Godbee at Colorado State University involved the use of urea versus soybean meal in the diets of growing horses, including weanlings, yearlings, 2-year-olds, and 3-year-olds (49). They found that the young weanlings fed soybean meal grew more (79 lb) than those fed the urea (24 lb). There was no difference in growth rate between the older groups of horses fed either urea or soybean meal. Thus, it appeared that horses are capable of utilizing urea more efficiently as they become older. Part of this may be due to the need for an adaptation period to urea feeding, such as occurs with ruminants. It usually takes 2–4 weeks for ruminants to adapt to urea feeding. Part of the explanation could also be that the older horse has a lower protein requirement than the young horse. The Colorado scientists also stated that it appears that horses are better able to utilize urea for minimal growth or maintenance than for maximum growth (weight gain).

A good many studies have been conducted with urea to show that it can be used to support nitrogen balance in mature horses fed low-protein diets (29–32). Theoretically, the increased nitrogen retention could be brought about through utilization of microbial protein (32,33) or endogenous protein synthesis in the liver from ammonia released by microorganisms (34). Data obtained in Ohio (31) strongly suggested that in the horse ammonia absorption occurs from the cecum. Studies in Colorado (35) showed that either urea or soybean meal are able to provide the ammonia pool needed for microbial protein synthesis. Studies in Cornell (36) indicated that significant quantities of urea are recycled and hydro- lyzed in the digestive tract of the horse. Studies in Maryland (37) suggested that the addition of yeast culture to the diet of thoroughbred yearlings acted by stimulating the conversion of recycled urea to microbial protein and amino acids. Another Maryland study (38) suggested that the decreased fermentability of the hemicellulose fraction of their straw-containing diet was responsible for the decreased microbial utilization of recycled urea and its metabolite, ammonia. This suggested that nitrogen intakes should be supplemented when mature horses are fed poorly fermentable feeds. Studies in California (39) showed that nitrogen retention was increased at each crude protein level (7, 10, 13, and 16%) by the addition of urea to corn gluten (an imbalanced protein source) diet. The response from urea decreased as the protein level in the diet increased. Studies in Wash- ington State (33) showed that biuret, a compound resulting from the condensa- tion of urea production, was slightly more beneficial than urea when added to a low-protein diet. Biuret is much less toxic than urea, since it is hydrolyzed more slowly, but is much more expensive. The slow release of ammonia provides safety against ammonia toxicity. The Washington State study (33) indicates that biuret can be used safely in horse diets. A field trial (40) also verified the safety of feeding biuret to horses.

In evaluating the use of NPN in horse diets, there is no adequate information to indicate how much of the increased nitrogen retention is due to dispensable amino acids synthesized in the liver of the horse or from the absorption of amino acids synthesized by the microflora in the digestive tract (41,42).

Two separate studies have shown that the horse can tolerate a fairly large amount of urea. Horses were fed from 0.5 to 0.55 lb of urea per day and did not develop any adverse effects. Scientists at Louisiana State University (43) fed horses levels of urea up to 0.5 lb per day for 4 weeks, and the horses exhibited weight gains, glossy hair coats, and good physical condition. In another study (44) horses were fed an average of 0.55 lb of urea daily for 5 months without any harmful effects. Thus, mature horses are not as susceptible to urea toxicity as some scientists once thought. Studies in Delaware (45) showed that mature ponies weighing 387 lb were fed a diet with 2.5% urea with no adverse effects. The average consumption of the ponies was 67.1 g of urea daily. An Abilene Christian University study (46) showed that weanling horses may safely be fed concentrate diets with 1% urea. Some investigators feel that young foals might be susceptible to urea toxicity. Therefore, one still needs to proceed with caution with regard to possible urea toxicity in horses. A Cornell study (47) showed that ponies weighing 286 lb fed 450 g of urea in one feeding died as a result of ammonia toxicity. The urea intake was equivalent to 25% of the diet.

One study (48) suggested that an inverse relationship between dietary crude fiber content and the apparent digestibility of nitrogen similar in magnitude to that for cattle may exist for the horse.

A study by L. M. Slade and R. G. Godbee of Colorado State University involved the use of commercial range protein blocks containing urea (49). They were fed to brood-mares during the last one-third of their gestation period. The mares consumed 5.5 lb of the protein block plus 17.6 lb of grass hay daily. The Colorado scientists concluded from their studies that the range protein blocks containing urea had no detrimental effects either to the fetus or the mare. This should decrease the concern of horse owners who run cattle and horses together and where the cattle are fed protein blocks containing urea.

Presently, urea is still not recommended for use in horse diets. There is still much to learn on whether it might be feasible and economical to use in certain horse diets. The age of the horse needs to be considered. The kind of diet used and the level and quality of the protein it contains may also influence the response obtained from urea feeding. There are still too many unanswered questions concerning urea use by the horse, and until this information is obtained, it is difficult to make a sound recommendation on its inclusion in horse diets. The person with young foals or with valuable performance horses should not feed urea in horse diets until more is known about its use and value.

The 1989 NRC report (1) states that there appears to be no beneficial effect of including nonprotein nitrogen (such as urea) in practical diets for horses. They also state that urea is well tolerated by the mature horse at levels of up to 4% of the total diet. They indicate that this tolerance is likely due to the high solubility of urea, which results in its absorption from the small intestine and subsequent excretion by the kidney.

Two studies indicate that growing horses cannot achieve maximum growth rates (49), and lactating mares do not give maximum milk production (50) when a major portion of the nitrogen requirement is provided by urea.

XIII. PROTEIN REQUIREMENT INFORMATION

The protein requirement depends on the amino acid content of the feeds used. If the total diet is balanced and is a high-quality protein diet, the protein level required will be lower than if the total diet fed is a low-quality protein diet. When protein of marginal quality is fed, the diet protein must be increased to provide the indispensable amino acids (51). Lysine is a limiting amino acid with linseed meal and with all the grains used in horse diets (see Table 7.2). Therefore, lysine supplementation may be needed for the young, growing horse unless other lysine-rich feeds are used. Weanling horses require 0.6–0.7% lysine and year-lings require 0.4% lysine in the diet (4,5).

A. Effect of a Protein Deficiency

A protein deficiency results in a depressed appetite, which leads to inadequate consumption of total feed. A poor hair coat and reduced hoof growth may also occur. A protein deficiency and a lack of energy often occur together. This results in a loss in weight in mature horses. In young horses, slow, inefficient growth and underdevelopment occurs (52,53). Other possible effects of a protein deficiency include reduced conception rate (54). Pregnant mares deficient in protein may have small, weak foals. Milk production is reduced with a lack of protein (55).

B. Research on Protein for Horses

The amino acid requirements for maintenance of the horse are not known (51). The body composition of the horse varies with age. Although the fat-free body of the horse tends to maintain a water content of 72%, a protein content of 22%, and an ash content of 6% (56,57), the whole body may vary from 2% fat at birth to 20% for a mature horse in good condition (58–60).

There is little or no increase in diet protein required above maintenance for work (51,61). A 1985 study (62) showed that exercise alone did not increase the protein requirement in growing horses. Another study (63) suggested that diets of exercising horses need not have protein at concentrations higher than that required for maintenance. A small amount of nitrogenous compounds, including protein, is lost in sweat (64). The energy requirements for work are met by extra feed intake. This results in a greater protein intake which provides enough protein to replace any nitrogen lost in the sweat (51,65). Beeley et al. (66) recently isolated a protein called latherin in horse sweat. They suggested that latherin may have a role in thermoregulation of the exercising horse. Sweat losses increase as activity levels increase, and losses as high as 5% of body weight have been estimated (67). Sweat may contain 1–1.5% nitrogen (67).

Nitrogen balances tend to be more positive in conditioned horses as compared to idle horses fed similar levels of nitrogen (68,69).

During the last 90 days of gestation the protein requirement increases considerably. The products of conception (fetus) contain 11.3% protein and total 10–

12% of the mare's body weight. It is assumed that 60–65% of the protein in the fetus is deposited during the last 90 days of gestation (1,70).

The protein concentration of mare's milk 5 days after parturition is about 3.1% and declines to about 2.2% in 2 months (71).

Studies on protein requirement of growing ponies and burros have varied from 14 to 15.8% (52,72–74). Diet constituents, energy levels, protein quality, amino acid availability, and many other factors affect the protein level required. It is to be expected, therefore, that variation will occur in levels of protein requirement as reported by different scientists using different conditions with ponies, burros, and horses.

C. Protein Requirements

Tables 6.5 and 6.6 give the levels of protein recommended by the 1989 NRC publication on nutrient requirements of the horse (1). These suggested requirements do not include a margin of safety. They are the suggested requirements based on the best judgement of the NRC committee for nutrient requirements of the horse. It is expected by the NRC committee that some safety factor may be added to those levels by those formulating horse diets to take care of the many factors that can affect protein requirements under various farm and management conditions.

University of Florida studies (75) showed that a daily intake of 848 g of crude protein gave maximum growth and bone development in thoroughbred and Quarter Horse weanlings.

A Texas A & M study (76) suggested that foals weaned at 3 months of age should be fed at least 20% protein to facilitate faster growth rates.

Limited information is available on the protein digestibility of individual feed ingredients for the horse. Since amino acids are mainly absorbed in the small intestine, and the predominant form of nitrogen absorbed from the hindgut area is ammonia, the site of absorption also greatly influences the efficiency of protein utilization (1). Therefore, diet formulation for horses should be based on crude protein (1) at the present time.

D. Suggested Protein Levels to Use

There are considerable differences in the quality of protein in the various diets fed to horses. The hay and pasture used varies a great deal in protein level as well as in quality of protein. Therefore, it is difficult to recommend one exact level of protein for each stage of the life cycle of the horse to be used on the farm. One needs instead to provide a small margin of safety to take care of the many factors that can affect protein requirements. Consequently, the protein requirements shown in Table 6.5 may need to be slightly increased. Another factor to consider is that too many different diets cannot be handled at the feed store or on the farm. There is not enough storage space at the feed store for the variety needed since feeds for other animals are handled as well. At the farm, storage space may be too limited for storage of many diets, and there may not be enough pastures to divide the horses into all the stages of the life cycle shown in Table 6.5. Moreover, it may be confusing for those feeding the horses to have all the different diets to meet the 19 life cycle categories shown in Table 6.5. Therefore, one needs to combine some of the categories into about five or six different diets. In doing so, there may be instances when some of the horses may be fed a higher level of protein than required. This is unavoidable, but is preferable to having a deficiency of protein.

It is assumed that the horse will consume on average about one-half concentrates and one-half-forage during all the various stages of its life cycle. However, this ratio of concentrates to forage (hay and/or pasture) will vary depending on the activity of the horse. For example, a hard working horse might consume 60–65% concentrates and 35–40% forage in its daily diet. A horse at light work or activity may consume only one-third concentrates and two-thirds forage. A young horse consumes more concentrates than forage. Table 6.5 gives the proportion of hay and concentrates during the various stages of the horse's life cycle. The older the horse, the more forage it consumes. If a low-quality forage is fed, the protein levels shown in Table 6.5 might need to be increased depending on the protein content in the forage. The person with quality horses who is pushing them for racing or high-level performance purposes usually insists on the use of high-quality forage at all times.

The protein needs of the horse decrease as it reaches mature size. Horses that are heavier at maturity usually have a higher protein requirement during early growth. All horses, however, have about the same protein requirement when they reach maturity. Protein requirements increase during the last one-third of pregnancy and during lactation. Most of this extra protein need is taken care of by feeding the mare more of the concentrate feed which gives her a higher total daily protein intake. In some cases, however, a higher protein level is needed in the concentrate mixture.

It is very important that the proper protein quality and level be used to ensure proper development of the foal. If the foal does not receive enough milk, a high-quality feed should be used to supplement the mare's milk. Some use creep feeds, calf starters, calf manna, special horse feeds, and other diets that contain dried skim milk (which closely resembles mare's milk) for supplemental feeding of the foal. Regardless of the feed used, the objective is to provide additional protein and other nutrients to ensure that the foal obtains adequate nutrition for rapid and proper growth. These feeds for the foal need to be palatable and well-balanced nutritionally.

It should be emphasized that the protein requirements shown in Table 7.3 are

TABLE 7.3

Recommended Protein Levels for Horses

Stage of life cycle	Percentage of protein in the total diet[a]
1. Creep feed and feed for foals	16–18[b]
2. Weaned foals to 12 months	14–16[c]
3. Yearlings, long-yearlings, and two-year-olds	12–14[d]
4. Mares (gestation and lactation) and breeding stallions	12–14[e]
5. Mature horses (idle and at work)	9–11[f]

[a]This is the protein in the total diet including the protein in the hay and/or pasture used. If very-poor-quality hay or pasture is used these levels would need to be increased and would justify the higher levels shown in the range of protein figures to be used.

[b]These feeds supplement the milk which the mare gives. The higher level protein would be used with poor milking mares.

[c]This is a critical time in getting a weaned foal started. A high-quality well-fortified diet with vitamins and minerals is needed.

[d]The faster growing and heavily trained animals would be given the higher level of protein and a higher level of vitamins and minerals.

[e]During early lactation, and with the heavy milking mares, the higher level of protein would be used.

[f]If the mature horses are idle, the lower level of protein is sufficient. With heavy work the higher level of protein would be used.

close to actual needs. It might be well to exceed these allowances a little with quite valuable horses for racing and high-level performance purposes in order to ensure an adequate level of protein. A little excess protein in the diet should have no harmful effect on the horse. The old thinking that excess protein was harmful to animals was shown to be incorrect when vitamin B12 was discovered. The harmful effects obtained were due to a deficiency of vitamin B12 with swine and poultry, which is needed for the utilization of protein and as one increases the level of protein one increases the need for this vitamin. Therefore, if one is feeding a well-balanced diet and vitamin B12 is adequate, there should be no harmful effect from the added protein in swine and poultry. It is possible the same may be true for the horse. The extra protein is broken down by the body and is used as a source of energy, just like carbohydrates or fats. However, since protein is a more expensive source of energy, it should be fed in excess only as insurance against a possible need for more protein or amino acids in the diet for valuable horses. This is especially the case when the hay quality is low and its protein quality and availability is also low.

The protein levels recommended by the writer are very close to those recommended by Cornell University scientists (77). They are a little higher than those recommended in the 1989 NRC report (1). This provides a small safety factor to take care of the many factors that may affect protein requirements.

XIV. CONCLUSIONS

Protein in the diet is not required as such. It is needed as a source of indispensable amino acids and of nitrogen for synthesis of dispensable amino acids. However, protein level is still used and is a good guide for meeting amino acid needs. An indispensable amino acid is one that cannot be synthesized or cannot be synthesized fast enough to satisfy the animal's need for it. Lysine is an indispensable amino acid for the young horse. Methionine and threonine may also be indispensable amino acids for the horse under certain conditions. The need for other indispensable amino acids is not known.

A balance of indispensable amino acids in the diet is very important. There are amino acid interrelationships with other nutrients which can affect amino acid requirement and that of other nutrients. But, little information is available in these two areas with respect to the horse.

The horse can tolerate some excess protein in the diet. However, excess protein should not be used unless it is a slight excess to ensure that protein needs are being met when hay protein level and quality is low and variable. High-energy diets increase protein needs since protein requirements are directly related to calorie level.

Adding urea to horse diets is not recommended, although the horse may make some use of it under certain circumstances, especially when on low-protein diets.

Suggested protein levels to use in horse diets are given. Signs of protein deficiency are also discussed.

REFERENCES

1. Ott, E. A., J. P. Baker, H. F. Hintz, G. D. Potter, H. D. Stowe, and D. E. Ullrey. "Nutrient Requirement of Horses." NAS-NRC, Washington, D. C., 1989.
2. Ott, E. A., R. L. Asquith, and J. P. Feaster. *J. Anim. Sci.* **49,** 620 (1979).
3. Ott, E. A., R. L. Asquith, and J. P. Feaster. *J. Anim. Sci.* **53,** 1496 (1981).
4. Breuer, L. H., and D. L. Golden. *J. Anim. Sci.* **33,** 227 (1971).
5. Hintz, H. F., H. F. Schryver, and J. E. Lowe. *J. Anim. Sci.* **33,** 1274 (1971).
6. Ott, E. A. *Proc. 30th Annu. Pfizer Res. Conf.* p.101 (1982).
7. Jordon, R. M., and G. Hackett. *Proc. Equine Nutr. Physiol. Symp., 5th,* p. 101 (1977).
8. Ott, J. P., R. L. Asquith, J. P. Feaster, and F. G. Martin. *J. Anim. Sci.* **49,** 620 (1979).
9. Rogers, P., W. W. Albert, and G. C. Fahey. *Proc. Equine Nutr. Physiol. Symp., 7th,* p. 73 (1981).
10. Ott, E. A., E. L. Johnson, R. L. Asquith, and M. D. Harrison. *Proc. Equine Nutr. Physiol. Symp., 11th,* p.113 (1989).

11. Glade, M. J. *Proc. Equine Nutr. Physiol. Symp., 11th*, p. 244 (1989).
12. Hintz, H. F., H. F. Schryver, and J. E. Lowe. *J. Anim. Sci.* **33**, 1274 (1971).
13. Breuer, L. H., and D. L. Golden. *J. Anim. Sci.* **33**, 227 (1971).
14. Borton, A., D. L. Anderson, and S. Lyford. *Proc. Equine Nutr. Physiol. Symp., 3rd*, p. 19 (1973).
15. Slade, L. M., R. Bishop, J. G. Morris, and D. W. Robinson. *Br. Vet. J.* **127**, 11 (1971).
16. Reitnour, C. M., J. P. Baker, G. E. Mitchell, Jr., C. O. Little, and D. D. Kratzer. *J. Nutr.* **100**, 349 (1970).
17. Potter, G. D. *Proc. Ga. Nutr. Conf.* p.49 (1985).
18. V. S. Serois, and R. M. Jordon. *Proc. Equine Nutr. Physiol. Symp., 8th*, p. 149 (1983).
19. Slade, L. M., L. D. Lewis, C. R. Quinn, and M. L. Chandler. *Proc. Equine Nutr. Physiol. Symp., 4th*, p.114 (1975).
20. Glade, M. J. *Equine Vet. J.* **15**, 31 (1983).
21. Glade, M. J., D. Beller, J. Bergen, D. Berry, E. Onder, J. Bradley, M. Cupelo, and J. Dallas. *Nutr. Rep. Int.* **31**, 649 (1985).
22. Hintz, H. F., K. H. White, C. E. Short, J. E. Lowe, and M. Ross. *J. Anim. Sci.* **51**, 202 (1980).
23. Frank, N. B., T. N. Meacham, and J. P. Fontenot. *Proc. Equine Nutr. Physiol. Symp., 10th*, p. 579 (1987).
24. Slade, L. M. *Proc. Equine Nutr. Physiol. Symp., 10th*, p. 585 (1987).
25. Pulse, R. E., J. P. Baker, and G. D. Potter. *Proc. Equine Nutr. Physiol. Symp., 3rd*, p. 34 (1973).
26. Hintz, H. F. *Feed. Int.* **3**(11), 41 (1982).
27. Meade, R. J. *J. Anim. Sci.* **35**, 713 (1972).
28. Tanksley, T. D., and D. H. Baker. *Pork Ind. Handb., Purdue Coop. Ext. Ser.* **PIH-5**, 4 (1984).
29. Hintz, H. F., and H. F. Schryver. *J. Anim. Sci.* **34**, 592 (1972).
30. Houpt, T. R., and K. A. Houpt. *Fed. Proc., Fed. Am. Soc. Exp. Biol.* **24**, 678 (1965).
31. Nelson, D. D., and W. J. Tyznik. *J. Anim. Sci.* **32**, 68 (1971).
32. Slade, L. M., D. W. Robinson, and K. E. Casey. *J. Anim. Sci.* **30**, 753 (1970).
33. Johnson, R. J., and J. W. Hart. *Nutr. Rep. Int.* **9**, 209 (1974).
34. Reitnour, C. M., and R. L. Salsbury. *J. Anim. Sci.* **35**, 1190 (1973).
35. Slade, L. M., M. L. Chandler, L. M. Lewis, G. M. Ward, and E. Keinholz. *J. Anim. Sci.* **42**, 1559 (1976).
36. Prior, R. L., H. F. Hintz, J. E. Lowe, and W. J. Visek. *J. Anim. Sci.* **38**, 565 (1974).
37. Glade, M. J., and L. M. Biesik. *J. Anim. Sci.* **62**, 1635 (1986).
38. Glade, M. J. *J. Anim. Sci.* **58**, 638 (1984).
39. Slade, L. M., and D. W. Robinson. *J. Anim. Sci.* **29**, 144 (1969).
40. Lynch, W. M., and R. W. Swart. *Proc. Nutr.* **8**, 13 (1974).
41. Coppock, C. E., A. D. Tillman, W. Burroughs, W. R. Featherstone, U. S. Garrigus, E. E. Hatfield, and R. R. Oltjen. "Urea and other Nonprotein Nitrogen Compounds in Animal Nutrition." *N.A.S.–N.R.C.*, Washington, D.C. (1976).
42. L. M. Slade, R. Bishop, J. G. Morris, and D. W. Robinson, *J. Anim. Sci.* **33**, 239 (1971).
43. Rusoff, L. L., R. B. Lank, T. E. Spillman, and H. B. Elliot. *Vet. Med. (Kansas City, Mo.)* **60**, 1123 (1965).
44. Ratliff, F. D., R. K. King, and J. B. Reynolds, *Vet. Med. (Kansas City, Mo.)* **58**, 945 (1963).
45. Reitnour, C. M., Delaware Agricultural Experiment Station, Newark (personal communication), 1970.
46. DuBose, L. E. *Proc. Equine Nutr. Physiol. Symp., 8th*, p.163 (1983).
47. Hintz, H. F., J. E. Lowe, A. J. Clifford, and W. J. Visek. *J. Anim. Vet. Med. Assoc.* **157**, 963 (1970).
48. VanderNoot, G. W., and E. B. Gilbreath. *J. Anim. Sci.* **58**, 638 (1984).

49. Godbee, R. G., and L. M. Slade. *J. Anim. Sci.* **53,** 670 (1981).

50. Gibbs, P. G., G. D. Potter, R. W. Blake, and W. C. McMullan. *J. Anim. Sci.* **54,** 496 (1982).

51. Hintz, H. F., J. P. Baker, R. M. Jordan, E. A. Ott, G. D. Potter, and L. M. Slade. "Nutrient Requirements of Horses." *N.A.S.–N.R.C.,* Washington, D.C. (1978).

52. Pulse, R. E., J. P. Baker, G. D. Potter, and J. Willard. *J. Anim. Sci.* **37,** 289 (1973).

53. Gill, R. J., G. D. Potter, J. L. Kreider, G. T. Schelling, and W. L. Jenkins. *Proc. Equine Nutr. Physiol. Symp., 8th,* p. 311 (1983).

54. Holton, D. W., and L. D. Hunt. *Proc. Equine Nutr. Physiol. Symp., 8th,* p. 107 (1983).

55. Gill, R. J., G. D. Potter, J. L. Kreider, G. T. Schelling, W. L. Jenkins, and K. K. Hines. *Proc. Equine Nutr. Physiol. Symp., 9th,* p. 84 (1985).

56. Mitchell, H. H. "Comparative Nutrition of Mare and Domestic Animals," Vol. 1. Academic Press, New York, 1962.

57. Robb, J., R. B. Harper, H. F. Hintz, J. E. Lowe, J. T. Reid, and H. F. Schryver. *Anim. Prod.* **14,** 25 (1972).

58. Julian, L. M., J. H. Lawrence, N. I. Berlin, and G. H. Hyde. *J. Appl. Physiol.* **8,** 651 (1956).

59. Pitts, G. C., and T. R. Bullard. *N.A.S.–N.R.C., Publ.* **1598** (1968).

60. Widdowson, E. M. *Nature (London)* **166,** 626 (1950).

61. Harvey, A. L., B. H. Thomas, C. C. Culbertson, and E. V. Collins. *Proc. Am. Soc. Anim. Prod.* p. 94 (1939).

62. Orton, R. K., I. D. Hume, and R. A. Leng. *Equine Vet. J.* **17,** 381 (1985).

63. Patterson, P. H., C. N. Coon, and I. M. Hughes. *J. Anim. Sci.* **61,** 187 (1985).

64. Freeman, D. W., G. D. Potter, G. T. Schelling, and J. L. Kreider. *Proc. Equine Nutr. Physiol. Symp., 9th,* p. 230 (1985).

65. Slade, L. M., L. D. Lewis, C. R. Quinn, and M. L. Chandler. *Proc. Equine Nutr. Physiol. Symp., 4th,* p. 114 (1975).

66. Beeley, J. G., R. Eason, and D. H. Snow. *Biochem. J.* **235,** 645 (1986).

67. Meyer, M. C., G. D. Potter, L. W. Greene, S. F. Crouse, and J. W. Evans. *Proc. Equine Nutr. Physiol. Symp., 10th,* p. 107 (1987).

68. Freeman, D. W. G. D. Potter, G. T. Schelling, and J. L. Kreider. *Proc. Equine Nutr. Physiol. Symp., 9th,* p. 236 (1985).

69. Freeman, D. W., G. D. Potter, G. T. Schelling, and J. L. Kreider. *J. Anim. Sci.* **66,** 407 (1988).

70. Bergin, W. C., H. T. Gier, R. A. Frey, and G. B. Marion. *Proc. 13th Annu. Conv., Am. Assoc. Equine Pract.* p. 179 (1967).

71. Ullrey, D. E., R. O. Struthers, D. G. Hendricks, and B. E. Brent. *J. Anim. Sci.* **25,** 217 (1966).

72. Yoakam, S. C., W. W. Kirkham, and W. M. Beeson. *J. Anim. Sci.* **46,** 983 (1978).

73. Jordon, R. M., and V. Meyers. *J. Anim. Sci.* **34,** 578 (1972).

74. Knapka, J. J., D. G. Brown, O. G. Hall, and L. L. Christian. *J. Anim. Sci.* **24,** 280 (1965).

75. Ott, E. A., and R. L. Asquith. *Proc. Equine Nutr. Physiol. Symp., 8th,* p. 39 (1983).

76. Craxton, D. E., G. D. Potter, and R. G. Haley. *Proc. Equine Nutr. Physiol. Symp., 6th,* p. 17 (1979).

77. Schryver, H. F., H. F. Hintz, and J. E. Lowe. *Cornell Coop. Ext. Serv. Inf. Bull.* **94** (1982).

8

Carbohydrates and Fiber

I. INTRODUCTION

Carbohydrates include the sugars, starch, cellulose, lignin, and related substances. The carbohydrates form the largest part of the horse's feed. Sugars and starches are readily digested and have a high feeding value for the horse. Cellulose, along with lignin, forms most of the skeletal portion of plants. Cellulose forms most of the fiber in horse diets and is poorly digested.

II. CLASSIFICATION OF CARBOHYDRATES

The carbohydrates contain carbon, hydrogen, and oxygen. Their classification is as follows.

1. Monosaccharides (contain one sugar molecule) are divided into pentoses, $C_5H_{10}O_5$ (five-carbon sugars), which include arabinose, xylose, and ribose, and hexoses, $C_6H_{12}O_6$ (six-carbon sugars), which include glucose, fructose, galactose, and mannose.

2. Disaccharides, $C_{12}H_{22}O_{11}$ (contain two sugar molecules), include sucrose, maltose, lactose, and cellobiose.

3. Trisaccharides, $C_{18}H_{32}O_{16}$ (contain three sugar molecules), include raffinose.

4. Polysaccharides (contain many sugar molecules combined) are divided into pentosans $(C_5H_8O_4)X$, such as araban and xylan, hexosans $(C_6H_{10}O_5)X$, such as dextrin, starch, cellulose, glycogen, and inulin, and mixed polysaccharides, such as hemicellulose, pectins, gums, and mucilages.

The monosaccharides, disaccharides, and trisaccharides are easily digested by the horse, as are starch and dextrin. The hemicelluloses contain many different sugars combined into a mixed polysaccharide which is easier to digest than cellulose but is less digestible than either starch or dextrin. Hemicelluloses are widely distributed in forage crops (such as hays or pastures). Cellulose is the principal constituent of the cell wall of plants and is fairly abundant in the more fibrous feeds. It is generally low in digestibility and it may also lower the digestibility of other nutrients. Cellulose is digested by microorganisms in the digestive tract, mostly in the cecum and colon of the horse. It requires a microbial enzyme, called cellulase, for digestion. The main end products of cellulose digestion are the volatile fatty acids (VFA), acetic acid, proprionic acid, and butyric acid. Lignin is not a true carbohydrate. It is found largely in very mature hays, straw, pastures, and hulls. Lignin is essentially not digestible by the horse and it also reduces the digestibility of the cellulose and other nutrients in forages. Lignin is of no nutritional value except in supplying bulk in the diet of the horse.

In feeds, the term fiber refers to the cellulose and related carbohydrates which form the cell walls of plants (along with lignin). The higher the fiber content of the feed, the less digestible it is. The fiber content of early cut hay is more digestible than the fiber of hay cut in late bloom when it is more mature. The fiber in lush, green pasture is much more digestible than in more mature dry pasture or most hays. The difference in digestibility is due to the carbohydrates becoming more complex as the plant matures. The easily digestible carbohydrates are replaced by hemicellulose and later by cellulose in the cell walls of the plant as it matures. The nutrients in the plant are inside the cell walls. As the plant matures, the cell walls become tougher to digest because they contain more complex carbohydrates. Therefore, as the plant matures not only is the fiber in the cell wall less digestible, but the tougher cell wall causes the other nutrients inside to be less available to the horse. Young growing horses, as well as race and high-level performance animals, need hay and/or pastures that are low in cellulose and high in the more digestible carbohydrates. This means the forages and other feeds in the diet need to be relatively low in fiber. Pastures should be short, lush, and green. Hays should be harvested early and cured properly so they will have plenty of leaves and a good, green color.

III. VOLATILE FATTY ACIDS

Most of the simple carbohydrates in the diet are digested in the small intestine and the end products that are absorbed are the simple sugars (1). Some of the simple carbohydrates and most of the complex carbohydrates (fiber) are digested by microorganisms in the cecum and colon and the end products are volatile fatty acids (VFA). In addition to the three already mentioned, a small amount of isobutyric, isovaleric, and valeric acids are also formed. VFA may supply 25% or more of the energy needed by the horse (2). These VFA are produced by microbial action in the digestive tract, mostly in the cecum and colon area. Microorganisms attack complex carbohydrates and produce volatile fatty acids which can be absorbed and utilized as sources of energy.

Studies in Kentucky (3) showed that an increase in cellulose digestion occurred with the addition of dehydrated alfalfa and distiller's dried grain solubles to oats and chopped straw diets. Another report (4) suggested that an unidentified substance in the fluid in the colon, in addition to bacteria, may enhance cellulose breakdown. The addition of oats to hay diets resulted both in an increase in the total number of bacteria and specific species of bacteria but no change in protozoa in the cecum of ponies (5). Lignin, as well as pectin, appears to be very resistant to digestion (6). A Kentucky study (7) showed that acetate and proprionate are absorbed from the equine cecum in a manner similar to their absorption from the stomach of ruminants. The butyrate absorption, however, differs from that in ruminants where it is largely converted into ketone bodies by the rumen epithelium (8,9). In the horse, the butyrate appears to be absorbed at a rate comparable to acetate and without conversion to ketone bodies (10). The total concentration of VFA in the intestinal contents and the percentage of acetate are decreased and the percentage of proprionate, isovalerate, and valerate increases when high-grain diets are fed (11,12). Studies at Cornell (13) indicate that the regulatory system for VFA and glucose utilization in ponies is markedly different from that of the ruminant. The rapid clearance of VFA from blood plasma is not attributed to insulin but is due to their uptake by the liver (7). Blood plasma glucose, on the other hand, appears to be under the control of the insulin produced by the pancreas.

Another Cornell study (14) showed that blood glucose levels were not influenced by the forage–grain ratio used. They reported that the total concentration of VFA in cecal fluid and the percentage of acetate were decreased and the percentages of proprionate, isovalerate, and valerate were increased when high-grain diets were fed.

IV. FORAGE DIGESTIBILITY

A Cornell report (15) indicated that horses digest fiber about two-thirds as efficiently as cows. It appears that digestion coefficients are similar for horses and ponies (16). This justifies using the pony in many horse research studies. Ruminants utilize dry matter and organic matter of grass hays more efficiently than horses. Crude protein and nitrogen-free extract are digested equally well by horses and ruminants (17). The ruminal bacteria digest dry matter and cell wall fractions of forages more efficiently than the cecal bacteria of the horse, even when they are exposed to the forages for the same length of time (18). Digestion of grass leaves by horses followed the same pattern as in ruminants, except that horses did not digest the leaf structures, which ruminants digest slowly (19). One explanation for the greater digestion of fiber by ruminants is that the rate of feed passage in the ruminant is slower than in the horse and so the fiber is exposed to microbial action for a longer period of time. But, other factors may also be involved (18).

A Maryland study (20) indicated that nitrogen intakes should be supplemented when mature horses are fed poorly fermentable feeds.

A New Jersey study (21) reported that crude fiber level was the best single predictor of all digestibilities in the horse. Another study (22) suggested that an inverse relationship between dietary crude fiber content and the apparent digestibility of nitrogen may exist for the horse, and this relationship may be similar in magnitude to that for cattle.

A USDA and University of Maryland study (5) reported that none of the types of cecal protozoa found in ponies were found in steers and vice versa. They also found that the cellulolytic bacteria per g of ingesta were similar in the pony's cecum and the steer's rumen.

A New Jersey study (23) reported that the legume forages with the greater portion of soluble carbohydrates were the most digestible for horses. It stated that the higher soluble carbohydrate content, along with the high apparent digestibility of soluble carbohydrates, is the distinct advantage that legumes have over grasses for horses.

A German study (24) reported that the fiber content of feed affected eating time, the amount of chewing, and the dry matter content of the swallowed bolus. They found an inverse relationship between the rate of eating and the fiber content of the feed. Another study (25) showed that the eating time was increased as the percentage of sawdust was increased in the diet of ponies. Moreover, the size of the meal increased and the meal frequency decreased with increasing amounts of sawdust. But, when the diet contained 50% sawdust, the feed intake was inadequate to maintain energy levels.

Another study (26) involved feeding a complete feed containing 66% hay along with concentrates, either unprocessed, ground or pelleted. Dry matter digestibility was not significantly affected by processing, but protein digestibility was higher when the diet was pelleted. Fiber digestibility was reduced after grinding and pelleting, and feed intake after pelleting was as high or higher than when the horses were fed unprocessed feed.

A Canadian study (27) involved ponies fed pelleted and unpelleted alfalfa and grass hays. Pelleting did not increase digestion or consumption of low-quality hays sufficiently to make them comparable to high-quality hays as a source of nutrients for ponies.

V. FORAGE LEVEL IN DIETS

Some fiber is needed in the diet of the horse, but the exact amount has not been definitely determined for horses under various conditions and at various stages of their life cycle. One report (28) indicated that horses should be fed at least 0.6 lb of forage per 100 lb body weight. Another study (29) reported that

0.4 lb of forage per 100 lb body weight was adequate. Many horse scientists recommend 0.5 lb of forage per 100 lb body weight as the minimum amount of forage needed. Many horse owners, however, prefer higher levels and will use from 0.5 to 1.0 lb of forage per 100 lb body weight as the minimum level of forage intake. Horses consume about 2 lb of feed, whereas the mature horse will consume less than 2 lb of feed per 100 lb body weight daily. This is an average figure throughout their lifetime. The young horse will consume less than 2 lb of feed per 100 lb body weight. Therefore, minimum forage intake, as long hay or pasture, is somewhere between 25 and 50% of the total diet on a dry matter basis.

Minimum forage (hay, pasture) intake needed will be influenced by the remainder of the diet fed. If a diet is bulky and is high in oats (which has 11–12% fiber as compared to 2% fiber in corn) and other high-fiber feeds (citrus pulp, which has 14% fiber and beet pulp which has 22% fiber), than the minimum forage intake can be decreased somewhat.

Older horses can utilize more forage than young horses. The microflora to digest fiber is not as well established in the young foal as it is in the older animal (30). Moreover, the young horse is growing rapidly and needs a higher energy diet than the more mature animals. The concentrate feeds contain approximately 50% more energy than the forage feeds on an equal dry matter weight basis. During periods of heavy energy demand (such as rapid growth, training, racing, performance) a higher level of concentrate feeds is needed in the diet. Moreover, the digestive capacity of the horse is not large enough to provide high energy requirements on high levels of forage in the diet. But, in the case of the mature horse, which has relatively low energy needs, forage can supply their energy requirements if the level of activity is not high.

In meeting the minimum forage requirements one needs to feed long hay for best results. Finely ground forage is not as good. In completely pelleted diets, 60–70% of finely ground forage is usually added to provide the forage requirement. If the forage is ground too fine in the pellet the horses may need to have some pasture, hay, or other high-fiber feed along with the pellets for best results.

VI. GLYCOGEN

Glycogen consists of many molecules of glucose. It is a carbohydrate reserve which exists in the liver and muscles. The liver contains the main supply of glycogen which can be as high as 10% of the liver. The liver glycogen is drawn upon when the blood glucose level is lowered. It is replenished when the blood glucose level increases beyond a certain level.

A Texas A & M report (31) indicated that diet and exercise can significantly affect muscle glycogen content, suggesting that glycogen is a limiting factor to the amount of aerobic work that can be performed. Horses reached fatigue at lower heart rates and lower blood lactic acid concentrations than horses with normal muscle glycogen concentrations. But, it was not demonstrated that glycogen overloading improved short-distance, explosive work. Another study (32) found that there was a significant use of muscle glycogen for energy production in the exercising horse and it was dependent on the muscle makeup of the horse. Horses with a high proportion of glycolytic white fibers used more stored muscle glycogen than those with more oxidative muscle fibers.

More detail on glycogen and its role in exercise physiology is given in Chapter 22.

VII. CONCLUSIONS

Carbohydrates form the largest part of the horses energy supply. Most of the simple carbohydrates are digested in the small intestine and are absorbed as simple sugars. Most of the complex carbohydrates are digested by microorganisms in the cecum and colon. The end products are volatile fatty acids which may supply 25% or more of the energy needs of the horse.

Horses digest crude fiber about two-thirds as efficiently as do ruminants. Digestion coefficients are similar for ponies and horses. Crude fiber appears to be the best single predictor of all digestibilities in the horse. The higher level of soluble carbohydrates in legumes is a major advantage which legumes have over grasses for horses. Horses need a minimum of 0.5 lb of forage per 100 lb body weight although a higher level is preferred by many horse owners. As plants approach maturity, they form complex carbohydrates in their cell walls called cellulose and lignin which are difficult for the horse to digest. This decreases digestibility and feeding value of forages.

Glycogen serves as a carbohydrate reserve in the muscles and liver. It is involved in exercise physiology which is discussed in Chapter 22.

REFERENCES

1. Crawford, B. H., Jr., J. P. Baker, and S. Lieb. *J. Anim. Sci.* **31**, 198 (1970).
2. Hintz, H. F., J. P. Baker, R. M. Jordan, E. A. Ott, G. D. Potter, and L. M. Slade. "Nutrient Requirements of Horses." *N.A.S.–N.R.C.*, Washington, D.C. 1978.
3. Leonard, T. M., J. P. Baker, and R. E. Pulse. *J. Anim. Sci.* **37**, 285 (1973).
4. Davis, M. E. *J. Appl. Bacteriol.* **31**, 286 (1968).
5. Kern, D. L., L. L. Slyter, J. M. Weaver, E. C. Leffel, and G. Samuelson. *J. Anim. Sci.* **37**, 463 (1973).
6. Trautman, A., and E. Hill. *Schweiz. Arch. Tierheilkd.* **95**, 286 (1953).
7. Lieb, S., J. P. Baker, and B. H. Crawford, Jr. *J. Anim. Sci.* **31**, 207 (1970).
8. Annison, E. F., J. K. Hill, and D. Lewis. *Biochem. J.* **66**, 592 (1957).
9. Cook, R. M., and L. D. Miller. *J. Dairy Sci.* **48**, 1339 (1965).

10. Giddings, R. F., and C. E. Stevens. *Proc. Equine Nutr. Physiol. Symp., 1st*, p. 15 (1968).
11. Stillions, M. C., S. M. Teeter, and W. E. Nelson. *Proc. Equine Nutr. Physiol. Symp., 2nd*, p. 21 (1970).
12. Hintz, H. F., H. F. Schryver, and J. E. Lowe. *J. Anim. Sci.* **33,** 1274 (1971).
13. Argenzio, R. A., and H. F. Hintz. *J. Nutr.* **101,** 723 (1971).
14. Hintz, H. F., R. A. Argenzio, and H. F. Schryver. *J. Anim. Sci.* **33,** 992 (1971).
15. Hintz, H. F. *Veterinarian* **6,** 45 (1969).
16. Slade, L. M., and H. F. Hintz. *J. Anim. Sci.* **28,** 842 (1969).
17. Vander Noot, G. W., and E. B. Gilbreath. *J. Anim. Sci.* **31,** 351 (1970).
18. Koller, B. L., H. F. Hintz, J. B. Robertson, and P. J. Van Soest. *J. Anim. Sci.* **47,** 209 (1978).
19. Harbers, L. H., L. K. McNally, and W. H. Smith. *J. Anim. Sci.* **53,** 1671 (1981).
20. Glade, M. J. *J. Anim. Sci.* **58,** 638 (1984).
21. Vander Noot, G. W., and J. R. Trout. *J. Anim. Sci.* **33,** 38 (1971).
22. VanderNoot, G. W., and E. B. Gilbreath. *J. Anim. Sci.* **58,** 638 (1984).
23. Fonnesbeck, P. V. *J. Anim. Sci.* **27,** 1336 (1968).
24. Meyer, H., M. Coenen, and C. Geerer. *Proc. Equine Nutr. Physiol. Symp., 9th,* p. 38 (1985).
25. Laut, J. E., K. A. Houpt, H. F. Hintz, and T. R. Houpt. *Physiol. Behav.* **35,** 549 (1985).
26. Raina, R. N., and G. V. Raghavan. *Indian J. Anim. Sci.* **55,** 282 (1985).
27. Cymbaluk, N. F., and D. A. Christensen. *Can. J. Anim. Sci.* **66,** 237 (1986).
28. Olson, N., and A. Ruudvere. *Nutr. Abstr. Rev.* **25,** 1 (1955).
29. Earle, I. P., N. R. Ellis, and C. Ŝ. Greer. *Army Vet. Bull. (Washington, D. C.)* **37,** 95 (1943).
30. Robinson, D. W., and L. M. Slade. *J. Anim. Sci.* **39,** 1045 (1974).
31. Topliff, D. R., G. D. Potter, T. R. Dutton, J. L. Kreider, and G. T. Jessup. *Proc. Equine Nutr. Physiol. Symp., 8th,* p. 119 (1983).
32. Valberg, S., B. Essen-Gustavsson, A. Lindholm, and S. Person. *Equine Vet. J.* **17,** 439 (1985).

9

Fatty Acids, Fat, Volatile Fatty Acids, and Energy

I. INTRODUCTION

There are many sources of energy in the diet. Carbohydrates are the main source of energy. Fat is also a source of energy and research is underway to determine the value of adding fat to horse diets. Excess protein is utilized as energy but is an expensive source of energy and is not recommended for energy use. This chapter will discuss all sources of energy and the horse's requirement for energy.

Owners of horses are very much concerned with the energy needs of their animals and are sometimes confused by the energy terms used in various reviews. The following scheme on energy terms may be helpful in understanding this subject.

Energy Breakdown during Digestion
Gross energy (GE) in feed
  ↓ (minus energy lost in feces)
  Digestible energy (DE)
    ↓ (minus energy lost in urine and gaseous products of digestion)
    Metabolizable energy (ME)
      ↓ (minus energy lost as heat of digestion)
      Net energy (NE)
          (used for maintenance and production)

Brief explanations on the various kinds of energy are discussed below.

1. GROSS ENERGY

The amount of energy a feed will yield is measured in calories. It is usually reported as kilocalories (kcal) or megacalories (Mcal). One kilocalorie contains 1000 calories (cal), and 1 Mcal contains 1000 kcal. Fat has 9.45 kcal/g whereas protein has 5.65 and carbohydrates 4.15 kcal/g.

2. Digestible Energy

This is the gross energy minus the fecal energy (which is the energy not digested by the animal in the intestinal tract). The fecal energy is the greatest energy loss that occurs in the animal. In the horse, the loss will amount to 35–40% of the gross energy in the feed.

3. Metabolizable Energy

If the energy that is lost in the gaseous products of digestion and in the urine is subtracted from the digestible energy, it gives the metabolizable energy. In general, urine energy losses will amount to about 2–5% of the gross energy in the feed for the horse. Very little experimental information is available on metabolizable energy levels in feeds for horses.

4. Net Energy

There is another energy loss that occurs before metabolizable energy is available for productive purposes by the animal. This loss is called the *heat increment* by the scientist. The heat increment is the increase in heat production that occurs from the process of digestion and the breakdown of nutrients in the body. When the heat increment is subtracted from the metabolizable energy the remainder is net energy. Some of the heat increment can be used to keep the body warm. The net energy is used by the animal for maintenance, growth, reproduction, and lactation. Very little information is available on net energy values in feeds for horses.

A calorie is defined as the amount of heat required to raise 1 g of water 1°C (1 degree centrigrade) from 14.5° to 15.5° C. A kilocalorie (1000 calories) is the amount of heat required to raise 1000 g (i.e. 1 kg or 2.2 lb) of water 1°C from 14.5° to 15.5°C).

5. Total Digestible Nutrients (TDN)

Most horsemen are familiar with the term total digestible nutrients, but it is being used less now in scientific investigations. TDN in feeds equals the percentage of digestible protein plus percentage of digestible nitrogen-free extract plus percentage of digestible fat times 2.25. One kilogram (2.2 lb) of TDN contains approximately 4400 kcal of digestible energy. This is the same as 2000 kcal of digestible energy per lb of TDN. Therefore, if one knows the TDN value of a feed, an estimated digestible energy value can be calculated from it. The 1989 NRC report states that energy values, such as TDN, DE, or NE, for feeds obtained from experiments with cattle may be significantly higher than those obtained from experiments with horses. Therefore, they state that such energy values should not be used directly in the formulation of horse diets, particularly if the feed contains a significant amount of fiber (1).

II. ENERGY USE BY THE HORSE

Many research studies during the past few years have shed light on how the horse makes use of the energy in feeds. The horse has the enzymes and mechanism for the digestion of starches and simple sugars, such as occur in feed grains, via the stomach and small intestine. These feeds are broken down to glucose, which is mainly absorbed through the small intestine. The breakdown and fermentation of fibrous feeds, such as hay, occurs via the active microorganisms in the cecum and colon. The complex carbohydrates in fibrous feeds are digested and absorbed as volatile fatty acids, and some glucose, in the large intestine. Feeds such as grains, with simple carbohydrates, are digested to a large extent before they reach the cecum, although some simple carbohydrates do get to the cecum. The fibrous feeds such as hay and pasture, which have complex carbohydrates, are digested primarily in the cecum and colon.

In the horse, the volatile fatty acids formed in the cecum are mainly acetic acid, proprionic acid, and butyric acid. A small amount of isobutyric, isovaleric, and valeric acids are also formed. They all serve as sources of energy for the body just like glucose does. However, they go through different pathways in their breakdown to energy. The volatile fatty acid concentration in the cecum and colon will vary with the kind of diet fed and with the ratio of grain to roughage used. Changing the diet from all forage to largely grain affects the glucose levels in the blood. With high-grain diets the total concentration of volatile fatty acids and the percentage of acetic acid decreases. The levels of proprionic, isovaleric, and valeric acids, however, increase.

Some of this information is complicated and difficult to understand by many horse people. Moreover, there is still much to be learned, although progress is being made on how energy in the feed is digested by the horse. Energy utilization is important to understand because an adequate energy intake is paramount to developing a winner at the race track or in high-level performance. Horses need a certain amount of roughage in the diet for best results. One needs to learn, therefore, how roughages can be used without slowing down the performance of the horse due to a lack of sufficient energy.

As studies determine how the horse utilizes volatile fatty acids most efficiently, methods will be found to vary the kind and level of each one formed in the digestive tract. Certain volatile fatty acids may be better sources of energy than others. The ratio of certain feeds to others alters the level and the kind of volatile fatty acids formed. When this subject is well understood, feeding programs will be modified to give the optimum amount of energy needed by the horse for the specific function it is performing. For example, at a low level of work or exercise, the mature horse may consume enough hay and/or pasture to meet its body needs for energy. As the level of work or exercise increases, however, the horse will need more grain, or other high-energy feeds, to provide

There is a limit to how much grain a horse can
ιt of roughage or hay a horse needs to prevent
ى other digestive disturbances. This explains why
ىorse's minimum needs for roughage while providing
ى energy intake required for the specific function assigned

ى scientists (2,3) studied energy utilization by the horse during
ι hey showed that muscle glycogen (from carbohydrates) and free fatty
ιfrom fat) play a dominant role in supplying energy to the muscles for work
ى both unconditioned and conditioned horses. The conditioned horse adapted
itself to utilize fat, in addition to glycogen, to meet the increased energy requirements of exercise and training. The unconditioned horse, however, was not able
to oxidize (or use) fat as efficiently as the conditioned horse. It used, instead,
more muscle glycogen for exercise. This is similar to the human. The untrained
human uses one-half as much free fatty acids for energy as the human who is
trained and in good condition. The New Jersey study showed that fat is made
available to the muscles in the form of free fatty acids. They are mobilized to
meet increased energy needs. The horse in good condition uses free fatty acids as
a source of energy for exercise or running. The muscle adapts itself to meet the
increased energy needs of exercise by storing more energy and developing more
efficient pathways to metabolize or make energy available. A thorough understanding is important since the secrets of muscle metabolism and its utilization of
energy and other nutrients determine the efficiency of the running horse. Until
this occurs, there is still much guess work involved in feeding and handling the
horse that is being developed for racing purposes. The breakdown and utilization
of fat, carbohydrates, and protein in the muscle involve very complex chemical
systems that consist of many steps and the need for many nutrients. Among them
are certain minerals and vitamins. Much research is needed before scientists can
better understand how to keep horses running at full speed longer without tiring
or slowing down as much. This is a challenging problem and one which can
provide much satisfaction when solved. Basic muscle metabolic studies are very
important and should be accelerated in horse research.

III. FATTY ACIDS

A. Unsaturated and Saturated Fatty Acids

Fats consist of esters of fatty acids and glycerol. Many refer to them as
triglycerides since they consist of a combination of glycerol and three fatty acids.
Some of the fatty acids are called "polyunsaturated fatty acids" which means
they have more than one double bond. The unsaturated fatty acids form unsatu-

rated fats, which are soft; some are liquid at room temperature. Other fatty acids are "saturated fatty acids," which means they have no double bonds in their molecule. They form saturated fats sometimes referred to as hard fats. The saturated fatty acids are butyric, caproic, caprylic, capric, lauric, myristic, palmitic, stearic, arachidic, and lignoceric. The unsaturated fatty acids are palmitoleic, oleic, linoleic, linolenic, arachidonic, and clupanodonic.

B. Rancidity

The unsaturated fats are more susceptible to rancidity. Both hydrolytic and oxidative changes are involved in rancidity. Various decomposition products occur. Peroxides are formed as intermediate compounds. These changes are accelerated by heat, moisture, and light. Rancid fats have disagreeable odors and flavors that can affect their desirability in horse diets. Moreover, rancidity results in the destruction of vitamin A, carotene, vitamin E, biotin, and other nutrients that are highly sensitive to rancid fats. Rancidity can be minimized by the use of proper antioxidants. Feeds should also be stored in cool, dry areas with good air circulation. In addition it is best to use feeds in a short period of time to avoid prolonged storage. If one is located in an area where conditions are likely to cause rancidity, it is good insurance to add a little extra vitamin A and E to the diet. This will help prevent a deficiency of these two very important vitamins.

C. Essential Fatty Acids

Linoleic acid is the main essential fatty acid, although arachidonic and linolenic acid are partially effective in replacing it. Arachidonic can be synthesized in the body from linoleic acid. The essential fatty acids are referred to as those the body is not able to synthesize, or at least not adequately, to meet body requirements. Linoleic and arachidonic acid are the most effective with linolenic being effective in certain animal species under certain conditions. Arachidonic acid is found only in animal fats and in relatively small amounts whereas linoleic acid is found in other feeds and at much higher levels. Linoleic acid, therefore, is the primary source of essential fatty acids and should be considered for animal diets. The chick, for example, cannot synthesize linoleic acid. Therefore, linoleic acid must be supplied in the diet. Arachidonic acid can be synthesized only from linoleic acid.

D. Requirement for Essential Fatty Acids

The requirement for essential fatty acids for the horse is not known (1). One report (4) suggests that polyunsaturated fatty acids are important in the development and maintenance of healthy skin and hair coat. In the pig, a level of linoleic acid between 0.03 and 0.22% of the diet is needed for optimum growth and skin development (5). A level of 0.03% linoleic acid is sufficient for normal growth, but 0.22% linoleic acid is needed for normal skin development. The 1989 NRC report (1) states that until more data are available dietary dry matter should include at least 0.5% linoleic acid in horse diets.

With other animals a lack of fat or an essential fatty acid causes a slower growth rate, an enlarged fatty liver, hair loss, dandrufflike dermatitis in the skin, harmful effects on reproduction and lactation, and, eventually, death. Therefore, an essential fatty acid deficiency can have quite harmful effects.

As far as is known, an essential fatty acid deficiency does not occur with practical, well-balanced diets for horses. Less fat is now left in protein supplements since many are solvent-extracted. For example, whole soybeans contain 9% linoleic acid, but solvent-extracted soybean meal contains only 0.4% linoleic acid. Corn contains 1.9% linoleic acid as compared to 0.85% in barley, 1.1% in milo, 1.5% in oats, and 0.6% in wheat. Therefore, diets with grains other than corn are lower in linoleic acid. Commonly used horse feeds should supply sufficient linoleic acid. Rancidity development, however, might complicate this matter. Linoleic acid, arachidonic acid, and linolenic acids are all unsaturated fatty acids and susceptible to rancidity. Therefore, rancidity should be avoided in handling horse diets.

IV. ADDING FAT TO DIETS

There has been considerable interest in the value of adding fat to horse diets. Normally, diets fed to horses already contain a certain amount of natural fat. This level usually ranges from 2 to 5% of the total diet. The kind of fat, its quality, and other attributes influence the level that can be used and the response obtained when they are added to horse diets.

A. Function of Fat

The fat in the diet serves a number of functions: (1) a source of essential fatty acids; (2) a source of concentrated energy; (3) a solvent that aids in the absorption of fat-soluble vitamins; (4) increases palatability of certain feeds; (5) reduces dustiness in feeds; and (6) aids pelleting by helping the passage of feeds through the dies by its lubricating effect.

High-level performance horses expend a great deal of energy in hunting, roping, cutting, and racing. Usually, these horses are fed high levels of grain to provide extra energy. If the grain level in the feed is too high it may cause compaction, founder, digestive disturbances, and possibly other problems. There is also a limit to the capacity of the digestive tract and the amount of feed that can be fed. Therefore, fats and oils, which contain about 2.25 times as much energy as carbohydrates, have been suggested as a possible concentrated source of energy for high-level performance horses. Body fat is mobilized during exercise, and free fatty acids are oxidized readily, especially during strenuous exercise (2,6). But, both of these studies reported that unconditioned horses do not appear to oxidize fat as efficiently as conditioned horses.

B. Effect of Added Fat

Horses accept fat additions to the diet quite readily if the fat is not rancid. A main reason for adding fat to the diet is to enable high-level performance horses to consume a high-energy diet that is highly digestible. Fat is readily digested by horses at fairly high levels. In one study (7) fat at the 20% level was 94% digested. A number of studies have shown that fats can be added to equine diets without causing palatability or digestibility problems (7–12).

Training has been shown to increase the utilization of fat as an energy source. The unconditioned horses use both fat and carbohydrates as energy sources but the conditioned horses exhibit greater utilization and a greater ability to oxidize fat (2). Increased fat utilization during exercise occurred in a Colorado study (12) as suggested by a progressive increase in blood fatty acids during exercise in trained horses.

There is interest in the utilization of fat as a means of increasing or sparing muscle glycogen. Additional detail on this is given in Chapter 22 on exercise physiology.

An excellent study with ponies was conducted at Virginia Polytechnic Institute to determine how much corn oil could be used in the diet (8). In a preliminary palatability trial, a diet containing 30% corn oil by weight was the upper level accepted. However, the consumption level of the 30% corn oil diet was very low, and it was felt that animals would not consistently eat a diet containing this level of corn oil. In their other studies they used 0, 5, 10, and 20% corn oil in the diet. The diets fed contained approximately 55% chopped alfalfa hay, 25% cracked corn, and 20% crimped oats fed in a loose form. Corn oil replaced 5, 10, and 20% of the total feed intake. The control diet (no fat added) provided a maintenance level of energy, with about 20% excess protein, calcium, and phosphorus. The additional protein and minerals were included to compensate for the amount replaced by the corn oil at the 20% level. The addition of the corn oil to the diet did not significantly affect the apparent digestibility of the crude protein or the true digestibility of fat. The digestibility of the corn oil, averaged over the three levels fed, was approximately 90%. The blood hemoglobin and hematocrit levels remained normal and were unaffected by the levels of oil used. The blood serum cholesterol levels increased the greatest with the 5% corn oil diet, but they were not significantly altered further with the higher levels of corn oil. The cholesterol values were 122, 144, 148, and 155 mg per 100 ml of blood serum according to the Virginia Polytechnic Institute scientists. The blood serum triglyceride levels were not affected by the addition of corn oil. There was no evidence of digestive disturbances in any of the animals fed the different levels of corn oil.

In a Cornell study (13), horses were fed a pelleted diet with and without 8% feed-grade animal fat. The control diet contained 40% alfalfa meal and 60% corn plus trace mineral salt fed free-choice. Both groups received 6 lb of timothy hay per horse daily. Both groups were conditioned for 9 weeks. After this period, four trail rides of 37 miles at 6 miles per hour were conducted. The rides were divided as follows: 6.8 miles ridden, 10 minute stop; 13.6 miles ridden, 1 hour lunch stop; 9.8 miles ridden, 10 minute stop; and then 6.8 miles ridden. After the first two rides, which were two weeks apart, the diets were switched and two more trail rides conducted after a 3-week adjustment period. They concluded that the performance of the horses was not affected by the added fat. They also indicated that further studies are needed to determine if horses ridden more than 37 miles would benefit from added fat. The study demonstrated that: (1) horses could utilize diets with 8% fat; (2) they required less total feed daily; and (3) they consumed the 8% fat diet with no difficulty. Furthermore, there was a partial protection against a decline in blood plasma glucose and a superior free fatty acid (FFA) to glucose ratio, which suggests that the fat probably was of some benefit. They also stated that preliminary work indicates that horses fed the 8% fat diet were less likely to founder than horses fed a high-carbohydrate diet. The Cornell study justifies more research to determine the value of fat and its effect on endurance in the horse. A recent Utah study (14) showed that the addition of a combination of 12% vegetable oil and 1000 IU of vitamin E resulted in the most desirable response in blood packed cell volume, hemoglobin, and glucose after exercise in endurance horses. Kentucky (15) added 15 and 30% corn oil to increase the energy content of horse diets. There is good reason for these fat studies, since it has been reported that the addition of fat improved the performance of pigeons (16) flying more than 200 miles and sled dogs (17) racing for 12 hours. Illinois scientists are working on methods to measure energy expenditure and physical condition in horses (18).

A VPI study (19) involved feeding mature ponies 0, 7.5, 10, and 15% of corn oil, peanut oil, inedible tallow, and animal vegetable blend no. 3. The addition of fat increased the digestible energy in the diet. The addition and kind of fat did not affect the apparent digestibility of acid detergent fiber or apparent absorption of minerals. Blood glucose levels were not significantly affected by level, kind, or addition of fat. Blood cholesterol levels were increased and higher at the higher fat levels but were in the normal range. Blood linoleic levels were not affected by the addition of fat but palmitic, stearic, and oleic acid were increased.

In another VPI study (20) 15% of either corn oil, inedible tallow, or blended fat (animal and vegetable fat mixture) was added to diets of mature gelding ponies. The addition of 15% fat increased the apparent digestibility of energy in the diet with the highest increase occurring with corn oil followed by inedible tallow and the blended fat. Total heat production increased with the addition of fat. The metabolizable energy was greater than the heat production increase due to the addition of fat. Thus, the energy balance for the 15% fat diets was approximately 20% greater than the control diet.

A Texas A & M study (11) found that feeding Quarter Horse mares 5% feed-grade-rendered fat in the diet resulted in shorter postpartum intervals (12 versus 14 days), fewer cycles per conception (1 versus 2 cycles), and a higher pregnancy rate (100 versus 89%), although the differences were not statistically significant.

A study at Texas A & M (21) with yearling Quarter Horses showed that 5 and 10% fat in the diet can be used for rapidly growing horses to replace carbohydrates with good results. There were no clinical or radiographic indications of skeletal abnormalities in the horses fed the added fat.

Texas A & M researchers (22) reported that adding 5 and 10% fat (feed-grade-rendered animal fat) to the diet of exercising mature Quarter Horse type geldings reduced the feed intake required to maintain constant body weight and condition. It appeared that adding fat to the exercising horse's diet had a sparing effect on muscle glycogen reserves.

A Colorado State University study (12) involved feeding four levels of soybean oil (4, 8, 12, and 16% of diet) to horses used for prolonged submaximal exercise. The four levels of fat had no adverse effects and proved a safe and efficient method of providing concentrated energy. Resting muscle glycogen increased 37% during the study as a result of conditioning, but there was no effect on liver glycogen at rest or after exercise. Increases in fat intake resulted in slight increases in liver glycogen at the resting level.

A Texas A & M study (23) showed that mature horses fed 10% fat in the diet provided more energy for work and may have increased muscle glycogen storage and subsequent mobilization of that glycogen for the anaerobic work typical of the cutting horse.

Arabian horses ridden for 36 miles required 15% less total feed when fed a diet containing 8% fat (12). Standardbred horses in a study in Sweden (9) fed 18% fat exhibited increased fatty acid metabolism and decreased lactic acid production when performing at 75% of their maximum capacity. A VPI study (24) showed no detrimental effects from feeding a 14% fat diet to horses during a period of conditioning.

A Texas A & M study (25) reported that feeding a moderately high fat diet (10% fat) increased muscle glycogen storage in fit horses, as well as increasing the amount of muscle glycogen utilized during an anaerobic workout. They stated that supplementing a performance horse's diet with fat will increase performance not only at aerobic intensities as has been previously shown, but also performance past the anaerobic threshold.

Another Texas A & M study (26) showed that adding 10% fat to the diet of weanling horses promoted greater efficiency and improved nutrient digestibility. No clinical or radiographic evidence of bone abnormalities occurred in any of the horses as has been reported by other scientists for young, rapidly growing horses.

A VPI study (27) involved feeding 10.5% fat to 2-year-old horses just starting exercise training. Their study indicated that the added fat appeared to have a muscle glycogen-sparing effect.

A Texas A & M study (28) showed that adding 10% fat to a concentrate diet improved the efficiency of digestible energy for maintenance of mature horses.

An Illinois study (29) concluded that growth in horses may be influenced by the dietary energy source (carbohydrate versus fat) and this effect may be mediated by the endocrine system and its regulatory effects on metabolic pathways.

A Texas A & M study (30) indicated that estimating digestible energy requirements without regard to season, ambient temperature, humidity, or body condition may result in substantial error in estimating feed requirements for the performance horse.

Another Texas A & M study (31) showed that excessive body fat stored during gestation in the multiparous mare may adversely affect milk production. The obese mares produced 3.08 lb less milk daily than mares with a moderate degree of body fat.

The addition of 5 or 10% fat in a Texas A & M study (32) to the diet of performance horses caused an overall decrease in heart rate, blood lactic acid, and blood total lipids, and an increase in packed cell volume. These workers felt that the lowered blood lipids might have been due to conditioning for more efficient fatty acid uptake by the working muscle.

A Kentucky study (33) reported on feeding Thoroughbred horses 0, 5, 10, or 20% corn oil in the diet. The horses were handled and trained in an effort to duplicate a race track environment. Blood glycerol levels were elevated above resting values throughout exercise and recovery indicating that fat was being mobilized. The corn oil supplemented diets proved to be very palatable and highly digestible.

As one adds higher levels of fat to the diet, the level of protein needs to be increased because protein requirements are directly related to the energy or calories in the diet. Certain minerals or vitamins may also need to be increased since they may also be related to the higher energy level in the diet.

V. BODY FAT

Cornell scientists (34) used ultrasonic measurements to estimate fat cover in horses and ponies. Ponies were fed in two lots, with one group being fed all the feed they could consume and the other being limited-fed. The full-fed group consumed feed at about 3% of body weight whereas the limited-fed animals were limited to 1.5% of body weight. The full-fed ponies had more fat over the rump, rib, and shoulder than the limited-fed group. Interestingly the limited-fed ponies had more fat over the shoulder than they had over the rump or ribs (which was different from full-fed ponies) as shown in the figures in the following tabulation (measurements made by ultrasound and given in cm).

	Rump	Rib	Shoulder	Actual rump fat at slaughter
Full-fed	1.30	0.30	0.83	1.53
Limited-fed	0.44	0.13	0.67	0.46

Twelve horses used by the Cornell Polo Team were studied for fat thickness over the shoulder, rib, and rump after 30 and 90 days of exercise (34). The fat in the shoulder and rump decreased after 30 days of exercise, but rib fat and body weight did not change. The rump fat was the thickest and the rib fat was the thinnest of the three areas measured.

Eight horses were used in a trial (34) to evaluate fat thickness and then slaughtered and analyzed for chemically extractable fat. Rump and shoulder fat were correlated with extractable fat. The study showed that extractable body fat could be accurately predicted from the rump fat measurement with ultrasound. The rump had the thickest fat cover followed by the shoulder and rib.

Eleven ponies were studied by the Cornell scientists (34); they showed the same patterns of fat deposits as the horse. The rump had more fat than the shoulder area and the rib area had the least amount of fat cover. The studies also showed that body fat can be accurately estimated in ponies from rump fat measurements. Ultrasonic measurements were made at 30 and 80 days on five exercised and six control (not exercised) ponies. The animals were slaughtered, and the extractable fat was determined. Values for rib and rump fat thickness, and extractable fat were lower for the exercised than for the nonexercised ponies. No differences were obtained between the control and exercised groups of ponies in estimated loin-eye area. Total body fat at slaughter for the ponies that were not exercised was 15.03%. For the ponies that were exercised, total body fat was 8.96%. The weights of the control and exercised groups of ponies at the initiation of the trial were 324.3 and 296.5 lb, respectively. At the time of slaughter after 80 days of exercise the weights of the control and exercised groups of ponies were 369.2 and 307.8 lb, respectively. The ponies that were not exercised gained 44.9 lb, whereas those that were exercised gained 11.3 lb. The difference in gain undoubtedly had some influence on the amount of body fat in the ponies.

The Cornell scientists concluded that fat cover in horses and ponies can be estimated by ultrasound measurements. They also concluded that the rump fat measurement by ultrasound is useful in predicting total body fat. This information can be used to determine: (1) the effect of different feeds and diet; (2) different levels of feeding; (3) different supplements and feed additives; and (4) the effects that different management programs have on body fat. This can also be correlated with different kinds and levels of exercise. Ultimately all of this information is needed to determine what effect fat may have on the performance of the horse. It could be that the level of body fat, if it goes beyond a certain point, may affect the ability of the horse to race, perform, or reproduce at its maximum. It is logical to assume that excessive fat may be detrimental to racing and performance. The unanswered question is how much body fat is detrimental and under what conditions would it be harmful.

Texas A & M research (35) has demonstrated that mares should be fed to a fatter condition than previously recommended to enhance reproductive performance. Thin mares gaining in weight at time of breeding were twice as likely to conceive as thin mares maintaining weight. In another Texas A & M study (36) it was shown that, in the multiparous mare, foaling was not impaired by an extremely high degree of fatness produced by overfeeding during gestation. Evidently horses can tolerate a higher level of fatness than can cattle or swine. But, milk production may be decreased in the obese mare (31). Table 18.1 gives a condition scoring system and Fig. 18.5 shows the areas of fat deposition in the horse as developed at Texas A & M (37). More detail on this subject is given in Chapter 18.

Hyperlipidema is an important clinical problem in small pony breeds (38). It is most common in mares in late gestation and lactation, and occurs when the animal is in negative energy balance because of inappetence, starvation, parasitism, pregnancy, lactation, or transportation. Common clinical symptoms include lethargy and dullness, progressing to severe depression, coma, and death. Fatty infiltration of the liver occurs in affected ponies, with concurrent hepatic dysfunction. Many hyperlipidemic ponies also exhibit pancreatitis and reduced insulin production.

VI. ENERGY REQUIREMENTS

A Canadian study (39) with young horses with *ad libitum* access to pelleted diets containing at least 60% forage ate 21–31% more dry matter and consumed 19–44% more digestible energy than horses fed diets limited to 1978 NRC (40) digestible energy guidelines for growth. The study also showed that 60% forage did not effectively restrict energy intake by young horses fed *ad libitum*. They indicated that, if energy intake is to be regulated in young horses, both concentrate and hay intake must be controlled.

A study at Texas A & M (41) showed that the digestible energy required to maintain the horse without forced exercise averaged 16.95 Mcal per day. A Cornell study (42) reported that the amount of energy expended by horses was exponentially related to speed and was proportional to the body weight of the riderless horse or the combined weight of the horse plus rider and tack. Another Cornell study (43) showed that the average digestibility of gross energy was 62.9 ± 0.8%. Urinary losses averaged 8.5 ± 0.32% of the digestible energy intake.

A Texas A & M study (44) with Belgian and Percheron horses showed that it apparently takes less feed per unit of body weight to maintain heavy draft horses than lighter horses. This is an observation that is frequently voiced by draft horse breeders. The Texas A & M scientists concluded that the lower energy needs may be due to a lower rate of activity and especially since draft horses do not appear to start and stop as suddenly as lighter horses when they move.

A deficiency of energy in the young horse results in a poor growth rate. In mature horses, it results in weight loss, poor condition, and performance. A lack of enough to eat is a common form of malnutrition with many horses.

The energy required for work is shown in Table 9.1. The data are based on two different studies (45,46). The information shown in Table 9.1 indicates the big difference in energy needs between various activities. For example, the energy expenditure for a strenuous effort, such as racing at full speed, requires considerably more energy than walking. This large difference indicates the need for an expanded research effort to understand thoroughly how to provide energy most effectively for optimum racing speed and endurance.

Estimates of energy needed for pregnancy during the last 90 days of gestation are 12% greater than for maintenance (40). The 1989 NRC report indicated that estimates of DE requirements for the ninth, tenth, and eleventh month of gestation were formulated by multiplying the maintenance requirements by 1.11, 1.13, and 1.20, respectively (1). The NRC committee (40) recognized that the

TABLE 9.1

Energy Requirements for Work[a]

	Digestible energy (DE) per hour[b]	
Activity	Per kg of body weight (kcal)	Per lb of body weight (kcal)
1. Walking	0.5	0.2
2. Slow trotting, some cantering	5.0	2.3
3. Fast trotting, cantering, some jumping	12.5	5.7
4. Cantering, galloping, jumping	23.0	10.5
5. Strenuous effort (polo, racing at full speed)	39.0	17.7

[a]Data from Hintz et al. (40).
[b]Above maintenance requirement.

voluntary intake of hay will decrease as the fetus increases in size. Therefore, the energy density of the diet should be increased during the last 90 days of pregnancy. The requirements for energy during the various stages of the horse's life cycle are shown in Tables 6.5 and 6.6.

The requirements for energy during lactation are shown in Tables 6.5 and 6.6. Mare of light breeds may produce as much as 52.8 lb of milk per day at peak lactation, which occurs at about 8 weeks, but the average milk production is probably within the range of 26.4–39.6 lb daily (40). It is estimated that ponies have an average daily milk production of 4 and 3% of body weight during early and late lactation, respectively (1,40). Mares of light breeds are estimated to produce milk equivalent to 3–2% of body weight during early lactation (1–12 weeks) and late lactation (13–24 weeks), respectively (1,40). The 1989 NRC report (1) used the assumption that it takes 792 kcal of DE per kg of milk produced which is the same figure used in the 1978 NRC report (40).

The nursing foal is assumed to be approximately 10% more efficient in utilization of digestible energy than are mature horses (40).

There is some variation in the energy needed for maintenance among horses (47–51). Some of the variation might be related to differences in the temperament of the horses or the breed involved (52).

Table 9.2 shows the digestible energy requirements for growth of foals. As the foal increases in age, it requires more energy per unit of body weight gain.

The 1989 NRC (1) assumed that for ponies and light horses (200–600 kg), light, medium, and intense work increased the energy requirements 25, 50, and 100% above maintenance, respectively. The energy requirements for draft horses depend on many factors such as size of load and type of work. Increasing maintenance requirements by 10% for each hour of field work should provide a reasonable guide (53) on energy needs.

Considerably more detail is given on energy requirements of the horse in the 1989 NRC publication on nutrient requirements of the horse (1).

TABLE 9.2

Digestible Energy Requirements for Growth
of Foals (1)

Age	Requirement (Mcal DE/kg of gain)
Weanling, 4 months	9.1
Weanling, 6 months	11.0
Yearling	15.5
Long yearling	18.4
2 years old	19.6

VII. CONCLUSIONS

Carbohydrates are the main source of energy in horse diets. Simple carbohydrates are broken down to glucose and absorbed through the small intestine. The fermentation of fibrous feeds by microorganisms in the cecum and colon results in volatile fatty acid production. They are absorbed in the large intestine and may provide about 25% of the energy needs of the horse.

Linoleic acid is an essential fatty acid, and it is assumed that well-balanced diets supply the horse's requirement for it. Rancidity in feeds, however, may destroy unsaturated fatty acids (linoleic acid is one of them) and so rancidity should be guarded against since it destroys other nutrients as well.

Levels of up to 20% fat can be added to horse diets with good results. But, most studies have involved adding 5–10% fat in the diet for exercising horses. Until more research is done, 5–10% fat in the diet of high-level performing horses should be the top limit used. Adding fat to their diet increases energy density and decreases diet volume, total feed intake, and digestive tract volume which should be helpful for them. It also increases the level of protein required and may increase the level of certain vitamins and minerals needed for utilization of the extra calories provided.

Excess feed intake increases body fat. Excess fat may decrease performance in high-level performance horses, but it may be beneficial for the mare during gestation and reproduction. Obesity should be guarded against, however, since it may decrease milk production.

Recent studies indicate that the energy requirements of the horse are higher than previously recommended by the 1979 NRC report (40). The new values are given in the 1989 NRC report and are discussed in this chapter (1).

REFERENCES

1. Ott, E. A., J. P. Baker, H. F. Hintz, G. D. Potter, H. D. Stowe, and D. E. Ullrey. "Nutrient Requirements of Horses." NAS-NRC, Washington, D. C., 1989.
2. Goodman, H. M., G. W. Vander Noot, J. R. Trout, and R. L. Squibb. *J. Anim. Sci.* **37,** 1 (1973).
3. Goodman, H. M., and G. W. Vander Noot. *J. Anim. Sci.* **33,** 319 (1971).
4. Mix, L. S. *VM/SAC, Vet. Med. Small Anim. Clin.* **61,** 598 (1966).
5. Cunha, T. J., J. P. Bowland, J. H. Conrad, V. W. Hays, R. J. Meade, and H. S. Teague. "Nutrient Requirements of Swine." *N.A.S.–N.R.C.,* Washington, D.C. (1973).
6. Anderson, M. G. *Equine Vet. J.* **7,** 27 (1975).
7. Bowman, V. A., J. P. Fontenot, K. E. Webb, Jr., and T. N. Meacham. *Proc. Equine Nutr. Physiol. Symp., 5th,* p. 40 (1977).
8. Bowman, V. A., J. P. Fontenot, K. E. Webb, and T. N. Meacham. *Va., Agric. Exp. Stn., Livest. Res. Rep.* **172,** 72 (1977).
9. Kane, E., J. P. Baker, and L. S. Bull. *J. Anim. Sci.* **48,** 1379 (1977).

10. Bowman, V. A., J. P. Fontenot, T. N. Meacham, and K. E. Webb, Jr. *Proc. Equine Nutr. Physiol. Symp., 6th,* p. 74 (1979).

11. Davison, K. E., G. D. Potter, L. W. Greene, J. W. Evans, and W. C. McMullan. *Proc. Equine Nutr. Physiol. Symp., 10th,* p. 87 (1987).

12. Hambleton, P. L., L. M. Slade, D. W. Hamar, E. W. Keinholz, and L. D. Lewis. *J. Anim. Sci.* **51,** 1330 (1980).

13. Hintz, H. F., M. Ross, F. R. Lesser, P. F. Leids, K. K. White, J. E. Lowe, C. E. Short, and H. F. Schryver. *Proc. Cornell Nutr. Conf.* p. 87 (1977).

14. Slade, L. M. Utah Agricultural Experiment Station, Logan (unreported data), 1979.

15. Kane, E., J. P. Baker, and L. S. Bull. *J. Anim. Sci.* **48,** 1379 (1979).

16. Goodman, H. M., and P. Greminger. *Poult. Sci.* **98,** 2058 (1969).

17. Kronfeld, D. S., E. P. Hammel, C. F. Ramberg, and H. L. Dunlap. *Am. J. Clin. Nutr.* **30,** 419 (1977).

18. Burke, D. J., and W. W. Albert. *J. Anim. Sci.* **46,** 1666 (1978).

19. Rich, G. A., J. P. Fontenot, and T. N. Meacham. *Proc. Equine Nutr. Physiol. Symp., 7th,* p. 30 (1981).

20. Snyder, J. L., T. N. Meacham, J. P. Fontenot, and K. E. Webb. *Proc. Equine Nutr. Physiol. Symp., 7th,* p. 144 (1981).

21. Scott, B. D., G. D. Potter, J. W. Evans, J. C. Reagor, G. W. Webb, and S. P. Webb. *Proc. Equine Nutr. Physiol. Symp., 10th,* p. 101 (1987).

22. Meyers, M. C., G. D. Potter, L. W. Greene, S. F. Crouse, and J. W. Evans. *Proc. Equine Nutr. Physiol. Symp., 10th,* p. 107 (1987).

23. Webb, S. P., G. D. Potter, and J. W. Evans. *Proc. Equine Nutr. Physiol. Symp., 10th,* p. 115 (1987).

24. Worth, M. J., J. P. Fontenot, and T. N. Meacham. *Proc. Equine Nutr. Physiol. Symp., 10th,* p. 145 (1987).

25. Oldham, S. L., G. D. Potter, J. W. Evans, S. B. Smith, T. S. Taylor, and W. S. Barnes. *Proc. Equine Nutr. Physiol. Symp., 11th,* p. 57 (1989).

26. Davison, K. E., G. D. Potter, J. W. Evans, L. W. Greene, P. S. Hargis, C. D. Corn, and S. P. Webb. *Proc. Equine Nutr. Physiol. Symp., 11th,* p. 95 (1989).

27. Greiwe, K. M., T. N. Meacham, G. F. Fregin, and J. L. Walberg. *Proc. Equine Nutr. Physiol. Symp., 11th,* p. 101 (1989).

28. Potter, G. D., S. P. Webb, J. W. Evans, and G. W. Webb. *Proc. Equine Nutr. Physiol. Symp., 11th,* p. 145 (1989).

29. Lawrence, L., J. Pagan, M. Pubols, J. Reeves, K. White, R. Douglas, and C. Gaskins. *Proc. Equine Nutr. Physiol. Symp., 11th,* p. 151 (1989).

30. Webb, S. P., G. D. Potter, J. W. Evans, and G. W. Webb. *Proc. Equine Nutr. Physiol. Symp., 11th,* p. 279 (1989).

31. Kubiak, J. R., J. W. Evans, G. D. Potter, P. G. Harms, and W. L. Jenkins. *Proc. Equine Nutr. Physiol. Symp., 11th,* p. 295 (1989).

32. Meyers, M. C., G. D. Potter, J. W. Evans, L. W. Greene, and S. F. Crouse. *Proc. Tex. A & M Agric. Ext. Serv., Horse Short Course* p. 20 (1987).

33. Durea, S., C. Wood, and S. Jackson. *Univ. Ky., Equine Line* Sept., p. EL-1 (1987).

34. Westervelt, R. G., J. R. Stouffer, H. F. Hintz, and H. F. Schryver. *J. Anim. Sci.* **43,** 781 (1976).

35. Henneke, D. R., G. D. Potter, and J. L. Kreider. *Theriogenology* **21,** 897 (1984).

36. Kubiak, J. R., J. W. Evans, G. D. Potter, P. G. Harms, and W. L. Jenkins. *Proc. Equine Nutr. Physiol. Symp., 10th,* p. 233 (1987).

37. Henneke, D. R., G. D. Potter, J. L. Kreider, and B. F. Yeates. *Equine Vet. J.* **15,** 371 (1983).

38. Jeffcott, L. B., and J. R. Field. *Vet. Rec.* **116,** 461 (1985).

39. Cymbaluk, N. F., G. I. Christison, and D. H. Leach. *J. Anim. Sci.* **67,** 403 (1989).

40. Hintz, H. F., J. P. Baker, R. M. Jordon, E. A. Ott, G. D. Potter, and L. M. Slade. "Nutrient Requirements of Horses." *N.A.S.–N.R.C.*, Washington, D.C. (1978).
41. Anderson, C. E., G. D. Potter, J. L. Krieder, and C. C. Courtney. *J. Anim. Sci.* **56,** 91 (1983).
42. Pagan, J. D., and H. F. Hintz. *J. Anim. Sci.* **63,** 822 (1986).
43. Pagan, J. D., and H. F. Hintz. *J. Anim. Sci.* **63,** 815 (1986).
44. Potter, G. D., J. W. Evans, G. W. Webb, and S. P. Webb. *Proc. Equine Nutr. Physiol. Symp., 10th,* p. 133 (1987).
45. Hintz, H. F., S. J. Roberts, S. Sabin, and H. F. Schryver. *J. Anim. Sci.* **32,** 100 (1971).
46. Kossila, V., E. Virtanen, and J. Maukoneu. *J. Sci. Agric. Soc. Finl.* **44,** 217 (1972).
47. Barth, K. M., J. W. Williams, and O. G. Brown. *J. Anim. Sci.* **44,** 585 (1977).
48. Hintz, H. F. *Proc. Cornell Nutr. Conf.* p. 47 (1968).
49. Hoffman, L., W. Klippel, and R. Schiemann. *Arch. Tierernaehr.* **17,** 441 (1967).
50. Stillions, M. C., and W. E. Nelson. *J. Anim. Sci.* **34,** 981 (1972).
51. Wooden, G., K. Knox, and C. L. Wild. *J. Anim. Sci.* **30,** 544 (1970).
52. Wolfram, S. A., J. C. Williard, J. G. Williard, L. S. Bull, and J. P. Baker. *J. Anim. Sci.* **43,** 261 (1976).
53. Brody, S. "Bioenergetics and Growth." Reinhold, New York, 1945.

10

Water Quality and Needs

I. INTRODUCTION

An adequate intake of water at all times is very important. Water is one of the most important nutrients needed by the horse, but it is frequently neglected (Fig. 10.1). A horse, or any other animal, will survive longer without feed than without water. An animal can lose practically all of its fat and over 50% of its protein and still live, but a loss of 10% of its body water results in disorders, and a 20% water loss results in death. A lack of water can lead to digestive disturbances such as colic and founder (1). Therefore, water should never be lacking in the diet. If a horse is given a free choice, the horse will consume enough water to maintain a proper water balance (Fig. 10.2).

A number of mineral elements, such as sodium, chloride, potassium, calcium, and others, are lost in the sweat and urine during physical activity. Electrolyte loss increases in line with the duration and degree of physical activity and with increasing temperature and humidity. The loss of these minerals causes decreased feed intake, fatigue, muscle weakness, and decreased ability to perform. To prevent this occurrence, the minerals which are lost should be replaced. The use of electrolytes and frequent watering is helpful, especially in endurance races or prolonged exercise or activity. After strenuous activity, however, the horse should be cooled before being allowed to drink all the water it wants.

The cecum is the main site of water absorption, but considerable amounts of water are also absorbed from the colon (2).

The horse obtains water by drinking it and from metabolic water which is formed in the body during the oxidation or breakdown of carbohydrates, protein, and fat. There are 41, 107, and 60 g of water formed per 100 g of protein, fat, and carbohydrates, respectively oxidized in the body (3).

II. WATER FUNCTION

There are many functions that water performs in the body. Some of the more important ones are as follows:

1. It is a carrier for various nutrients and other compounds absorbed and transported throughout the body.

2. It is a carrier of waste products removed from all cells and organs of the body. The urine contains the soluble products of metabolism that must be eliminated. The others are eliminated in the feces.

3. It facilitates many functions of absorption, digestion, and metabolic reactions.

4. It is involved in body temperature regulation. It helps cool the animal via sweating and evaporation from the skin and the breathing process.

5. It is needed to maintain proper water balance in the body. It is a constituent of all cells and body fluids. Surplus water is excreted via urine, feces, perspiration, and vapor from the lungs via breathing.

6. Saliva, which is very important in moistening the feed and in digestion, requires water. It is the first step in feed intake.

7. It helps lubricate joints via synovial fluid.

8. It cushions many body organs and systems.

Fig. 10.1 This is an example of a water tank which can be used with horses. The photo was taken at T-Cross Ranch, Peyton, Colorado. (Courtesy of Larry Slade, Utah State University.)

Fig. 10.2   Quarter Horse mares and foals in an excellent quality pasture near a water stream. Note the quality and excellent condition of the mares and their foals. (Courtesy of Don Treadway, *American Quarter Horse Association and Quarter Horse Journal.*)

III. FACTORS AFFECTING WATER REQUIREMENT

The body of the mare contains about 60% water, 17% protein, 17% fat, and 4.5% mineral matter (ash). This does not include the constituents inside the digestive tract. A Kentucky study (4) with mature pony geldings showed the empty body composition to be 65.87% water, 19.51% protein, 8.06% ether extract, and 5.37% ash. The empty body composition on a fat-free basis for the ponies was 71.87% water, 21.3% protein, and 5.84% ash. The empty body composition on a dry fat-free basis was 78.89% protein and 21.11% ash (4). The younger the horse, the more water its body contains. The tissue of young horses contains about 70–80% water which indicates the extra need for water during growth (5). The fatter the animal becomes, the less water its body contains. As a horse matures, it requires less water per 100 lb body weight. This is due to the horse consuming less feed per unit of body weight and more of the water in the body being replaced by fat.

The water intake of the horse will correlate with the dry matter or total feed intake. However, the composition of the dry matter or feed will influence the water needs. For example, in one experiment in New Jersey (6), the daily water intake of horses was 69 lb with an all-hay diet, whereas it was 38.5 lb with a hay–grain diet. German studies also indicate a higher water consumption when horses are fed hay (7). Water consumption will also vary with the kind of hay or grain that is fed. This is because the mineral, fiber, and other constituents in different feeds will affect water intake.

Studies by Dr. P. V. Fonnesbeck of New Jersey (6) showed that an average water intake was 3.6 lb per lb of hay when horses were fed only hay. Water consumption was 2.9 lb per lb of a hay–grain diet for horses. He also found that water intake per lb of dry matter did not correlate with protein content of the diet but that it correlated significantly with the ash and cell wall content of the diet (6). A review of a number of studies suggests that the horse needs about 2–4 lb of water per lb of feed. This is an approximate guideline to use if one wants to determine the water intake of horses. However, the climate or temperature in the area will also influence water needs. Warm temperatures will increase water requirements. A rise in the environmental temperature from 55 to 77°F increases the water requirement by 15–20% (5,8). The amount of water lost in the feces is also a factor. Diarrhea is one of the most common causes of dehydration, thus increasing water needs. The amount of riding, exercise, or work demanded of a horse will have considerable influence on water needs and could double that needed by a horse that is resting. Lactating mares may increase water needs by 50–65% depending on their level of milk production.

The more digestible the feed is, the less the amount of fecal residue that is left and the less water required for its excretion. With the most digestible feeds, therefore, the horse consumes less water per unit of dry matter. This is why less water is consumed with high-grain diets as compared to high-hay diets. Fecal water is highest in horses consuming hay. Thus, the quantity of feces is the main factor contributing to water excretion in the horse.

The amount of water excreted in the urine is correlated with the total amount of dry matter digested. This is because water is required for the absorption of digested nutrients from the intestinal tract. Therefore, the water absorbed with the nutrients from the digestive tract must be voided to maintain water balance in the body of the horse. This explains why there is more urinary excretion with legume forages as compared to grass forages. The horses consume more dry matter and more digestible dry matter with legume than with grass hays (this is assuming comparable quality and maturity of the hay). Therefore, the horse consumes more water, absorbs more of this water from the intestinal tract, and voids more of it in the urine. Under conditions of water scarcity, the horse may concentrate its urine to some extent by resorbing a greater amount of water than usual, thereby lowering its requirement for water (3).

Ponds and stagnant waters of any kind are likely to be polluted. Many disease outbreaks and parasite infestations are often traceable to contaminated water sources. Thus, one needs to exercise caution in watering horses from ponds.

Many horse owners use automatic waterers and have them available in all stalls and corrals (Fig. 10.3). This ensures a continuous supply of clean, fresh water. Whatever type waterers are used, they should have drains for easy and frequent cleaning (Fig. 10.4). In cold areas of the country, the water should be warmed to keep it from freezing, otherwise, the horses may not get enough water to drink. It is recommended that the water be heated to approximately 45°F in cold areas. Horses may suffer from a lack of water when it is carried or hauled to them because the caretaker may not have or take the time to keep them supplied with water at all times.

A lack of water lessens feed consumption, which decreases growth, feed efficiency, and productivity. Some horses can detect differences in water when

Fig. 10.3 One of the many types of automatic waterers available. This is a float-type waterer and located on the outside of the fence to avoid possibility of injury. (Courtesy of Professor Norman K. Dunn, Cal Poly University, Pomona.)

Fig. 10.4 Pasture waterer with pressure regulator and float-type valve. Concrete slab around water-er keeps area dry and avoids deep mud holes. Many older horses and especially weanlings will hesitate to drink due to deep mud. Tank located at Cal Poly University, Pomona. (Courtesy of Professor Norman K. Dunn, Cal Poly University, Pomona.)

they travel from one city to another on the show or racing circuit. This can affect their water and feed intake and throw them off feed. Some horse owners counter this by adding a small amount of cane molasses to the drinking water while on the road to mask any differences in taste. This is beneficial for the few horses that otherwise would not perform to their full potential. Sometimes paying attention to a small detail like this can mean the difference between having a champion or just another horse.

Within a few hours, hard working horses may lose more than 11 lb of sweat, which contains considerable amounts of electrolytes, per 220 lb body weight (9). A German study indicated that feeding 3–4 hours before an event, along with giving the horse free access to water, probably can help the horse during exercise to balance water and electrolyte metabolism (7).

Whenever possible, having water available near shade encourages horses to drink more frequently than if the water source is in a sunny, hot area.

Donkeys and ponies have less water in their feces than horses fed the same diet. There are indications that donkeys and mules are more tolerant of a lack of water than horses.

IV. SAFE MINERAL LEVELS IN WATER

The levels that the NRC recommends as being a safe level in water for a number of mineral elements for livestock are shown in Table 10.1 (3). Since specific data for the horse are not available, the levels shown for livestock can be used as a guide for the horse as well.

Water sometimes contains mineral elements at high enough levels to be toxic. Mineral elements such as chromium, copper, cobalt, and zinc seldom cause any problem with farm animals because they do not occur at high enough levels in soluble form or because they are toxic only in excessive concentrations. Also, these mineral elements do not appear to accumulate in meat, milk, or eggs to any extent that would constitute a problem in livestock drinking water under any but the most unusual conditions. Mineral elements such as lead, mercury, fluorine, and cadmium must be considered actual or potential problems, because they are occasionally found in waters at toxic levels.

A toxic level of a mineral element in water consumed for a short period of

TABLE 10.1

Mineral Elements in Water[a]

Item	Safe upper limit of concentration (mg/liter)[b]
Arsenic	0.2
Cadmium	0.05
Chromium	1.0
Cobalt	1.0
Copper	0.5
Fluoride	2.0
Lead	0.1
Mercury	0.01
Nickel	1.0
Nitrate-N	100.0
Nitrite-N	10.0
Vanadium	0.1
Zinc	25.0

[a] Data from Shirley *et al.* (3).
[b] Same as ppm (parts per million).

time may not have an effect that can be observed. However, consumption of the water for a long time can cause serious effects. Different species of animals may react differently to a toxic level of minerals in the water. The young and the healthy animal may not respond in the same manner as unthrifty or mature animals. The rate and level of water intake may also influence the toxicity effect. Different forms of the mineral element may result in a different toxicity effect on the animals. Therefore, many factors can influence the toxicity of a mineral element in the water. The levels shown in Table 10.1 should be the upper limit of how much can be allowed in water given to horses.

Water sources should be analyzed occasionally to make sure they are safe. It is especially important to check water levels when horses are raised under intensified conditions or near large cities where water contamination may occur. In almost all cases there should not be a problem, but checking the water is good insurance. The figures given in Table 10.1 can be used as a guide in determining whether a problem exists in the water supply.

Horses sweat a great deal in exercise and riding. Therefore, their level of water consumption can be quite high. Moreover, they may be exposed to different sources of water. In many areas, and especially near industrial plants, there may be some water contamination. Therefore, it is important that water quality be checked to make sure it is safe for drinking purposes. Water contains all the mineral elements that are needed as nutrients. Most of them are there at low levels and very seldom present a problem.

V. SALINE WATER

Many high-saline or -alkaline waters exist, and so it is important to know the limitations of salinity in water for horses. This is especially the case when horses are located near the coast or other areas where there is saline intrusion in the water. The ions most commonly involved in high-saline waters are calcium, magnesium, sodium, bicarbonate, chloride, and sulfate. Saline waters usually contain chlorides and sulfates of sodium, calcium, magnesium, and minor levels of carbonates, bicarbonates, and other ions. The damage that can be caused by high-saline water depends more on the total amount of minerals present rather than on any specific one. Usually the chlorides are less harmful than sulfates. Magnesium chloride appears to be more injurious than calcium or sodium salts. Table 10.2 gives the NRC recommendations on saline waters for horses and other animals.

Where possible, it is best to use water with a low-saline content. Sometimes the only water available is high in saline. Table 10.2 can serve as a guide on its use for horses and other livestock. Animals prefer water that is low in saline content, but they will get used to the saline water after a short period of time.

TABLE 10.2

A Guide to the Use of Saline Waters for Livestock[a]

Total soluble salts content of waters (mg/liter)[b]	Comments and discussion
Less than 1000	These waters have a relatively low level of salinity and should present no serious burden to any class of livestock
1000–2999	These waters should be satisfactory for all classes of livestock; they may cause temporary and mild diarrhea in livestock not accustomed to them, but should not affect their health or performance
3000–4999	These waters should be satisfactory for livestock, although they might very possibly cause temporary diarrhea or be refused at first by animals not accustomed to them
5000–6999	These waters can be used with reasonable safety for dairy and beef cattle, sheep, swine, and horses; it may be well to avoid the use of those approaching the higher levels for pregnant or lactating animals
7000–10,000	These waters are probably unfit for swine. Considerable risk may exist in using them for pregnant or lactating cows, horses, sheep, the young of these species, or for any animals subjected to heavy heat stress or water loss; in general, their use should be avoided, although older ruminants, horses, and even swine may subsist on them for long periods of time under conditions of low stress
More than 10,000	The risks with these highly saline waters are so great that they cannot be recommended for use under any conditions

[a]Data from Shirley *et al.* (3).
[b]Same as parts per million (ppm).

VI. CONCLUSIONS

It is very important for horses to have a supply of clean, fresh water at all times. A loss of 10% of a horse's body water can lead to digestive disturbances, and a 20% water loss results in death. After strenuous activity a horse should be cooled off before being allowed to drink all the water it wants. A horse needs about 2–4 lb of water per lb of feed. But, these water needs may be increased considerably by the duration and degree of physical activity, temperature, humidity, the stage of its life cycle, the kind and level of diet fed, and many other factors. Water should be analyzed since it sometimes contains certain minerals at levels high enough to be harmful or toxic. Saline or alkaline water may exist in certain locations. Guidelines are given on its use.

REFERENCES

1. Argenzio, R. A., J. E. Lowe, D. W. Pickard, and C. E. Stevens, *Am. J. Physiol.* **226,** 1035 (1974).
2. Argenzo, R. A., M. Southworth, and C. E. Stevens. *Am. J. Physiol.* **226,** 1043 (1974).
3. Shirley, R. L., C. H. Hill, J. T. Maletic, O. E. Olsen, and W. H. Pfander, "Nutrients and Toxic Substances in Water for Livestock and Poultry." NAS-NRC, Washington, D. C., 1974.
4. Elser, A. H., S. G. Jackson, J. P. Lew, and J. P. Baker. *Proc. Equine Nutr. Physiol. Symp., 8th,* p. 61 (1983).
5. Hintz, H. F., J. P. Baker, R. M. Jordan, E. A. Ott, G. D. Potter, and L. M. Slade, "Nutrient Requirements of Horses." NAS-NRC, Washington, D. C., 1978.
6. Fonnesbeck, P. V. *J. Anim. Sci.* **27,** 1350 (1968).
7. Coenen, M. and H. Meyer. *Proc. Equine Nutr. Physiol. Symp., 10th,* p. 531 (1987).
8. Caljuk, E. A., *Tr. Vses. Inst. Konevodstvo* **23,** 295 (1961); cited in *Nutr. Abstr. Rev.* **32,** 574 (1962).
9. Meyer, H., H. Perez, Y. Gomda, and M. Heilemann. *Proc. Equine Nutr. Physiol. Symp., 10th,* p. 67 (1987).

11

Relationships between Nutrition and Diseases[1]

I. INTRODUCTION

There are interrelationships between nutrition, disease, and performance. These relationships are important to understand since they affect the results obtained in developing an outstanding horse program.

It is estimated that diseases (including parasites) decrease animal productivity by about 15–20% in the U.S. and 30–40% in developing countries (1,2). Therefore, the control of certain diseases and the use of adequate health systems can greatly increase horse production efficiency throughout the world. Real progress will occur when diseases can be prevented rather than merely treated after they occur.

As intensified horse production systems increase in size and number and as the level of production rate increases, there is a greater need for better nutrition and health care since problems with diseases and parasites increase. Moreover, some of the problems that were minor become more important. As producers increase the number of horses in a farm, they may encounter disease problems that were not previously apparent. This is due to the animals being closer to each other, which increases the opportunity for diseases to transfer from one animal to another. Overcrowding, therefore, should be avoided as much as possible, and a good disease-prevention program should be followed. It is known that well-fed animals are more resistant to many bacterial and parasitic infections. Some of this may be due to better body tissue integrity, more antibody production, improved detoxifying ability, increased blood regeneration, and other factors. But it is felt by some that well-fed animals may be more susceptible to some virus diseases because certain viruses need a well-nourished body cell to grow and reproduce in. While this variation occurs on the effect of nutrition on the susceptibility to certain diseases, there is general agreement that good nutrition is beneficial to recovery from all diseases including those caused by viruses.

1 This chapter was reviewed by Gerald E. Hackett, Jr. DVM, M. S., Director of Equine Research, California State Polytechnic University, Pomona.

Fig. 11.1 Horses being shipped by plane from Ontario, California arriving at Bluegrass Field, Lexington, Kentucky. Studies are needed to minimize the effect of shipping and change of locations on horses. (Courtesy of John P. Baker, University of Kentucky.)

Dr. Ben Norman, Extension Veterinarian at the University of California, Davis, also has a Ph.D. in animal nutrition. In an address to veterinarians, he stated "I really would like for you to go back to your practice and think about the interaction between malnutrition and disease and remember that in 85% of the cases, nutrition will influence the outcome of the disease entity that you work with" (3).

II. DISEASE EFFECTS

A virus or other diseases can wipe out an entire group of animals in a short time. In a chronic state, diseases can debilitate animals and make them more costly to feed and prevent their potential production level from being reached. Therefore, a sound disease-prevention and control program is essential for a successful and profitable horse operation.

It should be noted that diseases are appearing in locations where they never existed before, or are reappearing where they had been eradicated years ago. No country is entirely secure from diseases in the 1980s (4). Rapid travel throughout the world means a person from a distant country may be visiting some horse farm the next day in another country either to purchase animals or observe new technological developments. Diseases can be transmitted via any laxity at airports and seaports and via the visitor to a horse unit. Many horse operations and research units now take extra precautions to minimize the possibility of disease contamination. But, even under the best of conditions, birds, other wildlife, farm equipment movement, and other means of disease transfer can intervene. This emphasizes the need to use a sound disease-prevention program with a horse operation.

III. DISEASE DEFENSE

Mucous membranes and the animal's skin are a first line of defense. Many nutrients are important in keeping epithelial tissue in a healthy condition. Protein, certain B-complex vitamins, and other nutrients are essential for the production of antibodies and phagocytes which serve as secondary defenders against infectious agents entering the animal's body. Adequate nutrition also enables an animal to respond properly to vaccination.

Certain diseases increase the need for various nutrients by animals. This can be due to: a reduced appetite, which causes an inadequate feed and consequent nutrient intake; fever; an infection which causes diarrhea or vomiting and thus a loss of nutrients from the intestinal tract; decreased absorption or utilization of nutrients; or other causes. A deficiency of nutrients can cause abnormality of the epithelial tissue and consequent penetration by various organisms. Additional nutrient supplementation may prevent this.

Diseases may destroy tissues, red blood cells, vital organs, and other parts of the body system. This increases the need for nutrients required for the repair and restoration of damaged cells. Therefore, extra nutrient supplementation may be helpful to the horse that is sick and in poor condition, provided that irreparable damage has not occurred in some vital area of the body system.

IV. COPING WITH INFECTIONS

An increasing amount of information is accumulating for many animal species to indicate that many nutrients are involved in improving the ability of animals to cope with infection. Magnesium, phosphorus, sodium, chloride, zinc, copper, iron, and selenium have been shown to be beneficial in this regard. Other mineral elements may also be found to be involved as more research occurs in this area. A deficiency of iron, zinc, copper, and selenium have resulted in lowered resistance to disease either through an impaired immune response or faulty leukocyte function (5). The leukocytes are divided into phagocytes and immunocytes which work together in protecting the body from foreign organisms.

A deficiency of zinc reduces thymus function and its important role in the immunological process (5). Pigs that received adequate iron survived a TGE (transmissible gastroenteritis) episode and gained weight more rapidly during the recovery period than anemic pigs (6). Numerous studies have linked selenium to immunocompetence (5). Vitamin E and selenium independently enhance the immune response in pigs and there is an additive effect of vitamin E and selenium in increasing hemagglutinin titers (7). Recent studies have shown that the mastitis–metritis–agalactia (MMA) syndrome, which has caused considerable loss in the swine industry, may be ameliorated by supplementation of the gestation–lactation diet with vitamin E and selenium (8). Recent studies indicate that sodium and chloride appear to influence antibody production and that 0.25% supplemental NaCl may not be adequate for maximum humoral immunity in the chick (9). This study indicates that growth rate is not always a good indicator of health and well-being of the chick. In research with salmonellosis it was found that a level of 2–3 times the dietary phosphorus level required for optimal growth increased the survival of animals infected with *Salmonella typhimurium* (10). A magnesium deficiency has a profound immunosuppressive effect in mice by significantly reducing the number of antibody-synthesizing cells and serum immunoglobulin concentrations (11). These are just a few of the studies on minerals and their relation to an animal's ability to cope with infection. It is apparent from these and other studies that the nutrient requirements for growth, feed efficiency, gestation, and lactation do not necessarily mean that those levels will be adequate for normal immunity and maximizing resistance to diseases. This points out the need for more sophisticated studies which include the horse's nutrient needs for normal immunity against diseases. Moreover, studies that also include histopathology data would give an even more complete evaluation of the horse's nutrient requirements. As one evaluates a few of the more sophisticated animal studies on nutrient requirements, it is apparent that higher nutrient levels than recommended by the NRC are sometimes needed for certain nutrients for optimum health of the animal.

Vitamins also play an important role in the immune response. In a recent review article it was pointed out that deficiencies of thiamin, riboflavin, niacin, pantothenic acid, pyridoxine, folacin, choline, and vitamin C decrease immune response in swine (12).

Protein and/or amino acid levels appear to also be involved in the immune response. Recent University of Wisconsin studies show that the dietary methionine levels that are adequate for growth may be inadequate for maintaining the chick's ability to mount an immune defense to challenging organisms (13). Some other information with poultry, where many more research data are available, is summarized in a recent review paper on the interaction of nutrition with stress or disease by Dr. L. S. Jensen of the University of Georgia (14). He indicated that a deficiency of essential fatty acids appears to greatly reduce the resistance of poultry to various diseases. He also cited research showing that adding selenium or vitamin E to the diet significantly reduced mortality in chicks exposed to malabsorption syndrome (15). The requirement of the chick for valine and threonine appears to be higher for antibody production than for optimum growth rate. He also cited evidence showing that a lack of vitamin A, pantothenic acid, riboflavin, vitamin B6, and ascorbic acid were concerned with immune response. Also, the immunization of chicks against coccidiosis was enhanced by selenium or vitamin E supplementation.

These animal studies indicate the importance of proper nutrition in coping with diseases. It also stresses the need for more and similar nutrition studies with the horse since it may react in a similar manner.

V. IMPORTANCE OF IMMUNITY

The pig is a good example to stress the importance of immunity in animal production since a considerable number of data are available. The baby pig's immune system is not developed enough to handle serious enteric disease challenge until 5–6 weeks of age. Before weaning, maternal immunoglobulins in milk afford the young pig protection against enteric disease. The immunoglobulin level in milk begins to decline 2 weeks after farrowing. At this time the young pig responds to infection by producing its own antibodies. But, this defense mechanism may not be fully developed if pigs are weaned early at 3–4 weeks of age. Consequently, outbreaks of diarrhea frequently occur with the early weaned pigs (16). This problem is increased when every effort is not made to prevent the introduction of new virulent strains of disease organisms from outside sources. This discussion about the pig stresses the need to produce maximum immune capability in the young foal by proper health care, proper nutritional supplementation, and excellent management to prevent costly foal losses. Dead or weak foals are a message that something is wrong, and it needs to be heeded.

In the horse there is no passage of maternal protective antibodies through the placenta. The foal is born lacking passive immunity and has to depend on receiving these antibodies via the colostrum. The absorption of antibodies from the colostrum and their transfer across the intestinal mucosa of the newborn foal is short lived, lasting approximately 24 hours. So, adequate intake of colostrum is very important to protect the foal from an environment filled with pathogenic antigens in the lag period after birth. It will take the foal at least 7–14 days to manufacture antibodies against antigens. The foal, therefore, is very susceptible to infectious diseases during the first few weeks of life. Proper nutrition of the mare and foal should help the foal during this period (17).

When a disease (due to a disease organism or a nutritional deficiency or both) occurs in a band of horses, some of the animals will be unaffected, some will be mildly affected, and some may become very sick and die. The level of immunity present is a very important factor in the ability of the horses to ward off disease organisms and how they respond to the challenge. The ability to ward off infection varies from individual to individual due partly to variation in the horse's immune system and partly to its total health status. Such resistance or susceptibility may be a genetic trait (18). It has been shown that some Arabian foals have a genetically defective immune system that renders them susceptible to many infections, including adenovirus (18). If the foal lacks a competent immune system to protect it from invasion of pathogenic organisms, the foal succumbs before 5 months of age to massive infection that is refractory to treatment, primarily of the respiratory tract (18).

VI. EFFECTS OF STRESS AND IMMUNOGENETICS

The effects of stress on nutritional requirements is an area where precious little information is available. Many scientists are of the opinion that higher nutrition levels are helpful under high-stress conditions. For example, hysteria, nervousness, and greater density stress in commercial hens has been helped by the addition of higher levels of niacin to the diet. It also increased egg production and hatchability (19,20). The writer has recommended higher nutrient levels for both "moderate stress" and "severe stress" conditions for the pig (21). Dr. Milton Scott of Cornell University has done the same for poultry.

Stress and subclinical disease levels increase as horse operations intensify. Any disease entity present in a band of horses kept under intensified conditions is more apt to infect horses nearby. Moreover, all horses are exposed more times to the diseases that may be present. But both stress and subclinical disease level can be handled with adequate nutrition, a good health care and disease-prevention program, and proper housing and management. Intensified horse operations, therefore, require more sophisticated management and the use of the latest technology available.

University of Missouri scientists believe that pigs from certain strains are more susceptible to stress than others (22). The same may be true for horses. Missouri is conducting studies to find criteria that can be measured to identify swine genetic lines that are the most resistant to stress (22).

Studies at Cornell University resulted in the development of two different strains of chickens that required 24 and 73 mg of niacin per kg of feed (23). Other examples of the effect of genetics on nutritional requirements are presented in a National Academy of Science publication (24). Unfortunately, this area lacks much research knowledge. This is especially the case for the horse. Many examples are available with poultry where a relatively short generation time and a large number of offspring from mating two individual birds make it easier to study the effect of genetic variation on nutritional needs and immunity. The field of immunogenetics is an area which needs considerable study. With horses, it is not surprising that a lack of knowledge exists since a large number of observations are needed which would require considerable time and expense. It is hoped, however, that some model may be developed in the future to adequately evaluate the role of genetics in the nutritional requirements and immune response of horses. Many scientists are of the opinion that genetics are involved in the ability of some horses to do well under low-level nutritional and relatively high disease level conditions. But research evidence is lacking and difficult to obtain with present technology and resources.

Growth and production suffer from heat stress. Usually the heaviest animals and the best producing females are those that are the most affected by the heat and high humidity. The first noticeable adverse effect is a decline in feed intake and thus less nutrient intake. This may necessitate some increase in nutrient level per unit of feed to make sure total nutrient intake is adequate to meet daily body needs. If the energy level of the diet is increased, certain nutrients in the diet need to be increased proportionally. Studies at the University of Florida and Texas A & M showed that heat-stressed dairy cows needed extra potassium in order to milk at an optimum level (25).

VII. SUBCLINICAL DISEASE LEVEL AND ANTIMICROBIALS

There are many explanations for subclinical disease level. A great deal of this information is available for the pig and so it will be discussed as a means of clarifying the subclinical disease situation. Many scientists feel that a subclinical disease level allows the pig to appear normal but prevents it from performing to its maximum genetic potential. A subclinical disease response is widely accepted as one reason for the beneficial effects from antibiotic usage. Antibiotics tend to suppress the microorganisms in the intestinal tract that cause a subclinical or a nonspecific disease effect. The evidence to support this is the fact that the response to antibiotics is greater: (1) under conditions of high levels of disease than under conditions of low levels of disease; (2) in a dirty versus a clean environment (in a new swine unit versus an older one); and (3) in young pigs, which are less tolerant of disease and stress than are older pigs (26–30).

Improvement in growth rate, feed efficiency, and survivability in pigs as a result of feeding antibiotics varies a great deal. The passive immunity that the pig acquires from colostrum drops to a relatively low level by 3 weeks of age and remains low until 6–8 weeks of age (28). As pigs become older, they develop greater immunological protection, so are better able to cope with disease-causing organisms in their environment. As a result, older pigs show a lesser response to antibiotics than younger pigs.

VIII. NUTRIENT SUPPLEMENTATION OF SICK ANIMALS

Extra nutrient supplementation may be helpful to the horse that is sick and in poor condition. Many producers use higher levels of nutrients with horses recovering from an attack of disease in the following manner.

1. Nutrient levels are used in the diet as high as 2–5 times the requirement level for a short time until the animal recovers. The increased level will depend on each nutrient involved. Some nutrients are used at lower levels than others to prevent exceeding the tolerance level for that nutrient which can cause harmful effects. There is evidence that nutrient needs for certain body measurements may differ. For example, University of Pennsylvania School of Medicine data showed that optimal vitamin A intake for maximal growth averaged 1.4 times the 1978 NRC recommendation; for liver secreted serum constituents 5.4 times; and for red blood cell criteria 10 times NRC requirements (31).

The Canadian Feed Industry Association Nutrition Council (32) indicates that vitamins A, D, and E should be used at levels in excess of the minimum levels for animal feeding under stress conditions and recommend the following as maximum levels to use: vitamin A, 10 times, vitamin D, 5 times, and vitamin E, 5 times the minimum level. These are maximum levels and are used only with severe stress conditions. Stress can be caused by diseases, nutritional deficiencies, environmental conditions (such as cold, wet, cramped, drafty, dusty, crowded, dirty quarters with poor air circulation), vaccination, castration, unusual and loud noises, handling, moving, and others.

2. If appetite is reduced, some of the nutrients may be added to the drinking water as a means of ensuring a higher daily intake. Horses will usually drink 2–4 lb or more of water per lb of feed intake.

3. In some cases, it is necessary to inject the nutrients as a means of getting them into the body system quickly and in effective doses to facilitate recovery.

4. In some cases, a combination of the suggestions given in 1, 2, and/or 3 may be used.

IX. DISEASE AFFECTS NUTRITIONAL NEEDS

A few examples of how diseases may affect the nutritional needs of horses are as follows (21,33).

1. Any disease that causes bleeding of the intestinal wall increases a need for the nutrients lost (via the bleeding) as well as those needed for tissue repair and for blood regeneration such as iron, copper, protein, certain vitamins, and other nutrients.

2. A disease that causes diarrhea or vomiting decreases intestinal absorption of nutrients and thereby increases their need. Electrolytes, extra protein, and

other nutrients may be beneficial in treating these conditions. Many animals with diarrhea die because of viral or bacterial invasion. In many cases, however, the precipitating cause of death is often electrolyte loss and dehydration.

3. A disease that causes reduced appetite and a decreased total feed intake increases the need for higher levels of nutrients in the feed consumed as a means of meeting the total daily body nutrient needs. It may also be necessary, in certain situations, to add certain nutrients to the water or to inject them to facilitate a more rapid response and recovery.

4. Carotene conversion to vitamin A occurs in the intestinal wall. Any disease that adversely affects the integrity of the intestinal wall may interfere with this conversion and thereby increase vitamin A needs. This may account for instances where beef cattle responded to vitamin A on lush, green pasture or on corn silage which supposedly had more than enough carotene to take care of their vitamin A needs. Since horses are quite susceptible to parasites and subsequent intestinal wall damage, one must be careful to watch for decreased carotene conversion to vitamin A by the intestinal wall. In the case of severe parasite damage, vitamin A supplementation may be warranted with the horse.

5. Recent studies show that vitamin D3 is changed to 25-hydroxy D3 in the liver and into 1,25-dihydroxy D3 in the kidneys. The 1,25-dihydroxy D3 is the hormonal form of D3 that is needed for proper calcium and phosphorus utilization in the body. This raises the question as to what occurs if the liver and/or kidneys become damaged and cannot convert vitamin D3 to 1,25-dihydroxy D3. This may account for situations where the writer has observed abnormal bone problems in various areas of the world with pigs on diets that supposedly contained the proper levels of calcium, phosphorus, and vitamin D.

6. The disease ketosis increases the need for niacin by dairy cows. It appears that about 50% of the cows in high-producing herds go through borderline ketosis during early lactation (34). A 6 g level of niacin daily appears beneficial with clinical and subclinical ketosis. The possibility that high milk producing mares may be affected in a similar manner needs research exploration.

X. FOAL LOSSES

The use of foster mothers, milk substitutes, proper nutrition, attention to proper disease prevention and control methods, attending the smaller or weak foals at birth and shortly thereafter, and proper housing and temperature control are some of the factors that are helpful in decreasing foal losses.

An excellent report from the University of Kentucky (34a) detailed the causes of mortality in foals that died at less than 6 months of age in central Kentucky. This area of Kentucky has about 9000 horses and a diagnostic laboratory that performs necropsies on most horses that die. A total of 516 necropsies performed in 1987 showed that half of these foals died in the first 7 days of life primarily from respiratory, cardiovascular, and musculoskeletal causes. The most common cause of death in neonates was perinatal distress (neonatal anoxia, delayed delivery) representing 19% of all deaths. Pulmonary causes contributed 36% of total deaths; musculoskeletal, 25%; gastrointestinal, 16%; cardiovascular, 15%; genitourinary, 3%; nervous, 2%; and no diagnosis, 3%. A total of 48 cases had gastric ulcers, with 23 of them perforated. There were 42 cases of trauma and fractures, many of which were due to mares kicking or stepping on their foals. There were 27 cases of contracted foal syndrome with extremely flexed limbs and scoliosis or curvature of the spine. Mares trying to deliver their foals often have dystocias and require veterinary intervention. There were nine cases of wobblers, or cervical vertebral malformation. There were nine cases of botulism resulting in shaker foal syndrome. The bacterium, *Clostridium botulinum,* is commonly found in soil and is ingested by the foal. The bacteria colonize the intestine where they produce neurotoxins which produce profound muscle weakness, potentially leading to death.

XI. THE NEED FOR CONTINUOUS GOOD NUTRITION

A Cornell University report (35) indicates that delayed growth may affect limb conformation in young horses. In one of their studies they used weanling and yearling horses. The weanlings were in excellent flesh and thrifty, whereas the yearlings were unthrifty, heavily parasitized, and small for their age. All the weanlings and yearlings were wormed and then fed all they would eat of a diet containing 65% TDN. The weanlings gained weight rapidly and few problems were encountered. The yearlings also gained weight rapidly, but within 1 or 2 months, four of the six horses developed very straight limbs in which the fetlock joint was more upright than normal. The condition became so severe in two of the horses that they knuckled over (the fetlock joint flexed forward during the later part of the support phase of gait causing great difficulty in walking). After this observation, the Cornell scientists thought this condition, which seemed to be identical to contracted tendons, may have been related to the growth restriction that was followed by rapid growth. They therefore set up a study where the growth of young foals was restricted and then followed by feeding a high-energy diet *ad libitum.*

They fed two groups of foals weaned at 4 months of age as follows. One group was fed all the feed it would consume for 8 months. The other group was fed a restricted-feed intake for 4 months, which was followed by 4 months of eating all the feed it would consume. They were fed a pelleted diet consisting of 35% alfalfa meal, 48% corn, 15% soybean meal, 1% trace mineralized salt, and 1% dicalcium phosphate. The diet contained 17% protein, 0.9% calcium, and

0.55% phosphorus. No skeletal abnormalities were observed in the foals fed all the feed they would consume for the 8-month trial. However, four of the six foals fed the restricted diet for 4 months developed the straight limbs similar to the "contracted tendons" within 1–3 months after being fed all the feed they would consume. The angle of the fetlock returned to normal after 8–9 months. The correction of the contraction coincided with the increase in exercise time. The foals were allowed more time for exercise in a dirt paddock in the late spring than during the winter when they were housed in concrete floors with wood shavings for bedding.

If one studies the scientific literature, there are a number of causes that have been proposed for contracted tendons. These have included: (a) types that result from inherited characteristics; (b) poor position of the fetus in the uterus; (c) a deficiency of calcium, phosphorus, vitamin A, or vitamin D; (d) injury that causes decreased use of the limb; (e) a late manifestation of congenital malformation of the bone which occurred while the foal was *in utero;* (f) a lack of sufficient exercise; and (g) other possible causes.

The Cornell study indicated that foals restricted in their feed intake for 4 months developed a condition that appeared to be like "contracted tendons." The foals developed the contracted tendons when they were rapidly gaining weight after the restricted feed intake period of 4 months. Their weight gains were no more rapid than the group of foals fed all the feed they wanted throughout the trial. It appears that the period of restricted feed intake had something to do with the contracted tendons developing, although, additional studies are needed to verify this.

This study, and its findings, reinforce the author's recommendation that high-quality horses used for racing or performance should not be allowed to experience periods of poor nutrition. For best results during growth and reproduction one should always feed a well-balanced diet with adequate energy, protein, minerals, and vitamins. It is hazardous to gamble with periods of time when certain nutrients are lacking in the diet, since one is never sure what effect it may have many months later. It is recognized with other animals that the diet that an animal is fed during growth can influence its ability to conceive, reproduce, and lactate many months later. It is also known that the diet a pregnant animal receives can influence the development of its young *in utero.* This, in turn, can influence how the newborn will fare during its growth period. Therefore, there is no period in the lifetime of the horse when one can relax and get by with feeding a poor diet without paying some penalty in terms of suboptimal performance later on. This is especially true with horses being developed for racing and high-level performance and when they are pushed for excellence at a young age. This added stress can result in breakdowns, poor performance, and problems in conception rate, reproduction, and lactation. The best advice is to feed horses a well-balanced diet, keep them in a thrifty condition (but not fat), and never neglect to supply the nutrients they require.

XII. RAISING REPLACEMENT ANIMALS

Using the pig as an example, where considerable data are available, the diet one feeds during growth may affect the animal's ability to function and to reproduce properly later on in life. Recently, Dr. D. C. Mahan of Ohio State University showed that the pig needs 0.1% more calcium and phosphorus in the diet for maximum bone development than is needed for its best rate of growth and feed utilization. A study by Dr. D. F. Calabotta indicated that 30–50% of swine breeding animals are culled each year for reasons related to leg weakness (36).

A number of studies indicate that young horses fed for rapid skeletal and body growth may develop metabolic bone disorders and lameness problems frequently attributed to mineral deficiencies or imbalances. A recent Kentucky study showed that the faster growing foals had a numerically smaller percentage of cortical area associated with the new bone growth (37). The study showed that young horses fed for rapid growth may not maximize bone deposition. They also found that creep feeding formulated to meet present nutrient requirements for growth was adequate to sustain a faster rate of skeletal and body growth in nursing foals (37).

As one decides which animals are to be used as replacements, it is suggested they be separated from those being developed for other purposes and fed a diet designed to ensure a normal reproductive tract which results in optimal breeding performance. Moreover, separating the young horses early in their development period and feeding them in a different area, lessens their exposure to disease outbreaks that may occur in the horses being used for horse shows, racing, rodeos, and other performance activities.

XIII. CONCLUSIONS

Many scientists argue as to whether proper nutrition or proper disease control is more important. Actually, both are very important and a good horse production enterprise needs to follow a good feeding and disease-prevention program in order to be successful. There are indications that certain diseases may increase the need for certain nutrients. Well-fed animals are more resistant to many bacterial and parasitic infections. There is general agreement that good nutrition is beneficial to recovery from all diseases including those caused by viruses.

REFERENCES

1. Cunha, T. J., K. H. Shapiro, J. M. Fransen, H. J. Hodgson, J. E. Johnston, W. H. Morris, R. R. Oltjen, W. R. Pritchard, R. R. Spitzer, and N. L. Van Demark. "World Food and Nutrition Study," Vol. 1, p. 141. National Academy of Science, Washington, D. C., 1977.

2. Cunha, T. J. *Int. Conf. Goat Prod. 3rd,* Univ of Arizona (1982).
3. Norman, B. *Calif. Agric. Ext. Leaflet. AABP* No. 51 (1977).
4. Mussman, H. C. *Int. Conf. Goat Prod. Dis. 3rd,* Univ. of Arizona (1982).
5. Miller, E. R. *J. Anim. Sci.* **60**(6), 1500 (1985).
6. Parsons, M. P., E. R. Miller, D. M. Bebiak, J. P. Erickson, M. V. Hogberg, D. J. Ellis, and D. E. Ullrey. *Mich. Agric. Exp. Stn., Rep.* **343,**10 (1977).
7. Peplowski, M. A., D. C. Mahan, F. H. Murray, A. L. Moxon, A. H. Cantor, and K. E. Ekström, *J. Anim. Sci.* **51,** 344. (1980).
8. Whitehair, C. K., O. E. Vale, M. Loudenslager, and E. R. Miller. *Mich., Agric. Exp. Stn., Res. Rep.* **456** (1983).
9. Pimentel, J. L., and M. E. Cook. *Abstr. Poult. Sci. Assoc. Meet.,* Ames, Iowa July–August (1985).
10. O'Dell, B. L. *Feed Manage.* **20**(1), 19 (1969).
11. Elin, R. J. *Proc. Soc. Exp. Biol. Med.* **148,** 620 (1975).
12. Blair, R., and F. Newsome. *J. Anim. Sci.* **60**(6), 1508 (1985).
13. Cook, M. E., University of Wisconsin, Madison (personal communication), 1985.
14. Jensen, L. S. *Proc. Ga. Nutr. Conf.* p. 46 (1986).
15. Colnago, G. L., T. Gore, L. S. Jensen, and P. L. Long. *Avian Dis.* **27,** 312 (1983).
16. Ahern, F. X. *Anim. Nutr. Health* **38**(3), 6 (1983).
17. Prades, M. *Feedstuffs* **56**(18), 55 (1984).
18. Smith, A. T. *J. Anim. Sci.* **51**(5), 1087 (1980).
19. Hansen, R. S. *Poult. Sci.* **49,** 1392 (1970).
20. Hinners, S. N., J. D. Ford, A. A. Marefi, and L. E. Strack. *Poult. Sci.* **54,**1773 (1975).
21. Cunha, T. J. *Squibb Int. Swine Update Rep.* **3**(1) 1984).
22. Ellersieck, J. R., T. L. Veum, T. L. Durham, W. R. McVickers, S. N. McWilliams, and J. F. Lasley. *J. Anim. Sci.* **48**(3), 453 (1979).
23. Scott, M. L., M. C. Nesheim, and R. J. Young. "Nutrition of the Chicken," 3rd ed. W. F. Humphrey Press, Inc., Geneva, New York, (1982).
24. Sunde, M. L., A. E. Freeman, R. H. Grummer, C. T. Hansen, M. C. Nesheim, C. E. Terrill, and E. J. Warwick. "The Effect of Genetic Variance on Nutritional Requirements of Animals." *N.A.S.–N.R.C.,* Washington, D.C. (1975).
25. Cunha, T. J. *Feedstuffs* **54**(42), 1 (1987).
26. Cunha, T. J. *Feed Manage.* **31**(2), 32 (1980).
27. Cunha, T. J. "Swine Feeding and Nutrition." Academic Press, New York, 1977.
28. Cromwell, G. L. *Anim. Health Nutr.* **38**(4), 18 (1983).
29. Braude, R., H. D. Wallace, and T. J. Cunha. *Antibiot. Chemother.* **3,** 271 (1953).
30. Cunha T. J., J. P. Bowland, J. H. Conrad, V. W. Hays, R. J. Meade, and H. S. Teague. "Nutrient Requirements of Swine." *N.A.S.–N.R.C.,* Washington, D.C. (1973).
31. Donoghue, S., D. S. Kronfeld, S. J. Berkowitz, and R. L. Coop. *J. Nutr.* **111,** 365 (1981).
32. Canadian Feed Industry Association Nutrition Council. "Recommended Minimum and Maximum Levels of Vitamins for Registration in Animal Feeds," CFIANC, 1978.
33. Cunha, T. J. *Feedstuffs* **57**(41), 37 (1985).
34. Cunha, T. J. *Feedstuffs* **54**(25), 20 (1982).
34a. Dwyer, R. M. *Univ. Ky., Coll. Agric., Coop Ext. Serv., Bull.* Equine Data Line, p. 1 March (1988).
35. Hintz, H. F., H. F. Schryver, and J. E. Lowe. *Proc. Cornell Nutr. Conf.* (1976).
36. Calabotta, D. F., E. T. Kornegay, H. R. Thomas, J. W. Knight, D. R. Notter, and H. P. Veit. *J. Anim. Sci.* **54,** 565 (1982).
37. Thompson, K. N., J. P. Baker, and S. G. Jackson. *J. Anim. Sci.* **66,** 1692 (1988).

12

Value of Feeds for Horses

I. INTRODUCTION

In developing diets for the horse, it is important to know the relative value of one feed to another (1,2). One needs to consider the protein, energy, vitamins, and minerals that feeds contain or are lacking in. One must select feeds that will not only result in a well-balanced diet but also an economical one. Unless the diet is economical, most horse owners will not use it. Moreover, the diet must be palatable—the horse must like it. Horses vary considerably in their likes and dislikes for certain feeds. This makes it more difficult to compound horse diets. To do all this successfully, one must be acquainted with the relative values of feeds, as well as any limitations they may have in horse diets.

In selecting feeds for diets, one needs to consider the purpose for which the feed will be used, such as growth, training, gestation, lactation, work, or maintenance. Also, it is important to know whether the diet will be fed on pasture or in partial or complete confinement. All of these factors, and many others, must be considered in selecting feeds to use in various horse diets. Unless the horse is fed properly, it will never develop to its full potential in speed, endurance, performance, or work.

II. AVOID MOLDS

The horse is the most susceptible class of livestock to moldy feeds. Therefore, one needs to avoid buying spoiled or moldy feeds. Moreover, the feeds purchased should be stored properly to avoid molds forming.

When molds (or fungus) grow on grains, hay, or other feeds, a group of compounds called mycotoxins develop. One of these is called aflatoxin. Mycotoxins can be very harmful to horses.

Hay is probably the biggest offender in having molds. So, horse owners should be careful in avoiding the purchase of hay which contains moldy areas inside the bale. Usually, this hay will be warm inside these areas or if the molds have already formed will have a musty odor. If one has hay which appears to be moldy, it should not be fed to valuable animals. If it needs to be used for horses at all, it might first be fed to a few of the least valuable horses to make sure the mold present is not harmful.

Improper storage and handling of feeds can cause molds to develop. The following are some suggestions to help in this regard.

1. Provide adequate moisture and humidity control in storage. Good air circulation in the storage barn is very helpful.

2. Avoid feeds sweating in storage. Protection from the rain helps since it prevents wet spots which can later become moldy.

3. Remove mold-damaged feeds before processing, before storage, and when found in storage.

4. Keep processing facilities as well as the storage area clean to avoid contamination of feeds.

5. Be careful when purchasing feeds from humid areas to make sure molds are not present.

III. DEFICIENCIES IN CEREAL GRAINS

The proteins in cereal grains are low in nutritional quality and lacking in lysine. Table 12.1 gives lysine levels for the cereal grains. Because all cereal grains are low in lysine and below the level of at least 0.60–0.70% lysine which weanling horses require in the diet (3,4), their use requires supplementation with other feeds. Corn and the sorghum grains have the lowest level of lysine. Both are used extensively in horse feeding. The protein supplements used to balance cereal grains and other high-energy feeds must not only supply enough protein but also protein with a good balance of the indispensable amino acids. Since corn contains the lowest amount of protein of the cereal grains (see Table 12.1), more protein supplement is needed to balance a corn diet than one containing wheat, barley, oats, or grain sorghums. This needs to be considered in deciding which grains to use at specific prices.

All the high-energy feeds shown in Tables 12.2 and 12.3 are lacking in vitamin D, salt, and calcium (except for beet pulp, citrus pulp, and sugarcane molasses, which are high in calcium.). Grains are fair sources of phosphorus, but only yellow corn contains appreciable amounts of carotene (provitamin A). Tables 12.1, 12.2, and 12.3 give more detail on the nutrients contained in each feed. These data are more applicable to the horse since they were developed by the 1989 NRC committee on the nutrient requirements of the horse (4).

More detailed information on the value of feeds for horses is provided in the 1989 NRC publication on nutrient requirements of horses (4). Additional information on feeds not contained in the 1989 NRC publication are available in a 1982 NRC publication (5) on U.S.–Canadian tables of feed composition.

TABLE 12.1

Relative Value of Grains[a]

Grain	Weight in		Digestible energy kcal/lb	Percentage of						
	Bushel lb	Gallon[b] lb		Dry matter	Crude fiber	Crude protein	Calcium	Phosphorus	Lysine
Corn	56	6.8	1540	88	2.2	9.1	0.05	0.27	0.25
Barley	48	6.0	1490	88.6	4.9	11.7	0.05	0.34	0.40
Sorghum	56	6.8	1460	90.1	2.6	11.5	0.04	0.32	0.26
Wheat	60	7.6	1560	88.9	2.5	13.0	0.04	0.38	0.40
Oats	32	4.0	1300	89.2	10.7	11.8	0.08	0.34	0.39
Rye	56	6.8	1530	87.5	2.2	12.0	0.06	0.32	0.41

[a] Adapted from 1989 NRC report on "Nutrient Requirements of Horses" (4).
[b] These figures indicate the need to feed by weight and not by volume. For example, corn fed by volume would have 70% more weight per gallon and about 86% more digestible energy per gallon than oats.

TABLE 12.2

Some Nutrients in Feeds Commonly Used in Horse Diets[a]

Feed	Dry matter (%)	Digestible energy (Mcal/lb)	Crude protein (%)	Lysine (%)	Ether extract (%)	Fiber (%)	Ash (%)
Alfalfa							
Hay, sun cured, early bloom	90.5	1.02	18.0	0.81	2.6	20.8	8.4
Meal, dehydrated, 17% protein	91.8	0.98	17.4	0.85	2.8	24.0	9.8
Alyce clover							
Hay, sun-cured	89.7	0.75	10.9	—	1.6	36.2	5.7
Bahiagrass							
Hay, sun-cured	90.0	0.79	8.5	—	1.8	28.1	5.7
Barley							
Grain	88.6	1.49	11.7	0.40	1.8	4.9	2.4
Hay, sun-cured	88.4	0.81	7.8	—	1.9	23.6	6.6
Beet, pulp	91.0	1.06	8.9	0.54	0.5	18.2	4.9
Bermudagrass, Coastal							
Hay, sun-cured, 29–42 days growth	93.0	0.89	10.9	—	2.4	28.0	6.2
Bluegrass, Kentucky							
Hay, sun-cured	92.1	0.72	8.2	—	3.0	29.9	5.4
Brome, Smooth							
Hay, sun-cured, mid-bloom	87.6	0.85	12.6	—	1.9	28.0	9.5
Canaraygrass, Reed							
Hay, sun-cured	89.3	0.81	9.1	—	2.7	30.2	7.3
Canola							
Seed meal, solvent extracted	90.8	1.28	37.1	2.08	2.8	11.0	6.4
Carrot							
Roots, fresh	11.5	0.20	1.2	—	0.2	1.1	1.0
Citrus, pulp	91.1	1.16	6.1	0.2	3.4	11.6	6.0

Clover, Alsike							
Hay, sun-cured	87.7	0.78	12.4	—	2.4	26.2	7.6
Clover, Ladino							
Hay, sun-cured	89.1	0.89	20.0	—	2.4	18.5	8.4
Clover, Red							
Hay, sun-cured	88.4	0.89	13.2	—	2.5	27.1	6.7
Corn, Dent, yellow							
Grain	88.0	1.54	9.1	0.25	3.6	2.2	1.3
Distiller's grains	92.0	1.46	27.8	0.81	6.6	11.3	3.1
Cotton							
Seed meal; solvent extracted	91.0	1.25	41.3	1.68	1.5	12.2	6.5
Fats and oils							
Fat, animal, hydrolyzed	99.2	3.61	—	—	98.4	—	—
Oil, vegetable	99.8	4.08	—	—	99.7	—	—
Fescue, Kentucky							
Hay, sun-cured	91.9	0.86	11.8	—	5.1	23.9	7.6
Fish, Anchovy							
Meal, mechanically extracted	92.0	1.25	65.5	5.03	4.2	1.0	14.7
Fish, Menhaden							
Meal, mechanically extracted	91.7	1.33	62.2	4.74	9.8	0.7	18.9
Flax, Common							
Linseed meal, solvent extracted	90.2	1.25	34.6	1.16	1.4	9.1	5.9
Lespedeza, Common							
Hay, sun-cured	90.8	0.88	11.4	—	2.3	26.2	4.5
Lespedeza, Kobe							
Hay, sun-cured	93.9	0.89	10.0	—	2.8	26.2	3.8
Meadow, Plants, Intermountain							
Hay, sun-cured	95.1	0.73	8.2	—	2.4	31.2	8.2
Milk							
Skimmed, dehydrated	94.1	1.73	33.4	2.54	1.0	0.2	7.9
Millet, Pearl							
Hay, sun-cured	87.4	0.61	7.3	—	1.8	32.2	8.9

(continued)

TABLE 12.2

(Continued)

Feed	Dry matter (%)	Digestible energy (Mcal/lb)	Crude protein (%)	Lysine (%)	Ether extract (%)	Fiber (%)	Ash (%)
Molasses and syrup							
Beet, more than 48% invert sugar, more than 79.5 degrees	77.9	1.20	6.6	—	0.2	0.0	8.9
Citrus, syrup	66.9	1.03	5.7	—	0.2	0.0	5.1
Sugarcane, dehydrated	99.4	1.46	9.0	—	0.8	7.1	12.0
Sugarcane, more than 46% invert sugar, more than 79.5 degrees brix (blackstrap)	74.3	1.18	4.3	—	0.2	0.4	9.9
Oats							
Grain	89.2	1.30	11.8	0.39	4.6	10.7	3.1
Groats	89.6	1.40	15.5	0.55	6.1	2.5	2.0
Hay, sun-cured	90.7	0.79	8.6	—	2.2	29.1	7.2
Orchardgrass							
Hay, sun-cured, early bloom	89.1	0.88	11.4	—	2.6	30.2	7.6
Pangolagrass							
Hay, sun-cured 29–42 days growth	91.0	0.74	6.7	—	1.8	29.5	7.3
Pea							
Seed meal	89.1	1.40	23.4	1.65	0.9	5.6	2.8
Peanut							
Seed meal, without coats, solvent extracted	92.4	1.36	48.9	1.45	2.1	7.7	5.8
Hay, sun-cured	90.7	0.79	9.9	—	3.3	30.3	8.2
Prairie plants, Midwest							
Hay, sun-cured	91.0	0.67	5.8	—	2.1	30.7	7.2

Redtop							
Hay, sun-cured, midbloom	92.8	0.83	11.1	—	2.4	29.0	6.0
Rice							
Bran with germs	90.5	1.19	13.0	0.57	13.6	11.7	10.4
Grain, ground	89.0	1.54	7.5	0.24	1.6	8.6	5.3
Mill run	91.6	0.28	6.3	0.26	5.2	28.9	15.7
Rye							
Grain	87.5	1.53	12.0	0.41	1.5	2.2	1.6
Ryegrass, Italian							
Hay, sun-cured, late vegetative	85.6	0.71	8.8	—	2.1	20.4	9.4
Sorghum, grain	90.1	1.46	11.5	0.26	2.7	2.6	1.7
Sorghum, Johnson-Grass							
Hay, sun-cured	90.5	0.68	6.7	—	2.0	30.4	7.7
Soybean							
Seed meal, solvent extracted, 44% protein	89.1	1.43	44.5	2.87	1.4	6.2	6.4
Seed meal without hulls, solvent extracted	89.9	1.53	48.5	3.09	1.0	3.5	6.0
Sunflower, common							
Seed meal without hulls, solvent extracted	92.5	1.17	45.2	1.68	2.7	11.7	7.5
Timothy							
Hay, sun-cured, early bloom	89.1	0.83	9.6	—	2.5	30.0	5.1
Trefoil, Birdsfoot							
Hay, sun-cured	90.6	1.19	14.4	—	1.9	29.3	6.7
Wheat							
Bran	89.0	1.33	15.4	0.56	3.8	10.0	5.9
Grain, hard, red winter	88.9	1.56	13.0	0.40	1.6	2.5	1.7
Grain, soft, red winter	88.4	1.55	11.4	0.36	1.6	2.4	1.8
Hay, sun-cured	88.7	0.76	7.7	—	2.0	25.7	7.0
Mill run, less than 9.5% fiber	89.9	1.42	15.6	0.57	4.1	8.2	5.1

(*continued*)

TABLE 12.2

(*Continued*)

Feed	Dry matter (%)	Digestible energy (Mcal/lb)	Crude protein (%)	Lysine (%)	Ether extract (%)	Fiber (%)	Ash (%)
Wheat, soft, white Winter Grain	90.2	1.61	10.6	—	1.5	2.2	1.5
Wheatgrass, Crested Fresh	28.5	0.33	6.0	—	0.6	6.2	2.9
Whey Dehydrated (cattle)	93.2	1.72	13.1	0.94	0.7	0.2	8.7
Low lactose (dried whey product), cattle	93.7	1.53	16.8	1.40	1.0	0.2	16.0
Yeast, Brewer's Dehydrated	93.1	1.40	43.4	3.23	1.0	3.2	6.7

[a]Adapted from 1989 NRC report on "Nutrient Requirements of Horses" (4).

TABLE 12.3

Some Vitamin and Mineral Values for Commonly Used Feeds in Horse Diets[a]

Feed	Percent in feed							I.U. per kg	
	Dry matter	Ca	P	Mg	K	Na	S	Vitamin A equivalent[b]	Vitamin E
Alfalfa									
Hay, sun-cured, early bloom	90.5	1.28	0.19	0.31	2.32	0.14	0.27	50,608	23.5
Meal., dehydrated, 17% protein	91.8	1.38	0.23	0.29	2.40	0.10	0.22	29,787	81.9
Alyce Clover									
Hay, sun-cured	89.7	—	—	—	—	—	—	—	—
Bluegrass									
Hay, sun-cured	90.0	0.45	0.20	0.17	—	—	—	—	—
Barley									
Grain	88.6	0.05	0.34	0.13	0.44	0.03	0.15	817	23.2
Hay, sun-cured	88.4	0.21	0.25	0.14	1.30	0.12	0.15	18,571	—
Beet pulp	91.0	0.62	0.09	0.26	0.20	0.18	0.20	88	—
Bermudagrass, coastal									
Hay, sun-cured, 29–42 days	93.0	0.30	0.19	0.11	1.58	—	—	—	—
Bluegrass, Kentucky									
Hay, sun-cured	92.1	0.24	0.25	—	1.40	—	—	—	—
Brome, Smooth									
Hay, sun-cured, midbloom	87.6	0.25	0.25	0.09	1.74	0.01	—	—	—
Canarygrass, Reed									
Hay, sun-cured	89.3	0.32	0.21	0.19	2.60	0.01	0.12	6,762	—

(continued)

241

TABLE 12.3

(*Continued*)

Feed	Dry matter	Ca	P	Mg	K	Na	S	I.U. per kg Vitamin A equivalent[b]	Vitamin E
Canola									
Seed meal, solvent extracted	90.8	0.63	1.18	0.55	1.22	0.01	1.23	—	—
Carrot									
Roots, fresh	11.5	0.05	0.04	0.02	0.32	0.06	0.02	31,160	6.9
Citrus pulp	91.1	1.71	0.12	0.16	0.70	0.08	0.07	85	—
Clover, Alsike									
Hay, sun-cured	87.7	1.14	0.22	0.39	1.95	0.40	0.17	65,285	—
Clover, Ladino									
Hay, sun-cured	89.1	1.20	0.30	0.42	2.17	0.12	0.19	57,475	—
Clover, Red									
Hay, sun-cured	88.4	1.22	0.22	0.34	1.60	0.16	0.15	9,727	—
Corn, Dent, Yellow									
Grain	88.0	0.05	0.27	0.11	0.32	0.03	0.11	2,162	20.9
Distiller's grains	92.0	0.10	0.41	0.06	0.17	0.09	0.42	1,104	—
Cotton									
Seed meal, solvent extracted	91.0	0.17	1.11	0.54	1.30	0.04	0.26	—	—
Fats and oils									
Fat, animal, hydrolyzed	99.2	—	—	—	—	—	—	—	—
Oil, vegetable	99.8	—	—	—	—	—	—	—	—
Fescue, Kentucky									
Hay, sun-cured	91.9	0.37	0.27	0.14	1.76	0.02	—	—	—

Fish, Anchovy Meal, mechanically extracted	92.0	3.74	2.47	0.25	0.72	0.88	0.72	—	5.0
Fish, Menhaden Meal, mechanically extracted	91.7	5.01	2.87	0.15	0.71	0.41	0.53	—	5.8
Flax, Common Linseed meal, solvent extracted	90.2	0.39	0.80	0.60	1.38	0.14	0.39	—	—
Lespedeza, Common Hay, sun-cured	90.8	1.07	0.17	0.22	0.94	—	—	—	—
Lespedeza, Kobe Hay, sun-cured midbloom	93.9	1.11	0.32	0.27	0.89	—	—	—	—
Meadow, Plants, Intermountain Hay, sun-cured	95.1	0.58	0.17	0.16	1.50	0.11	—	12,744	—
Milk Skimmed, dehydrated	94.1	1.28	1.02	0.12	1.60	0.51	0.32	—	9.1
Millet, Pearl Hay, sun-cured	87.4	—	—	—	—	—	—	—	—
Molasses and syrup Beet, more than 48% invert sugar, more than 79.5 degrees	77.9	0.12	0.02	0.23	4.72	1.16	0.46	—	4.0
Citrus, syrup	66.9	1.18	0.09	0.14	0.09	0.28	0.14	—	—
Sugarcane, dehydrated	99.4	1.03	0.14	0.44	3.39	0.19	0.43	—	5.2
Sugarcane, more than 46% invert sugar, more than 79.5 degrees brix (blackstrap)	74.3	0.74	0.08	0.31	2.98	0.16	0.35	—	5.4

(continued)

TABLE 12.3

(*Continued*)

Feed	Dry matter	Percent in feed						I.U. per kg	
		Ca	P	Mg	K	Na	S	Vitamin A equivalent[b]	Vitamin E
Oats									
Grain	89.2	0.08	0.34	0.14	0.40	0.05	0.21	44	15.0
Groats	89.6	0.08	0.42	0.11	0.36	0.03	0.20	—	14.8
Hay, sun-cured	90.7	0.29	0.23	0.26	1.35	0.17	0.21	10,792	—
Orchardgrass									
Hay, sun-cured, early bloom	89.1	0.24	0.30	0.10	2.59	0.01	0.23	13,366	—
Pangolagrass									
Hay, sun-cured, 29–42 days growth	91.0	0.42	0.21	0.14	1.27	—	—	—	—
Pea									
Seed meal	89.1	0.12	0.41	0.12	0.95	0.22	—	285	3.0
Peanut									
Seed meal, without coats, solvent extracted	92.4	0.29	0.61	0.15	1.18	0.03	0.30	—	2.9
Hay, sun-cured	90.7	1.12	0.14	0.44	1.25	—	0.21	12,618	—
Prairie Plants, Midwest									
Hay, sun-cured	91.0	0.32	0.12	0.24	0.98	—	—	—	—
Redtop									
Hay, sun-cured, midbloom	92.8	0.58	0.32	—	1.57	—	—	1,856	—
Rice									
Bran with germ	90.5	0.09	1.57	0.88	1.71	0.03	0.18	—	85.3
Grain, ground	89.0	0.07	0.32	0.13	0.44	0.06	0.04	—	14.0
Mill run	91.6	0.15	0.46	0.10	0.52	—	0.18	—	5.3

Rye									
Grain	87.5	0.06	0.32	0.11	0.45	0.02	0.15	35	14.5
Ryegrass, Italin									
Hay, sun-cured, late vegetative	85.6	0.53	0.29	—	1.34	—	—	99,287	—
Sorghum, grain	90.1	0.04	0.32	0.15	0.37	0.01	0.13	468	10.0
Sorghum, Johnson Grass									
Hay, sun-cured	90.5	0.80	0.27	0.31	1.22	0.01	0.09	14,102	—
Soybean									
Seed meal, solvent extracted, 44% protein	89.1	0.35	0.63	0.27	1.98	0.03	0.41	—	3.0
Seed meal without hulls, olvent extracted	89.9	0.26	0.64	0.29	2.12	0.01	0.44	—	3.3
Sunflower, Common									
Seed meal without hulls, solvent extracted	92.5	0.42	0.94	0.65	1.17	0.02	0.31	—	11.1
Timothy—									
Hay, sun-cured, early bloom	89.1	0.45	0.25	0.11	2.14	0.01	0.12	18,719	11.6
Trefoil, Birdsfoot									
Hay, sun-cured	90.6	1.54	0.21	0.46	1.74	0.06	0.23	52,250	—
Wheat									
Bran	89.0	0.13	1.13	0.56	1.22	0.05	0.21	1,048	14.3
Grain, hard, red winter	88.9	0.04	0.38	0.13	0.43	0.02	0.13	—	11.1
Grain, soft, red winter	88.4	0.03	0.36	0.12	0.35	0.01	0.13	—	15.6
Hay, sun-cured	88.7	0.13	0.18	0.11	0.88	0.19	0.19	30,304	—
Mill run, less than 9.5% fiber	89.9	0.10	1.02	0.47	1.20	0.22	0.17	—	31.9
Wheat, Soft, White Winter									
Grain	90.2	0.06	0.30	0.10	0.39	0.02	0.12	—	18.0

(continued)

TABLE 12.3

(Continued)

Feed	Dry matter	Ca	P	Mg	K	Na	S	Vitamin A equivalent[b]	Vitamin E
				Percent in feed				I.U. per kg	
Wheatgrass, Crested									
Fresh, early vegetation	28.5	0.12	0.09	0.08	—	—	—	49,377	—
Whey									
Dehydrated (cattle)	93.2	0.85	0.76	0.13	1.16	0.62	1.04	48.8	0.2
Low lactose (dried whey product), cattle	93.7	1.50	1.11	0.22	2.86	1.45	1.07	—	—
Yeast, Brewer's									
Dehydrated	93.1	0.14	1.36	0.24	1.68	0.07	0.44	—	2.1

[a]Adapted from 1989 NRC report on "Nutrient Requirement of Horses" (4).

[b]The vitamin A equivalent was calculated as carotene × 400 except for value for dehydrated whey, which represents actual amounts of vitamin A. CA = calcium; P = phosphorus; Mg = magnesium; K = potassium; Na = sodium, and S = sulfur in this table.

IV. EVALUATING GRAINS

The cereal grains and their by-product feeds make up a large part of the concentrate feeds used for horses.

Many of the grains are sold on a price per bushel. Therefore it is important to know what the weight of the various grains are per bushel (Table 12.1). Grain weight per bushel can vary. Some of the grains may be higher in fiber than others, which will influence weight per bushel. High-moisture grains will usually weigh more per bushel. The moisture level of grains is important to know since the feeding value of a grain for horses depends to some extent on its dry matter content.

There is also the problem of storing grains. If the moisture is high, then the grain may heat up and develop molds. Therefore, one needs to develop expertise in buying grains at the correct moisture level, which usually should be below 14–15%. If grains heat up too much there is some loss in the nutritional value, especially of vitamins A (carotene) and E. There is considerable variation in the protein level of the various grains. This is important since lower protein grains will require higher levels of protein supplements, which are more expensive than grain. Wheat, for example, may have from 10 to 17% protein.

V. PROCESSING GRAINS

Some grains can be fed whole without any processing. Some horse owners feed whole oats and shelled corn, although many prefer to process the corn and oats, especially with horses used for racing or high-level performance purposes. One should not feed whole grains to horses with poor teeth. Wheat, milo, or barley should be processed for maximum utilization.

Grains should not be finely ground. This causes dustiness and can result in digestive disturbances and heaves. It usually reduces the palatability of the grains as well. In addition, dustiness can cause a significant amount of grain loss in grinding, handling, and in feeding (unless it is fed as a pellet). If grains are ground, it should be a coarse rather than a fine grind.

The most popular method of processing grain for horses is steam rolling. This results in a grain that is less dusty and more palatable. Moreover, the steam-rolled grain results in a bulkier product, which is important in many situations. Steam rolling may also increase the feeding value of the grain. Horses relish steam-rolled grains, and many people are convinced that horses perform better when they are used in the diet.

Some crimping and micronizing of grains is used. In the micronizing process the grain is dry heated and then crimped. Studies at Texas A & M (6) showed that crimping or micronizing grain sorghums was of value for the horse. Horses fed crimped grain sorghum gained weight slightly more slowly than those fed crimped oats or micronized oats or micronized grain sorghums. Micronizing either oats or grain sorghums tended to improve the performance of the horses. In another Texas A & M study (7) micronizing grain sorghums improved prececal starch digestion by over 20% over that of crimped sorghum. They felt that this minimizes hindgut disturbances from hindgut starch overloads.

Buying, handling, storing, processing, and feeding grains is not a simple matter if one wants to obtain optimum performance and maximum returns of each dollar spent.

VI. FEEDING VALUE OF GRAINS

A. Oats

Oats are considered the standard by which other grains are evaluated, and are the most widely used and the most popular grain for horses. Ponies preferred oats over other grains in a feed preference trial at Cornell University (8). Corn was the next preferred grain. But, Dr. H. F. Hintz pointed out the intake of one grain may be satisfactory if fed alone as compared to giving the horse a choice of two or more grains. Heavy, bright oats, which contain a small percentage of hull, are preferred. They are obtained by screening out the lighter oats from the heavier oats. Oats average 30% hulls, with poor-quality oats having 50% hulls and the best-quality oats may contain only 25% hulls (9). Musty oats should never be used since they may cause colic or heaves. Oats are bulky, which is helpful in preventing impaction or founder. Hulled oats are especially valuable for young foals. They are called oat groats and are obtained from oats from which the hull has been removed. Oats may be fed whole, crimped, or rolled. Crimping or rolling oats improves digestibility by about 5% (10). Many prefer to feed steam-rolled oats. It is best to roll or crush oats for horses with poor teeth or for young foals. Micronizing oats improved prececal starch digestion by 14% over crimped oats in horses (7). An Oklahoma study (10a) showed that the energy and dry matter digestibility of wheat are higher than oats, while apparent protein digestibility is similar.

B. Barley

Barley is a popular grain for horses and is used heavily in some areas of the country, especially in the western states where it is the leading grain used. Barley is also popular in some foreign countries. Since barley is hard, it should always be coarsely ground or preferably steam-rolled. The outer hull adheres tightly to the kernel; thus, processing increases its digestibility (11). Some horse breeders feel that barley may cause colic when fed alone. Mixing barley with more bulky feeds such as 15% or more wheat bran or 25% oats may minimize the occurrence of colic. Most horse owners prefer to feed barley with oats, brewer's dried grains, citrus pulp, beet pulp, or other bulky feeds even though barley is higher in fiber than corn and grain sorghums. Barley varies considerably in weight per bushel which makes its feeding value variable.

C. Corn

Corn is widely used for horses and ranks second to oats in use. Corn may be fed to horses whole, coarsely ground, or steam-rolled. One should not feed whole corn to horses with poor teeth or to young foals. Most horse people prefer to grind coarsely, or steam flake, or roll the corn for horse feeding. Corn is high in energy and low in fiber. Many horse owners feed corn during the cool weather and take it out of the diet or decrease it to lower levels during warm weather. This is due to the fact that heat produced during digestion increases as the level of fiber in the feed increases. Finely ground corn should not be fed to horses since it may cause colic or founder. Corn is well utilized by the horse if it is balanced with other feeds to offset its deficiencies. Its protein and lysine level is low. Many prefer to feed corn combined with oats and/or bulky feeds such as wheat bran, brewer's dried grains, citrus pulp, or beet pulp. Sometimes the whole ear corn is coarsely ground and fed to mature horses. The cob weighs about 20% of the whole ear of corn and supplies bulk in the diet. An Illinois study (12) with Quarter Horses showed that substituting corn for oats decreased diet and intestinal volume and improved racing time. There is more interest in the use of corn in horse diets than was the case previously.

D. Sorghum Grains (Milo)

Sorghum grains are similar to corn in composition and in feeding value although they vary more in protein content. They are used heavily in the Southwest. Sorghum grains should be coarsely ground, micronized, or steam-rolled for horses. Texas A & M studies (6) showed that sorghum grains could substitute for corn but the best results were obtained when 45.5% of sorghum grain was combined with 15.5% oats and 6.5% corn. In certain situations it would be well to combine the grain sorghums, which are a heavy feed, with bulky feeds such as oats, wheat bran, brewer's dried grains, citrus pulp, or beet pulp. Texas studies (6) showed that micronizing grain sorghums increased grain intake by 1.08 lb daily and the horses required 0.6 lb less feed per lb of gain. Micronizing grain sorghum also decreased prececal starch digestion (7). In the micronizing process the grain is dry heated and then crimped.

E. Wheat

Wheat is not used very much for horse feeding. Wheat contains gluten, which is a sticky substance. When wheat is ground, it is rather doughy and tends to ball up with moisture and produces palatability problems. If used, it is best to steam roll and mix it at a low level with bulky feeds such as oats, brewer's dried grains, beet pulp, and citrus pulp. Wheat should be used in small amounts of 10–20% in the concentrate diet. Wheat is usually too expensive to feed to horses since it is used mainly as human food.

F. Rye

Rye is frequently contaminated with a fungus called ergot. The ergot makes rye even more unpalatable. Ergot may cause abortion and lactation failure in the sow. It is not known whether the mare is affected in a similar manner. Rye is not as palatable as other grains. It is not recommended that rye be fed to valuable horses. If used, it should be fed at no higher a level than 10–20% of the total diet and mixed with palatable feeds such as molasses. If rye is contaminated with ergot, it should not be used.

G. Triticale

Triticale is a hybrid obtained by crossing wheat and rye. Certain varieties of triticale may become infested with ergot. For the pig, triticale has 90–95% the value of corn, but it is less palatable than corn. No information is available to make a sound recommendation on using triticale for horses. The same precautions discussed with rye should be used with triticale.

VII. FEEDING VALUE OF OTHER HIGH-ENERGY FEEDS

A. Wheat Bran

Wheat bran is valuable for its mild laxative effect, its bulky nature, and its palatability. It is generally used at levels of 5–20% of the diet. Wheat bran is made from the hard outer coating of wheat. Many horse owners feed wheat bran as a means of preventing constipation and maintaining correct bowel movements. Some will prepare a wet bran mash by adding warm water and feeding it to horses that are constipated, sick, or foaling.

B. Wheat Middlings

Wheat middlings contain about 20% protein and 5% fat. It is a heavy feed and floury, which means it is dusty. If used in the diet, it should be pelleted to avoid problems with dustiness. The use of 5–10% of sugarcane molasses would help the dustiness problem. Wheat middlings are a good energy feed but should be limited to 5–20% of the concentrate diet.

C. Brewer's Dried Grains

Brewer's dried grains are a good source of protein and higher than the cereal grains in lysine content. They contain 27% protein, 7% fat, and 16% fiber. Brewer's dried grains are used primarily as a source of energy, protein, and bulk. Since their palatability may be a little low, it is best to mix them with more palatable feeds or to pellet the diet. The use of 5–10% of sugarcane molasses in the same diet will increase their palatability. Studies in Florida (13) have shown that brewer's dried grains are a good feed for horses. In a digestion trial the brewer's dried grains were fed at a 20 and 40% level as a substitute for oats and soybean meal in the diet. No differences were obtained in digestibility of dry matter, crude protein, energy, cell wall constituents, or soluble carbohydrates when either level of brewer's dried grains were added to the diet.

In a second study, 24 foals about 8 months of age were fed 20% brewer's dried grains. The foals fed 20% brewer's dried grains decreased their feed intake and rate of gain slightly and sorted out some of the ingredients in the diet (which was not pelleted). There were indications that the diet with 20% brewer's dried grains may have required lysine supplementation.

A third study at Florida (13) was conducted with yearlings averaging 299 days of age. The diets used were pelleted, which prevented sorting of ration ingredients. Levels of 10 and 20% brewer's dried grains were used. With 20% brewer's dried grains, levels of 0.15 and 0.30% lysine were added, whereas levels of 0.10 and 0.20% lysine were added to the diets with 10% brewer's dried grains. The results obtained indicated that the growth of foals and yearlings fed pelleted concentrate diets containing brewer's dried grains as the only supplemental source of protein was only slightly less than that obtained with soybean meal. The addition of lysine to the brewer's dried grain diet resulted in a slight improvement in weight gain, indicating that the lysine level had been too low for young growing horses to obtain maximum growth. Since the lysine-supplemented brewer's dried grain diet resulted in weight gains above those on the soybean meal diet, it is possible that the amino acid composition of the lysine-supplemented brewer's dried grain diets was superior to the amino acid composition of the soybean meal diet.

The source of protein in the diets (brewer's dried grains versus soybean meal) did not appear to influence hoof composition or the amino acid levels in the blood plasma. The increased hoof growth rate in the yearlings on the brewer's dried grain diets, plus lysine supplementation, suggests that hoof growth rate may be influenced by amino acid intake. These studies indicate that brewer's dried grains are a good feed for horses. They should also be considered as another source of bulk in the diet. For young growing horses a maximum of 10–20% of brewer's dried grain intake should be used. If used at a 20% level, lysine supplementation or high-lysine feeds should be provided (see Table 12.2). It would be best to use soybean meal as the protein supplement if 20% brewer's dried grains are used with the young growing horse. Older horses can use higher levels of 20–40% of brewer's dried grains in the diet. Attention should be paid to the lysine level if levels above 20% are used.

Wet brewer's grains were fed at a level of 0, 66, or 81% of the dry matter in the diet (14). Their dry matter digestibility was 66% and the protein digestibility was 84%. These researchers suggested that, owing to the instability of the wet grains, they should only be fed to horses living near the breweries.

D. Sugarcane Molasses

Sugarcane molasses (also called cane molasses or blackstrap molasses) is an excellent feed for horses. It increases palatability and reduces dustiness in the diet. It also adds moisture, which is helpful in the pelleting of feeds. Molasses contains 4.3% protein, but no information is available as to how digestible the protein is for the horse. Some of it is in the form of nonprotein nitrogen (NPN) compounds, which have limited value for the horse. Molasses is usually added at a level of 5–15% in the concentrate feed. Therefore, it would be contributing only a small amount of protein to the total diet. In hot, humid areas, molasses should be limited to between 5 and 10% of the diet. Higher levels might cause wetness in the feed and result in mold growth.

E. Beet Molasses

Beet molasses can be used for the horse in the same manner as sugarcane molasses. Many horse people prefer sugarcane molasses, but either one is satisfactory and both have the same feeding value per pound of dry matter. Since both beet and sugarcane molasses can vary considerable in dry matter content, this should be considered when purchasing molasses.

F. Citrus Pulp

Citrus pulp has been used successfully in the diet by horse producers. Little research information is available on its use in horse feeding. It contains about 6.1% protein and about 3.4% fat. The fat level will vary depending on how much citrus seed (which is high in oil) is contained in the citrus pulp. Citrus pulp contains about 1.71% calcium, which is added in the processing. The calcium level needs to be kept in mind in balancing horse diets. The phosphorus level is about 0.12%. Citrus pulp is a bulky feed and an energy source. Sometimes the citrus pulp also contains citrus molasses, which is readded during processing. Citrus pulp can be used at a level of 5–15% of the diet (15). The lower level would be used with younger horses and the higher level with older, more mature horses. It might help to add 5–10% of sugarcane molasses to diets containing the higher levels of citrus pulp, which will increase the palatability.

Citrus pulp consists of the pulp and residue of oranges and grapefruit being processed for juice. Sometimes lemons, limes, and tangerines are contained in citrus pulp in small amounts. The palatability and feeding value of citrus pulp varies with the processing conditions and the percentage of the various citrus fruits it contains. Care should be taken to obtain high-quality citrus pulp for horses. A Florida study (15) showed that dried citrus pulp is a suitable feed in pelleted diets for mature horses at levels up to 15% as a replacement for oats. The Florida study showed that six of eight horses refused to eat the coarse grain concentrate mix with 30% citrus pulp and consumed only 8.6% of that offered. But, they consumed more of the 30% citrus pulp diet if it was pelleted. More studies are needed on citrus pulp fed as a pelleted feed.

G. Beet Pulp

Beet pulp can be used in the same manner as citrus pulp for feeding horses. It might be limited to 15% of the concentrate diet until more research on it is obtained.

H. Rice Bran

Sometimes rice bran is considered as a feed for horses. It contains about 11.7% crude fiber, 13.0% protein, and 13.6% fat, which is highly unsaturated and easily becomes rancid. If rice bran is used for horses, it should be fed fresh. Because of the high risk of rancidity, rice bran should not ordinarily be used with valuable horses. If used, rice bran should be limited to 5–10% of the concentrate diet and the possibility of rancidity closely monitored.

I. Hominy Feed

Hominy feed is a by-product of the process used for producing corn meal for human use. It consists of a mixture of corn germ, corn bran, and a part of the starchy portion of the kernels. It resembles corn in feeding value. Hominy feed contains about 6.9% fat, which is highly unsaturated and becomes rancid easily. Hominy feed should be fed fresh and kept in a cool, well-ventilated storage area. It should be fed at levels no higher than 20–25% of the concentrate diet. Because of the rancidity problem, it is not usually recommended that hominy feed be used

TABLE 12.4

Relationship of Crude Fiber to Expected
Digestible Energy in Mixed
Concentrate Feeds (10)

Crude Fiber in Feed (%)	Digestible Energy in Feed (Mcal/lb)
2.0	1.60
4.0	1.55
6.0	1.45
8.0	1.35
10.0	1.25
12.0	1.15

with valuable horses. If it becomes rancid, it should not be fed to any horses since rancidity destroys nutrients and causes digestive disturbance.

As one evaluates energy feeds it might be well to keep in mind that over the range of 2–12% crude fiber there is a close inverse relationship between crude fiber content of a feed and its expected energy value (10). This is shown in Table 12.4

VIII. PROTEIN SUPPLEMENTS

Protein supplements should be selected on the basis of total protein level, balance of amino acids, and value in correcting protein and amino acid deficiencies in the energy feeds used in horse diets. One should also consider their mineral and vitamin content when balancing diets. Consideration must also be given to the cost of the protein supplement per unit of protein. Tables 12.2 and 12.3 give information on the protein level, lysine, and other nutrients contained in protein supplements and other high-protein feeds. Protein supplements that contain a high level of the indispensable amino acids, especially lysine, are preferred for young growing horses and may benefit older horses too. Plant protein concentrates are used rather than animal protein concentrates for horses. Small levels of animal protein concentrates, such as fish meal, meat meal, meat, and bone meal, are sometimes used in horse diets. They are good sources of lysine but are usually more expensive than plant protein supplements. They also vary more in quality of protein. Dried skim milk, however, is an excellent quality source of animal protein for young horses.

Table 12.2 gives data on the more commonly used protein supplements.

A. Soybean Meal

Soybean meal is an excellent source of protein, especially for the young growing horse. It contains the highest level of lysine (Table 12.2) of the plant protein concentrates used in horse feeding. It is even higher in lysine than dried skim milk, which itself is an excellent source of protein for the young foal. Soybean meal contains either approximately 44% or 48% protein; the 48% soybean meal is prepared by removing the hulls. Soybean meal is the most widely used protein supplement in horse feeding with excellent results. In practice, however, it is usually fed in combination with other protein sources. Properly processed soybean meal by the hydraulic, expeller, or solvent method has about the same feeding value. Overheating soybean meal destroys lysine whereas underheating soybean meal makes its methionine less available. For maximum nutritional value, therefore, soybean meal needs to be properly heat treated and processed.

Whole uncooked soybeans should not be fed to horses since they contain a trypsin inhibitor which prevents the enzyme trypsin from digesting protein (10). There may be other factors in uncooked whole soybeans which require heat treatment before use by horses.

B. Cottonseed Meal

In volume used, cottonseed meal ranks next to soybean meal in feeding horses. It also is available in many areas of the world. High-quality cottonseed meal low in free gossypol is a satisfactory protein supplement for horses. Gossypol in excess is toxic to pigs. It is not known what effect it may have on the horse. The symptoms of gossypol toxicity in the pig are as follows: excessive fluid in the pleural and peritoneal cavities; flabby and enlarged hearts; congested and edematous lungs; a general congestion of other organs such as the liver, spleen, and lymph glands; and finally death. Since it is not known if these symptoms occur in the horse, they might be used as a guide until data are obtained for horses. Mature horses have been fed 1.0–1.5 lb of cottonseed meal per day per 1000 lb live weight without any harmful effect from gossypol (16). It is best to use cottonseed meal in combination with other protein supplements for best results since cottonseed meal contains approximately half the lysine level of soybean meal. Texas A & M studies (17–19) showed that direct-solvent or prepared-solvent-extracted cottonseed meal, containing 0.2% or less free gossypol, can be used as a protein supplement for young horses if either synthetic lysine or lysine-rich feeds (e.g., soybean meal, fish meal, etc.) are added to the diet. No gossypol toxicity symptoms were reported in any of the studies. Another study at Cal Poly, Pomona (20), showed that weanling fillies fed diets for 6 months containing 0, 10, or 20% low-gossypol (0.04% free gossypol) cottonseed meal plus lysine gave comparable results to the controls. They stated that low-gossypol cottonseed meal with added lysine is an acceptable source of supplemental protein for young horses.

C. Linseed Meal

Linseed meal is a very popular protein supplement for horse feeding. Most horse owners feel that linseed meal contains something (probably mucins or fat) which produces bloom and luster in the hair coat. The solvent-extracted linseed meal is rather fine and dusty. Consequently, many horse feeders prefer to use it in a pelleted form. Some horse people feel that the solvent-extracted linseed meal is too low in oil and does not have as good an effect on the bloom and hair luster. They prefer the old process linseed meal which has more oil, is more palatable, and has more effect on bloom and hair luster. Linseed meal is low in lysine content and should not be used as the only protein supplement with foals or young rapidly growing horses since it could result in a lysine deficiency (3). Linseed meal also has a laxative effect which limits it being used in high amounts in the diet. It is best to use linseed meal in combination with other protein supplements or other feeds, which are high in lysine.

D. Peanut Meal

Peanut meal is the product that remains after the extraction of oil from peanuts. In some foreign countries it is called groundnut meal. Expeller-processed peanut meal contains about 5% fat. The fat in peanut meal is unsaturated and can develop rancidity easily. It is best to use solvent-extracted peanut meal, which contains about 2.1% fat. If peanut meal is used for horses, it should be fed fresh and stored only for short periods of time in a cool, well-ventilated area. Because of the rancidity possibility and its low level of lysine and methionine, peanut meal should not be fed to valuable foals or young rapidly growing horses. It is best to use it at about one-fourth to one-third of the protein supplement added to concentrate diets for older horses. Peanut meal is a very palatable protein supplement.

E. Cull Peas

Cull peas contain 23.4% protein and can be used as a grain or protein substitute. Cull peas fed to pigs and sows on pasture and in dry lot with wheat and barley diets produced similar results to soybean meal and meat meal (21,22). Cull peas are low in lysine and methionine and, therefore, should not be fed to foals or young, rapidly growing horses unless proper amino acid supplementation is used. It is best to use them as one-fourth to one-third the protein supplement added to the concentrate diets for older horses. They can also be used to replace part of the grain or energy feeds in the diet. However, their low level of lysine and methionine would limit their use to a rather low level except to mature horses.

F. Sunflower Meal

Sunflower meal is produced from the seed of the sunflower plant. It is low in lysine which is about half that of soybean meal. The meals with the highest protein levels are those that have the largest quantity of seed hulls removed. The 45.2% protein solvent-extracted meal has 11.7% crude fiber. Extrapolating experimental work with swine, sunflower meal should replace no more than about 20% of the soybean meal in the diet and be fed only to older animals (23). Until more data for the horse are obtained, it would be best not to feed sunflower meal to foals unless it is supplemented with lysine or fed at about one-fourth to one-third the level of protein supplementation with soybean meal. A Texas Tech University study (24) compared sunflower meal to soybean meal in weanling horse diets. The diets were formulated to contain equal amounts of protein, fiber, and TDN. Overall performance of the horses was similar for the two diets.

G. Dried Skim Milk

Dried skim milk is very palatable and highly digestible. It contains lactose, which is the best source of carbohydrates for the young foal. It is an excellent source of high-quality protein and is also a good source of vitamins and minerals. Therefore, it is an excellent feed for the horse. It is too expensive, however, to be used in large amounts in protein supplements for horses. It is used primarily in foal diets to supplement the mare's milk or as part of a milk replacer to reinforce diets fed to foals after weaning to give them a good start.

H. Fish Meals

There are many fish meals on the market. Properly processed fish meal is an excellent protein supplement. It is exceptionally high in lysine and methionine (Table 12.2). If fish meal contains too large an amount of bone and fish heads, its feeding value decreases because of the lower nutritive value of bone as compared to the flesh of the fish. Fish meal is expensive, which limits its use. It is used at low levels in the diet, however, in many cases to provide part of the protein supplement and to supply extra lysine and methionine, and to help balance the amino acids in the diet.

I. Meat Meal and Meat and Bone Meal

Both of these animal products are sometimes used as a small portion of the protein supplement added to concentrate diets for horses. They are good sources of lysine, being considerably higher than the plant protein supplements (Table 12.2). They are also very high in calcium and phosphorus (Table 12.3). They also contain a good supply of vitamins and minerals. There is considerable variation in the nutritional value of these meat products depending on how much meat and internal organs they contain. Quality of product, therefore, needs to be carefully observed.

J. Rapeseed Meal and Canola Meal

Rapeseed meal contains about 36% protein and 2.1% lysine. It is high in fiber with about twice the level of 44%-protein soybean meal. Studies by H. F. Hintz at Cornell University showed that the protein of rapeseed meal was digested at about the same rate as that of soybean meal. He encountered no feed-acceptability problems in diets containing up to 30% rapeseed meal. Canadian studies found no difference in feed intake, daily gain, or feed efficiency in growing horses fed diets containing 15% rapeseed meal or 15% soybean meal.

In the last few years, the term canola meal has been used for varieties of rapeseed meal that are low in erucic and glucosinolate, the toxic substances present in high amounts in the original varieties. The canola meal has been found to be superior to rapeseed meal for swine and poultry feeding. It would appear that canola meal might be preferable to rapeseed meal for use in horse diets until more research is available to differentiate their relative feeding values for horses, especially young, high-level performance horses.

K. Urea

Urea and its use in horse diets is discussed in Chapter 7 on protein requirement of horses.

IX. FEEDING VALUE OF HAYS

Forages, which include hay and pasture, are very important in the horse diet. There is a minimum level of forage needed for best horse performance. Some scientists feel the minimum forage requirement is 0.5% of the horse's body weight. This means the horse should consume 0.5 lb of forage per 100 lb of body weight. However, many horse owners prefer to feed at least 1 lb of forage per 100 lb of body weight. They feel this higher level is safer and more apt to result in less problems with digestive disturbances, colic, and other disorders.

There is considerable variation, however, in the best level of forage for the individual horse. Horses vary in stomach capacity and this will influence the level of forage needed. A horse with limited stomach capacity requires less forage. The level of activity of the individual horse is also important. Horses being developed for racing or high-level performance purposes should be fed no more forage than is necessary. They need as much energy as possible in their total daily diet. This means a relatively low level of forage should be used. Many performance horses are fed too much forage. This results in a distended digestive tract, which causes discomfort when they are exercised or worked heavily. It may also cause quick tiring and labored breathing. As a result, many prefer to feed one-half or more of the forage allowance in the evening after the horse is retired for the day. The remainder of the forage used is divided between the other feedings during the day when the horse is active. Sometimes it may be necessary to muzzle greedy horses to keep them from eating the hay and often times the bedding when their forage allowance is restricted.

The kind of forage used is also important. Some forages are more easily digested than others. Some are eliminated from the digestive tract more rapidly. The kind and quality of the forage used is of importance in determining the level to use in the total diet.

Table 12.5 gives the suggested range of forage to use in horse diets. The levels shown in the table can vary depending on individual horses. It is important to emphasize that horses being developed for top performance will vary in the level of forage and/or concentrates needed for best results. The feeder or trainer needs to study each individual animal and determine the level it prefers in order to obtain successful results. This can be the difference between developing a winner or just another horse.

A study of Table 12.5 shows that nursing foals, weanlings, and performance (intense work) horses consume the least amount of forage as a percentage of their diet. These animals require high-energy levels, and their diet must contain a higher level of concentrates and less forage. Idle horses require the least amount of concentrates. If need be, they can do well on high-quality forage alone. The pregnant mare can do well on about half forage and half concentrates. During the last one-third of gestation, the level of concentrate needs to be increased to take care of the foal's development in the uterus. During lactation, the percentage of concentrate in the diet needs to be increased even further to take care of milk production.

The condition of the horse also influences the level of concentrates and forage used. The horse's condition should be used as a guide. A level of feeding that causes one horse to get fat may keep another in a thrifty condition. This may be due to the activity of the horse or to its ability to utilize the feed it consumes. Some horses are more efficient than others in this respect.

The level of activity or performance required of the horse will also influence

TABLE 12.5

Expected Feed Consumption by Horses (4) (% Body Weight)[a]

	Forage	Concentrate	Total
Mature horses			
Maintenance	1.5–2.0	0–0.5	1.5–2.0
Mares, late gestation	1.0–1.5	0.5–1.0	1.5–2.0
Mares, early lactation	1.0–2.0	1.0–2.0	2.0–3.0
Mares, late lactation	1.0–2.0	0.5–1.5	2.0–2.5
Working horses			
Light work	1.0–2.0	0.5–1.0	1.5–2.5
Moderate work	1.0–2.0	0.75–1.5	1.75–2.5
Intense work	0.75–1.5	1.0–2.0	2.0–3.0
Young horses			
Nursing foal, 3 months	0	1.0–2.0	2.5–3.5
Weanling foal, 6 months	0.5–1.0	1.5–3.0	2.0–3.5
Yearling foal, 12 months	1.0–1.5	1.0–2.0	2.0–3.0
Long yearling, 18 months	1.0–1.5	1.0–1.5	2.0–2.5
Two year old (24 months)	1.0–1.5	1.0–1.5	1.75–2.5

[a]Air-dry feed (about 90% dry matter).

the energy level needed and, therefore, the ratio of concentrates to forages fed. The greater the physical activity, the less forage relative to concentrates used. One needs to feed the horse according to: (1) its individual needs; (2) its level of activity; (3) its response to the diet used; and (4) the condition of the horse. The data shown in Table 12.5 can be used as a guide on the ratio of forage to concentrates to use. These levels can be altered or modified to meet the specific needs of the individual horse involved.

A. Form of Hay Used

There are many ways in which hay can be fed. Baled hay is the most widely used today. Some pelleted or wafered hay is also fed. In buying or making hay, it should be free of dust and molds which can cause respiratory difficulties including heaves. Wafer size should be such that it breaks readily on consumption and will not cause choking. Horses usually consume more hay fed as wafers or cubes than when it is fed as baled hay.

Pellets are also used and the size usually is about $\frac{1}{2}-\frac{3}{4}$ inch pellet for the more mature horses and a smaller pellet for the young horses (usually about a $\frac{1}{4}-\frac{1}{2}$ inch pellet). Pellets should not be too hard. A softer pellet is easier to chew and consume. But, pellets should not be so soft that they break up and produce fines. The same applies to wafers.

One of the big advantages to pellets and wafers is that they reduce dust. They also save hay loss since 8–20% of long hay may be wasted. A Canadian study (25) showed that using a feeder for hay resulted in daily gains in horses which were 18.3% higher than those in horses fed an equal amount of hay on the ground. The use of a feeder resulted in an 18.5% improvement in feed efficiency because of reduced feed wasted. In addition, the use of pellets or wafers saves storage space, reduces transportation costs, and increases hay or forage intake by the horse. The use of pellets allows the addition of minerals, vitamins, and other nutrients. It allows adaptation to automation when it is feasible. One disadvantage is that more wood chewing will occur. This can be minimized by increasing the fiber length of the ground hay in the pellet or by feeding some hay or using pasture.

B. Hay Quality Varies Considerably

Dr. C. G. Depew of Louisiana State University (26) developed the information shown in Table 12.6. It demonstrates the tremendous difference which may occur in the protein and TDN value of 5 different hays which are either excellent or poor in quality. The vitamin, mineral, other nutrients, and the palatability would also decrease in close proportion to the decrease in protein and TDN shown in Table 12.6. The data in the table demonstrate that poor-quality alfalfa is lower in nutritional value than the other four hays which are excellent in quality. This stresses the importance of using quality as the number one priority when purchasing hay. It also indicates there may be more difference within a specific hay than between different kinds of hay.

The best indication of hay quality is how much of it the horse consumes. The

TABLE 12.6

Nutrient Differences in Various Types and Qualities of Hay (26)

Quality	Type	Total protein (%)	Total digestible nutrients (%)
Excellent	Alfalfa	20.0	60.8
Poor	Alfalfa	8.0	36.9
Excellent	Bermuda	16.0	56.3
Poor	Bermuda	6.0	40.7
Excellent	Timothy	12.0	50.0
Poor	Timothy	5.0	35.0
Excellent	Bahia	11.8	55.8
Poor	Bahia	4.9	34.6
Excellent	Ryegrass	17.3	61.3
Poor	Ryegrass	3.7	37.8

higher the hay quality, the more of it the horse will eat. This is what scientists refer to as "voluntary forage intake." A quick, easy chemical test needs to be developed which correlates the voluntary forage intake of the horse with the quality of the hay. There are many indicators of quality which can be looked for in physically evaluating hay if chemical analyses are not available.

The green color in hay is a very important indicator of quality. It indicates that the hay was harvested, cured, and stored before prolonged exposure to the sun. It also indicates that the hay was not rained upon. Rain leaches out many of the nutrients from hay and lowers its nutritional value. The green color also indicates the forage was harvested while it was immature and of high quality.

The percentage of leaves in hay is also an indication of quality since leaves have the highest nutritional value. So, high-quality hay should contain a high proportion of leaves relative to stems.

Evaluating the hay for stem size, pliability, and length of stem may indicate how mature the hay was when harvested. Large and numerous stems result in low-quality, less-palatable hay. Foreign materials, such as stubble, weeds, sticks, stones, manure, dirt, metal objects, etc., should not be present in hay. If mature seeds are present it is an indication of the degree of maturity at harvest. Mature plants make stemmy, low-protein, high-fiber hay with poor digestibility.

The hay should have a clean, fresh crop odor, which indicates that it was properly cured and stored with good air circulation. Moldy, musty, or dusty odors in hay are undesirable.

Proper hay making requires that hay be baled as quickly as possible after it is dry enough. Then it should be stacked or put under a roof as soon as possible to avoid sun and rain. It should also be stored in a way to allow good air circulation, which keeps it from heating up or becoming moldy or musty. Sometimes hay is windrowed and then put into stacks. The same suggestions as were recommended for baling hay would also apply to stacked hay.

Hay should be used before it becomes too old. The longer hay is stored the lower it becomes in quality and feeding value. Most forages can be stored successfully as hay if allowed to dry to 20% or lower in moisture content (4). At higher moisture levels, the forage is subject to molds, heating, and spoilage. When a forage is harvested for hay at higher moisture levels, preservatives can be applied to prevent molding and spoilage. One preservative product consisting of proprionic acid (80%) and acetic acid (20%) has been shown to successfully preserve alfalfa baled at moisture levels of up to 35%, and this hay has been fed to horses with excellent results (27–29).

With forages harvested at higher moisture levels and stored as haylage or silage, the risk of the presence of substances toxic to horses is much greater (4,30). A serious outbreak of botulism from horses eating big bale silage has been reported (31).

C. Alfalfa

Alfalfa is increasingly being used in well-balanced horse diets. The prejudice that used to exist against alfalfa for horses has gradually disappeared as research evidence and experience with alfalfa use indicate that it is an excellent feed for horses. Many horse owners use it exclusively with good results.

Dr. James H. Bailey of Pampa, Texas, fed brood mares with as high as 5–10 lb of 17%-protein dehydrated alfalfa meal (dehy) pellets daily in place of hay while the horses were on dry grass during the winter (32). The number of heat periods required per conception were decreased with the horses fed alfalfa. He also observed that foals from mares receiving dehy were much above average in physical condition with no contracted tendons, crooked legs, or other abnormalities. This trial showed that mares can be fed as high as 5–10 lb of dehy daily with beneficial results.

Studies at Texas A & M by Dr. L. R. Breuer, Jr., and others in 1970 showed that a 20% level of dehy replacing 20% of milo had a slightly higher energy value for maintenance of mature horses. In another group of horses, 40% of dehy replaced 20% of milo and 20% of sorghum hay and also had a slightly higher energy value than the two feeds it replaced. In fact, the diet with 40% dehy was observed to be the most efficient for the maintenance of the mature horses used. They studied 800–1200 lb mature horses, which were fed completely pelleted diets ($\frac{3}{8}$ inch pellet) for a 189-day trial period. The Texas study showed that 40% of dehydrated alfalfa meal could be successfully fed to mature horses. They also showed that dehydrated alfalfa meal could serve as the only roughage in a completely pelleted horse feed. They, like other scientists, obtained no evidence of harmful effects from feeding alfalfa to horses. They observed wood and tail chewing, which may be related to the completely pelleted diet. Coprophagy (eating of feces) was observed in several horses. This has been reported by other scientists feeding a completely pelleted diet with no access to hay and/or pasture.

A study at Washington State University involved studying alfalfa fed as a cube versus baled hay for horses (33). Two of the horses did not readily accept alfalfa cubes right away, but did within a short time. The horses fed the alfalfa cubes consumed on average 18.7 lb of alfalfa cubes plus 4.3 lb of concentrates daily. The horses fed the baled alfalfa hay consumed 18.6 lb of alfalfa plus 5.8 lb of concentrates per day. During the 65-day trial the horses fed the alfalfa cubes gained 31 lb, whereas the horses fed the baled alfalfa lost 19 lb. This indicated that the horses utilized the alfalfa more efficiently when it was fed as a cube, since they consumed less total feed daily but gained instead of losing weight. The Washington State scientists reported that the cubes presented no observable problems in chewing or swallowing and no evidence of choking. They also reported no evidence of increased wood chewing when the cubed alfalfa hay was used.

This study indicates that cubed alfalfa is a satisfactory method of feeding alfalfa to horses. It also presents additional evidence that horses can do well on high levels of alfalfa. Mature horses from the Washington State University Horse Show Team were used in this experiment. The scientists were not trying to feed for maximum gain but rather to maintain the horses in a proper condition. When the trial terminated they placed three of the cubed alfalfa-fed horses on a free-choice feed for 2 weeks. This resulted in a rapid weight gain when the horses were allowed to eat all the cubed alfalfa they wanted.

Scientists from the University of Delaware fed Thoroughbred mares and their nursing foals $\frac{3}{4}$-inch alfalfa pellets as the only feed for 1 year continuously (34). The mares were in excellent physical condition during the trial and the foals grew as well as those on a conventional feeding system. The only problem that arose was wood chewing. This occurs with some completely pelleted diets and can be helped by feeding some long hay, providing some pasture, or feeding a wafered hay. The Delaware scientists found that horses consumed 17% more wafers and 24% more pellets than loose hay. The nutritive value index was 32% higher for alfalfa pellets as compared to loose alfalfa hay. The nutritive value index was about the same for pellets as for wafers. Using pellets or wafers reduces bulkiness in transporting hay. For example, pelleting or wafering reduces the bulkiness of baled hay by as much as 75% (34).

Studies at the University of Kentucky showed that dehydrated alfalfa meal increased cellulose and energy digestibility when 18.2% dehydrated alfalfa meal was added to a completely pelleted diet (35).

These studies, and others, indicate that alfalfa can be used successfully in horse diets. Alfalfa is a good source of protein, vitamins, minerals, and other nutrients that are beneficial for the horse.

G. L. Ward and co-workers (36) concluded that 20–30% of the calcium in alfalfa is in the form of oxalate and is apparently unavailable to ruminants. Studies at Cornell University (37) found no differences in apparent or estimated true digestibility of calcium, phosphorus, and magnesium in alfalfa pellets with varying calcium to oxalic acid ratios. They felt further work is needed to clarify the differences in the effect of oxalic acid on calcium utilization between horses and ruminants. They also stated that several tropical grasses such as buffel, pangola, and kikuyu contain levels of oxalic acid that significantly interfere with mineral utilization in horses.

D. Timothy Hay

Timothy hay has been the preferred hay for many horse people for many years. It is low in protein as compared to many other hays and may require more protein supplementation (Table 12.2). However, it is usually grown in combination with red clover or alsike clover. The clovers are high in protein and increase the level of protein in the timothy–clover hay combinations. It is best to cut timothy at the prebloom stage in order to increase its protein content, decrease its fiber level, and increase its palatability and digestibility. Mature, late-cut timothy is a poor-quality hay. Timothy hay is usually free of molds and mildew.

E. Oat Hay

Oat hay is used a great deal in horse feeding. It is low in protein and more protein supplementation is needed; or, it can be fed with alfalfa or other legume hays to increase the total protein level in the diet. Oat hay should be cut in the soft dough stage when it is highly nutritious and lower in fiber. There is usually less shattering of the kernels when the oats are cut for hay at the soft dough stage. If the heads in oat hay are lost, its feeding value decreases.

F. Other Cereal Grain Hays

Other small grain hays are used for horses. Horses usually prefer oats, barley, wheat, and rye hay in that order (10,38). They should all be cut in the soft to stiff dough stage.

G. Other Grass Hays

In addition to timothy and other small grain hays there are many other grass hays used for horse feeding. These hays include the following: bahiagrass, coastal bermudagrass, sorghum-type hays, bluegrass, smooth bromegrass, reed canarygrass, tall fescue, orchardgrass, pangolagrass, praire, and others. These hays are all low in protein as compared to the legume hays. The nutritional value of these grass hays varies depending on the stage of growth they are cut, how much exposure they get to the sun after cutting, whether they are rained on, and how well they are handled, stored, and fed. As the grass, and legumes as well, increase in maturity before cutting for hay, the fiber level increases and the protein level, palatability, digestibility, and feeding value decreases. Good-quality grass hays should be leafy, soft, and pliable to the touch, have no or comparatively few seed heads, and should be free of mold, dust, and weeds.

H. Other Legume Hays

In addition to alfalfa, there are other legume hays used for horses. These legume hays include the following: alsike clover, crimson clover, ladino clover, red clover, soybean, birdsfoot trefoil, lespedeza, and cowpea hay. In recent years some varieties of perennial peanuts have shown promise for hay production (4). These legume hays are higher in protein than the grass hays. In many cases both a grass hay and a legume hay are fed together to horses as a means of balancing the protein level provided by both. One hay may be fed in the morning and the other in the evening.

Chapter 13 on pastures for horses gives additional information on hays for feeding horses. It also discusses limitations and problems which may occur with certain hays.

X. FEEDING VALUE OF SILAGE

Silage is not usually fed to horses. Silage is highly susceptible to molds and sometimes spoils because of too little moisture, exposure to oxygen, and other reasons. Horses are very susceptible to spoiled or moldy silage. It can cause digestive upsets and even death. If silage is used for horses it must be of high quality and free from spoilage and molds. Unfortunately, it is difficult to produce mold-free silage.

Corn silage, sorghum silage, and grass–legume silages can be successfully fed to horses. Silage should not be used without some hay also being fed. Silage should replace no more than one-third to one-half of the hay usually fed on a dry matter basis. Usually 1 lb of hay is equivalent to about 3 lb of silage.

Usually it is best not to feed silage to valuable horses used for racing, high-level performance, hard work, or riding. Silage is high in moisture and distends the digestive tract if too much is fed. Many who use silage feed it to mature idle horses, mares not in foal, and horses not used extensively for work or riding.

Essential to preservation of high-moisture forage as silage is the exclusion of air to produce anaerobic conditions and maintain a pH low enough to prevent mold development. The optimum pH ranges from 3.5 to 4.5 (4). A patented procedure for vacuum packing high-moisture forage ("Horse Hage") was developed in England and recently introduced to the United States: it is now being marketed to the horse industry (4).

XI. OTHER FEEDS

There are many other feeds that can be used in horse diets, but there is so little information available on them for the horse that it is difficult to make recommendations on them. A few will be listed, however, with some information provided.

A. Grain Straws

Straw from grains has very little feeding value for the horse. They are all high in fiber (about 37%), but they can be used as a part of the roughage in a completely pelleted diet as a means of adding bulk, provided that adequate supplemental feeds are used. The level used should be limited to about 10% of the diet in order not to decrease the feeding value of the pellet too much. Straws are usually dusty and sometimes contain dirt and other materials. Only high-quality grain straw as free as possible from dust, mold, dirt, and other matter should be used for horse feeding. Ryegrass straw fed at 51% of the diet gave poorer results than fescue hay when fed in a completely pelleted, isonitrogenous, and isocaloric diet in studies in Oregon (39). Another Oregon study (40) showed that ryegrass straw fed at 50 or 68% of the total diet to mature pony mares and geldings maintained weight and did not adversely affect feed intake when fed in combination with concentrates. But, the energy and fiber digestibility levels may be decreased. Therefore, one needs to be careful in using different straws and limit the level at which they are used. A number of studies have shown that straw treated with ammonia, sodium hydroxide, or acid followed by yeast inoculation is more digestible than untreated straw (41–43).

B. Rice Hulls

Rice hulls are a very poor-quality feed. They are high in fiber (40.7%) and their use in horse diets might be as a roughage substitute to add bulk to a completely pelleted diet. Rice hulls should be used at a level no higher than 10% of the diet as a source of bulk in order not to decrease the feeding value of the pellet. Some feel the sharp edges of the rice hulls may irritate the mouth or digestive tract of the horse. No research information is available on this matter. This problem may be eliminated by grinding the rice hulls before using them in the pellet. Only high-quality rice hulls as free as possible from mold, dust, and other material should be used for horse feeding.

C. Oat Hulls

Oat hulls are high in fiber (29.2%) and low in feeding value. They can be used, however, as part of the roughage in a completely pelleted diet as a means of adding bulk. The level used should be limited to about 10% of the diet in order not to decrease the feeding value of the pellet appreciably. Only high-quality oat hulls as free as possible from dust, mold, or other materials should be used in horse feeding.

D. Dried Bakery Product

Dried bakery product is produced from reclaimed bakery products that have been blended and standardized to give a feed containing about 9.5% protein and 13% fat. With pigs, the dried bakery product has compared favorably to dried skim milk in starter diets. Dried bakery product would have maximum value in creep diets, or milk-replacer diets for the foal. However, only a high-quality dried bakery product that is blended, with the fat protected by an antioxidant, and that is standardized to give a constant product should be used. Some bakery products on the market should not be used for feeding foals because they are not properly processed and standardized.

E. Feather Meal

Feather meal contains 85–87% protein. It needs to be hydrolyzed for maximum feeding value. If used in horse diets, it should be limited to a level no higher than 2.5% in the concentrate diet and no higher than one-half the protein supplement used.

F. Distiller's Dried Solubles

Distiller's dried solubles are valuable as a source of B-complex vitamins. If used in horse diets, the level used should be no higher than 2.5–5.0% of the concentrate diet. They are low in lysine and tryptophan.

G. Distiller's Dried Grains with Solubles

They vary in composition because of the various kinds of distiller's dried grains available. They are excellent sources of the B-complex vitamins but are low in lysine. If used in horse diets, they should be limited to a level no higher than 2.5–5.0% in the concentrate diet.

H. Corrugated Paper

Studies at Cornell (41) showed that corrugated paper boxes and computer paper were efficiently utilized by horses when ground and added to pelleted diets. Diets containing as much as 50% paper have been fed. Paper products, such as newsprint, which contain high levels of lignin, would probably not be effectively utilized by horses since even ruminants do not utilize newsprint effectively. Moreover, the ink used in newsprint may create problems. Corrugated paper boxes and computer paper contain very low levels of protein, minerals, and vitamins but do contain a high level of digestible cellulose. If used, the paper must be free of heavy metals such as lead. Polychlorinated biphenyls (PCBs) and other potentially toxic materials should be monitored. Paper is a potential source of feed but more studies are needed. If used, it would be best to feed it to mature horses, idle horses, and other less valuable animals used for periodic riding or work.

I. Whole Corn Plant

Scientists at Oregon State (44) fed the whole corn plant (contains 6.2% protein) as a pellet to mature horses. They concluded that the mature horse could be maintained on the whole corn plant as the major energy source if it is supplemented with adequate protein, minerals, and vitamins. A 3-month maintenance trial indicated that the mature horse could be maintained by feeding 13.64 lb per animal daily (1.2% of body weight) of the whole corn plant. However, considerable appetite depravity and excessive coprophagy (eating of feces) was observed among all the horses. Supplementing the whole corn plant with soybean meal to increase the protein content to 10% eliminated all coprophagy within 5–7 days. Subsequent removal of the supplemental protein resulted in coprophagy again within 7–10 days.

A Minnesota study (45) showed that pony mares fed an equal amount of bromegrass hay with pelleted corn plant increased mare weight by 15%. The pellets were readily consumed.

J. Dried Poultry Waste

Studies at Cal Poly University, Pomona (46) showed that no differences were obtained in height, weight, or heart girth measurements of 7-month-old horses fed 0, 5, or 10% dried poultry waste in a pelleted diet. This preliminary trial showed that at least 10% of dried poultry waste can be safely fed in the diet of young growing horses. Until more studies are conducted, it would be wise not to include dried poultry waste in the diet of high-level performance horses.

K. Vegetables

Fruits, carrots, turnips, and other root crops are often used as treats for horses. They are used as a treat like some people put cubes of sugar in their hand for horses to nibble on or like a human would give candy to a child. In some cases, higher levels of vegetables and other crops are given to mature or idle horses in some foreign countries. These crops are quite high in water, however, and so there is a limit to how much one can feed to horses.

L. Gelatin

Some horse owners and trainers feel that gelatin may be helpful in hoof development. Studies at Cornell University by Dr. H. F. Hintz and at the University of Kentucky by Dr. John P. Baker have shown no beneficial effect on hoof development from gelatin supplementation. Levels of 0.2–0.25 lb of gelatin were used. If gelatin supplementation has an effect on the hoof of the horse, other research is needed to verify it since these two well-conducted studies showed no value for gelatin use.

M. Live Yeast Culture

One report showed that adding live yeast culture to diets of growing Thoroughbreds enhanced nitrogen retention and hemicellulose digestibility (47). Others (48,49) had previously shown yeast culture to benefit horses. Another report (50), however, found no influence on nutrient digestibility or nitrogen retention in horses fed yeast culture. In another study (51) the data suggested an enhanced conditioning effect from chronic yeast culture supplementation, which was evidenced by increased aerobic capacity. The report also indicated that dietary yeast culture supplementation of horses entering into conditioning programs may well enhance athletic training. In a Northwestern University report (52) the findings suggested that supplemental yeast culture may influence dietary nitrogen retention by improving the effective biological value of feed protein. It may also be associated with positive effects on both protein synthesis and protein degradation in young growing horses. A Kentucky study (53) reported the lack of a significant difference between yeast culture supplementation and digestibility which suggested that either not enough yeast culture was fed or that the yeast culture was digested and utilized as a nutrient source, rather than for enhancing the fermentation in the lower gut. Also, rice hulls furnished the major portion of the diet and because of their indigestible nature, they may have contributed to the lack of a significant difference from yeast culture supplementation.

N. Sunflower Hulls

A Minnesota study (45) showed that a level of 65 and 80% pelleted sunflower hulls in place of grass hay decreased rate of gain in yearling ponies. They did provide bulk and appetite satiation. They can be used at lower levels to add bulk to the diet.

O. Soybean Hulls

A Minnesota study (45) showed that diets containing 50% soybean hulls resulted in gains comparable to ponies full-fed diets containing 75% alfalfa. Weanling and pony mares ate less of a diet containing 75% soybean hulls as compared to 50% soybean hulls. They can be used at lower levels to add bulk to the diet.

P. Cottonseed Hulls

Cottonseed hulls contain about 50–70% as much digestible energy as average-quality grass hay (4). Despite their low nutritive value, cottonseed hulls can be used to a limited extent to provide fiber and bulk in the diet.

Q. Peanut Hulls

They are rarely fed to horses since they are susceptible to invasion by aflatoxins (4). But, they can be used to provide bulk and fiber in the diet.

R. Rice Mill Feed

It is a by-product of rice milling and consists of rice grain and hulls. It contains over 32% fiber and is used in some horse feeds.

S. Ground Corncobs

They contain approximately 50–70% as much digestible energy as the average grass hay (4). They can be used as a source of fiber and bulk in horse diets.

XII. CONCLUSIONS

The relative value and limitations of various feeds used for horses are discussed in this chapter. Feeds need to be combined properly into diets that are well balanced nutritionally, palatable, and economical for the work activity the horse will perform. Grains, other high-energy feeds, protein supplements, hays, silage, and other feeds are discussed in detail. Recommendations are given on their value and use in horse diets. Information is also given on the level of important nutrients in certain feeds which serve as a guide on how best to use them in horse diets.

REFERENCES

1. Hintz, H. F., J. P. Baker, R. M. Jordon, E. A. Ott, G. D. Potter, and L. M. Slade. "Nutrient Requirements of Horses." *N.A.S.–N.R.C.*, Washington, D.C. (1978).
2. Sunde, M. L., J. R. Couch, L. S. Jensen, B. F. March, E. C. Naber, L. M. Potter, and P. E. Waibel. "Nutrient Requirements of Poultry." *N.A.S.–N.R.C.*, Washington, D.C. (1977).
3. Breuer, L. H., and D. L. Golden. *J. Anim. Sci.* **33,** 227 (1971).
4. Ott, E. A., J. P. Baker, H. F. Hintz, G. D. Potter, H. D. Stowe, and D. E. Ullrey. "Nutrient Requirements of Horses." *N.A.S.–N.R.C.*, Washington, D.C. (1989).

5. Conrad, J. H., C. W. Deyoe, L. E. Harris, P. W. Moe, R. L. Preston, and P. J. Van Soest. "U.S.-Canadian Tables of Feed Composition." NAS-NRC, Washington, D. C., 1982.

6. Householder, D. D., G. D. Potter, R. E. Lichtenwalner, and J. H. Hesby. *J. Anim. Sci.* **43**, 254 (1976).

7. Householder, D. D., G. D. Potter, and R. E. Lichtenwalner. *Proc. Equine Nutr. Physiol. Symp.*, *5th*, p. 44 (1977).

8. Hintz, H. F. *Proc. Cornell Nutr. Conf.* (1980).

9. Schryver, H. F., H. F. Hintz, and J. E. Lowe. *Cornell Univ. Ext. Serv. Publ. Feeding Horses* (1982).

10. Householder, D. D., and G. D. Potter. *Tex. A & M Univ. Mimeo Rep.* (1988).

10a. Topliff, D. R., and D. W. Freeman. *Proc. Equine Nutr. Physiol. Symp.*, *11th*, p. 157 (1989).

11. Ott, E. A. "Effect of Processing on the Nutritional Values of Feeds," p. 373. NAS-NRC, Washington, D. C., 1972.

12. Burke, D. J., W. W. Albert, and P. C. Harrison. *Proc. Equine Nutr. Physiol. Symp.*, *7th*, p. 197 (1981).

13. Ott, E. A., University of Florida, Gainesville (personal communication), 1978.

14. Walter, R., J. P. Vellette, and A. Daste. *Ann. Zootech.* **32**, 497 (1983).

15. Ott, E. A., J. P. Feaster, and S. Lieb. *J. Anim. Sci.* **49**, 983 (1979).

16. Morrison, F. B. "Feeds and Feeding," 22nd ed., p. 481. Morrison Publ. Co., Ithaca, New York, 1956.

17. McCall, M. A., G. D. Potter, and J. L. Kreider. *Proc. Equine Nutr. Physiol. Symp.*, *7th*, p. 82 (1981).

18. Potter, G. D., and J. D. Huchton, *Proc. Equine Nutr. Physiol. Symp.*, *4th*, p. 19 (1975).

19. Wirth, B. L., G. D. Potter, and G. A. Broderick. *J. Anim. Sci.* **43**, 261 (1976).

20. Moise, L. L., and A. A. Wysocki. *Proc. Equine Nutr. Physiol. Symp.*, *7th*, p. 85 (1981).

21. Cunha, T. J., E. J. Warwick, M. E. Ensminger, and N. K. Hart. *J. Anim. Sci.* **7**, 117 (1948).

22. Colby, R. W., T. J. Cunha, and M. E. Ensminger. *Wash., Agric. Exp. Stn., Circ.* **153** (1951).

23. Seerley, R. W., D. Burdick, W. C. Russom, R. S. Lowery, H. C. McCampbell, and H. E. Amos. *J. Anim. Sci.* **38**, 947 (1974).

24. Heird, J. C., D. H. Hurley, R. C. Albin, and C. R. Richardson. *Proc. Equine Nutr. Physiol. Symp.*, *6th*, p. 84 (1979).

25. Coleman, R. J., J. D. Milligan, and L. D. Burwash. *Proc. Equine Nutr. Physiol. Symp.*, *11th*, p. 164 (1989).

26. Depew, C. G. *La., Agric. Ext. Serv., Mimeo Horse Nutr.* (1986).

27. Hintz, H. F., J. E. Lowe, and W. F. Miller. *Proc. Equine Nutr. Physiol. Symp.*, *8th*, p. 1 (1983).

28. Lawrence, L., K. J. Moore, H. F. Hintz, E. H. Jaster, and L. Weschover. *Can. J. Anim. Sci.* **67**, 217 (1987).

29. Battle, G. H., S. G. Jackson, and J. P. Baker. *Nutr. Rep. Int.* **37**, 83 (1988).

30. Smith, G. R., and L. G. Murray. *Vet. Rec.* **114**, 75 (1984).

31. Ricketts, S. W., T. R. C. Greet, P. J. Glyn, C. D. R. Ginnett, E. P. McAllister, J. McCraig, P. H. Skinner, P. M. Webbon, D. L. Frape, G. R. Smith, and L. G. Murray. *Equine Vet. J.* **16**, 515 (1984).

32. Bailey, J. H. *Feedstuffs* **42**, 22 (1970).

33. Johnson, R. J., and I. M. Hughes. *Feedstuffs* **46**, 31 (1974).

34. Haenlein, G. F. W., R. D. Holdren, and Y. M. Yoon. *J. Anim. Sci.* **25**, 740 (1966).

35. Leonard, T. M., J. P. Baker, and J. G. Willard. *J. Anim. Sci.* **39**, 184 (1974).

36. Ward, G., L. H. Harbers, and J. J. Blaha. *J. Dairy Sci.* **62**, 715 (1979).

37. Hintz, H. F., H. F. Schryver, J. Doty, C. Lakin, and R. A. Zimmerman. *J. Anim. Sci.* **58**, 939 (1984).

38. Bradley, M., and W. H. Pfander. *Univ. Mo., Mimeo Feeds Light Horses* (1978).

39. Schurg, W. A., and R. E. Pulse. *J. Anim. Sci.* **38,** 1330 (1974).
40. Schurg, W. A., and D. W. Holton. *Proc. Equine Nutr. Physiol. Symp., 6th,* p. 86 (1979).
41. Hintz, H. F., and H. F. Schryver. *Proc. Cornell Nutr. Conf.* p. 27 (1978).
42. Hintz, H. F. "Horse Nutrition—A Practical Guide." Arco Publ., New York, 1983.
43. Schurg, W. A. *Proc. Equine Nutr. Physiol. Symp., 7th,* p. 8 (1981).
44. Schurg, W. A., D. L. Frei, P. R. Cheeke, and D. W. Holtan. *J. Anim. Sci.* **45,** 1317 (1977).
45. Jordon, R. M. and G. Kosmo. *Proc. Equine Nutr. Physiol. Symp., 6th,* p. 77 (1979).
46. Wysocki, A. A., A. S. Brown, and J. A. Grimmett. *Calif. State Poly Univ., Pomona, Mimeo Ser.* No. 76-1 (1976).
47. Glade, M. J., and L. M. Biesik. *J. Anim. Sci.* **62,** 1635 (1986).
48. Godbee, R. Research Bulletin. Clemson University, Clemson, South Carolina, 1983.
49. Mason, T. R. Research Bulletin. McNeese State University, Lake Charles, Louisiana 1983.
50. Webb, S. P., G. D. Potter, and K. J. Massey. *Proc. Equine Nutr. Physiol. Symp., 9th,* p. 64 (1985).
51. Campbell, M., and M. J. Glade. *Proc. Equine Nutr. Physiol. Symp., 11th,* p. 72 (1989).
52. Glade, M. J. *Proc. Equine Nutr. Physiol. Symp., 11th,* p. 119 (1989).
53. Hall, R. R., S. G. Jackson, and J. P. Baker. *Proc. Equine Nutr. Physiol. Symp., 11th,* p. 124 (1989).

13

Pasture for Horses

I. INTRODUCTION

Good horses and good pastures go together. By instinct, horses are forage consumers. They are also particular about the pasture plants they like and eat. Pastures are not only a source of excellent-quality feed but also provide a means of exercise and help in bone and body development. Most horse people feel there is no substitute for a high-quality pasture. Many of them feel the other feeds used are a supplement to the pasture rather than vice versa. Good pastures are an excellent source of protein, vitamins, minerals, and other nutrients. Moreover, they save considerably on the feed bill. So, a good horse-breeding farm program should be built around an excellent-quality pasture program. Pastures are ideal for young animals and breeding stock. Pastures are used in many ways. Three of the most important are as follows: (1) to provide all or most of the feed; (2) to provide a large portion of feed needs, the remainder coming from supplements and possibly some energy source; (3) to provide very little feed but to be used primarily as a holding area or for exercise purposes.

Unfortunately, what many horse people use as a pasture is more of an exercise area rather than a source of feed. Therefore, if pastures are going to provide a significant amount of the feed supply, they need to be high quality and managed properly. It is becoming increasingly more difficult to have a good pasture program in suburban areas where the greatest numbers of horses exist since there is a lack of sufficient land available for this purpose.

II. PASTURE HELPS REPRODUCTION

Pastures are ideal for young animals and breeding stock. Many horse owners who have difficulty getting their mares to breed, have observed that a lush, green pasture is beneficial in getting them to conceive. In some areas, the conception rate does not start to improve until pastures grow and become green in the spring. This is one reason why the greatest conception rates occur in March, April, and May. Whether something in the pasture causes the increase in conception rate or whether some other factors are involved is not definitely known. It is possible that some environmental factor (temperature, sun's rays, length of daylight, humidity, etc.) could be involved. Or, the pasture could be supplying some nutrients that increase conception rate. But regardless of what is involved, most horse owners prefer to have as much high-quality pasture as possible. Unfortunately, many of the top horse farms producing high-level performance and race horses are located near large cities where land is scarce and high priced and many have to settle for a limited pasture acreage. In some areas, the weather is such that pastures cannot be raised during the cold months which also limits the amount of pasture available during the year. The next alternative is to use high-quality hay and/or high-quality dehydrated alfalfa meal as substitutes for limited pasture availability. Some horse farms have put in temporary pastures such as oats, rye, wheat, barley, or others for green grazing during the winter months. In certain localities, irrigation is needed for these temporary pastures to do well. Since horses have a birthday each January 1, it is important to have them foaled as close to this date as possible. Therefore, the use of green pasture at breeding time can be of help in getting mares bred earlier.

Fig. 13.1 Pasture renovator originally developed at the University of Kentucky. Pastures need to be renovated periodically in order to aerate the soil and improve pasture production. (Courtesy of John P. Baker, University of Kentucky.)

III. MANAGING PASTURES

A good-quality pasture should be kept lush, green, and not allowed to get too short or too tall. If allowed to get too short, the horses will eat the pasture too close to the ground. This not only increases the danger of parasite infestation, but also may be harmful to the pasture since the horses can overgraze it too close to the roots. If the pasture gets too tall, it becomes too mature and decreases in digestibility and feeding value. For good parasite prevention and good pasture management, therefore, the horses should be rotated between pastures periodically in order to prevent overgrazing, to rest each field regularly, and to allow it to recover from grazing.

Usually, horses will consume the immature, short pasture in preference to the more mature pasture (1). This can result in horses concentrating in one area and keeping it grazed closely while other areas grow untouched. This can be counteracted by dividing pastures into smaller areas and stocking the pastures with enough horses to keep the pastures short, lush, and green. One needs to guard against too many horses in a pasture, which can result in overgrazing and aggravation of the parasite problem.

Fig. 13.2 Pasture sprinklers. Permanent sprinkler systems along fence lines reduce labor and damage of sprinkler pipes from hauling and installation. Sprinklers should be set 8 ft above ground level to avoid damage by horses. Pasture pictured at Cal Poly University, Pomona. (Courtesy of Professor Norman K. Dunn, Cal Poly University, Pomona.)

Pastures require proper fertilization and management in order to keep them highly productive, nutritious, and palatable. Soil tests should be made periodically to determine the level of major and minor mineral elements, as well as pH (acidity), in the soil. These tests should be properly evaluated by the farm operator or by soil and pasture specialists. The advice of the County Agent or Farm Advisor is very helpful since they are acquainted with fertilizer needs in the county or area. The timing of the fertilizer applications can be very helpful in regulating the amount of forage available for grazing. The fertilizer formulation is also very important. Factors that are important in determining fertilizer needs include the following: (1) soil type; (2) kind of forages used; (3) soil fertility; (4) moisture level; (5) season of the year; (6) plant maturity; (7) plant health; and (8) climate as well as other factors.

Grass, legume, or grass–legume pastures can be used for horses. During certain periods of the year, some pastures may be too laxative for horses that are exercised a great deal. The laxative effect can be alleviated by feeding a dry grass hay, in addition to the pasture, or by letting the horses run into another field which has dried pasture. Brood mares and young colts should best be kept in pastures that give a laxative effect. The horses used for performance or racing purposes should be kept on pastures that keep the horses in good condition as far

Fig. 13.3 A portable sprinkler system which separates in sections of three links. Easily moved by one person and is suited for small pastures. Sprinklers located on McCoy's Arabian Farm, Chino, California. (Courtesy of Professor Norman K. Dunn, Cal Poly University, Pomona.)

as their bowels and sweating are concerned. If necessary, one can also restrict the horses to a certain number of hours daily on the pasture in order to prevent too much of a laxative effect with heavily exercised or worked horses.

Many horse owners turn their animals out on pasture at night as well as on idle days. This gives the horses an opportunity to relax and exercise at will (Fig. 13.4). It also decreases the amount of grain and hay that is fed (Fig. 13.5).

It is important to avoid having stumps, pits, poles, holes, and any sharp obstacles in the pastures that can injure the horses. Many good horses have been ruined by these hazards in pastures. It is helpful to place mineral boxes in a corner of the pasture to avoid horses running into them. A plastic rather than wooden mineral box will also lessen the possibility of an injury.

Some horse owners use cattle in the rotation of pastures and have them graze the surplus forage. They keep the horses in the pastures when the forage is short and when it starts to mature, the cattle are used, and the horses are moved to another pasture with short, lush, green forage. Cattle graze taller forages than horses. Beef cattle and horses can also be grazed together. They do well once they become accustomed to each other. The cattle used also bring in additional income from grazing the surplus forage (2).

Good horse pastures do not just happen. They require considerable planning

Fig. 13.4 Pastures afford young horses an opportunity for necessary exercise as well as providing a quality roughage. Pasture located at Cal Poly University, Pomona, California. (Courtesy of Professor Norman K. Dunn, Cal Poly University, Pomona.)

Fig. 13.5 Good horses and good pastures go together. Pastures are an excellent source of feed. They also provide an area for romping and exercise. It is excellent for mares and their foals. (Courtesy of T. J. Cunha, Cal Poly University, Pomona.)

and management to be successful. Following are some helpful management hints for use in developing and maintaining quality pastures.

1. Spread manure droppings with a chain harrow to eliminate their accumulation in numerous manure piles (3).

2. Try not to graze pastures during very wet weather to avoid damage to the pasture turf. Horses are generally more destructive to pasture than cattle because of the tearing action of their hooves while galloping (4).

3. Avoid dropping wire, nails, and similar sharp objects which might be hazardous to horses.

4. Eliminate overgrazing or overstocking pastures since it may decrease desirable pasture species, cause a decline in forage production, and also increase weed growth.

5. If possible to do so, it is best to graze young horses separately from other horses.

6. Horses seldom bloat but they may founder when first turned into lush

spring pasture. To minimize the possibility of founder, give the horses plenty of hay before they go out to the pasture for the first time. Allowing them only a few hours of grazing during the first few days will also help. Founder or laminitis is a condition that can cause lameness or inflammation of the feet.

7. Plan the pasture system so it consists of a number of smaller pastures instead of just a few larger pastures. This will facilitate rotational grazing and having pastures exclusively available for mares and foals, young horses and other categories of horses which the size of farm justifies.

8. No single forage plant meets all the criteria to provide as much year-round grazing as possible. So the pasture program should involve some forages which can stand cold weather, others that do well during warm weather, clovers if they do well in the area, and temporary forages which can be used to lengthen the pasture season if it is necessary to do so.

9. In most areas, the maturity date among the various grasses differs. Many horse owners, prefer one grass species along with one or two legumes in the mixture. But, a different grass species might also be used in another pasture with one or two legumes. This procedure is followed since one grass may grow early in the year whereas the other comes in later. This difference in maturity of each grass species in the same pasture causes differences in palatability and nutritional value and causes the horses to graze the most palatable grass first and let the other grow tall. Therefore, many horse owners use a number of different grass species but prefer each one in a separate pasture. But sometimes there are certain grass species which can be used in combination with each other in certain areas of the country.

10. Legumes are helpful along with grasses in various forage combinations. Legumes are high in nutritional value and add nitrogen to the soil. Since horses seldom bloat, there is usually no problem with a grass–legume pasture where there is a proper balance between grasses and legumes (2). Sometimes an excessive amount of legumes in a pasture mixture will cause slobbering. If the legumes can be kept at 30–35% of the mixture, slobbering is minimized or eliminated (5). Some of the legumes used in horse pastures are: white and ladino clover, alfalfa, red clover, birdsfoot trefoil, alsike clover, and lespedeza.

11. If horses are given supplemental feed on pastures it should be done gradually. It is best to start with a small amount of feed and gradually increase it to the desired level in a week to 10 days (6).

12. Horses on pasture should have access to plenty of clean water. It is best to provide water free choice so they can drink it whenever they want. If the horses need to be hand watered it should be supplied at least twice and preferably three times daily.

13. Lactating mares, weanling horses, and hard working horses should usually be fed some supplemental feed even if grazing excellent pastures. Failure to

provide the mare with enough total feed during lactation may lead to diminished milk production and poorly developed foals. The young horses may need additional supplemental energy and protein on pasture for the proper development of bone and muscle.

14. If horses are fed on pasture, feeding at regular times prevents digestive upsets. Horses are creatures of habit and appreciate being fed at regular times.

15. Horses on pasture should not be overfed since excessive fat can reduce performance and reproductive efficiency. Sometimes horses grow faster with supplemental feed, but the extra gain may be largely fat which should be avoided.

IV. SUPPLEMENTATION ON PASTURE

In many cases, pastures supply only part of the total feed needs of the horse. This is especially the case with high-level performance horses that are exercising, growing, or producing a foal (6,7). Texas A & M showed that yearling stock horses on well-managed bermudagrass and rye–ryegrass pastures plus trace mineralized salt can attain similar or greater average daily gains than projected by the 1978 NRC report (8). How much supplementation is needed therefore depends on many factors which include the following: (1) quality and quantity of pasture; (2) the condition of the horses; (3) how much is demanded of the horse in work or performance; (4) the stage of the life cycle of the horse (such as foal, yearling, 2 year old, mare, or stallion); (5) how well the owner wants the horses to look; and (6) many others.

Trace mineralized salt should be used when pasture makes up a significant part of the diet. This is because plants do not need selenium or iodine and can grow normally and produce optimum yields even though they contain an insufficient amount of these two minerals to meet animal requirements. Two studies in Missouri showed that selenium supplementation resulted in a trend toward improving reproductive efficiency in mares grazing fescue pasture (9,10). Cobalt is only needed by the nitrogen-fixing microorganisms in the nodules of the roots of legume plants. Moreover, certain plants grow normally and produce optimum yields even though they contain less iron, zinc, manganese, copper, and cobalt than is required by horses and other animals. Therefore, a pasture may look lush and green and is yielding at a maximum level but the horses grazing it can be suffering from a lack of certain trace minerals, hence the reason for self-feeding trace mineralized salt on pastures for horses (11).

It is rare for plants to suffer from a deficiency of sodium and chloride. Therefore, the commonly used feeds do not contain enough sodium to meet animal body needs (11). Most concentrates and some forages are also low in chloride. Therefore, sodium and chloride are supplemented by the use of salt. This is another reason for trace mineralized salt being self-fed to horses via mineral boxes located in each pasture. Moreover, the mineral boxes should be inspected periodically and not allowed to remain empty for more than a few days at a time. Horses need minerals daily. They also sweat a great deal which increases their need for salt since sweat contains about 0.7% salt. Calcium, phosphorus, and other minerals may also be needed depending on the pastures used and the remainder of the total feed program. Mineral supplements are available which take into account the forages available and the requirements of the horses involved.

Protein and energy supplementation may also be needed by horses on pastures. This will especially be the case when the quantity of forages is low or when the forage becomes too mature. Dr. C. G. Depew at Louisiana State University demonstrated what happens to forage quality as it matures (Table 13.1).

The level of protein and TDN (total digestible nutrients) decreases as the forage matures and the fiber level increases. Moreover, the nutrients in the forage become less available to the horse as the forage matures. The consumption of the forage decreases as it matures to a point where less than one-half is consumed by the horses. This indicates that mature forages may need protein, energy, mineral, and vitamin supplementation in order for horses to do well. More detail on supplementation with vitamins, minerals, protein, and energy are given in the chapters dealing specifically with these subjects (Chapters 5–9, respectively).

The information shown in Table 13.1 verifies the need to utilize forages when they are lush, green, and have a high nutritional level. The high nutritional value may be maintained by rotational grazing, clipping, hay production, or using cattle to utilize the surplus, more mature forage. Rotational grazing is an excellent method since forage quality can be maintained by proper stocking rate and rotating horses from one pasture to another before forage maturity sets in. Clip-

TABLE 13.1

The Effect of Cutting Time on Quality and Consumption of Bermudagrass Hay by Horses

Cutting time (weeks)	Total protein (%)	TDN (%)	Fiber (%)	Consumption (lb)
4	16.9	55	29.5	17.8
8	10.4	51	34.8	15.0
12	7.7	44	38.0	12.3
16	3.4	36	42.3	7.4

ping the tall forage will entice the horses to graze the smaller, younger, and faster growing forage areas. If large pastures cannot be divided into smaller ones, then part of the pasture which is getting close to maturity can be clipped and made into hay. This will result in always having some short, lush forage available for grazing with the surplus forage made into hay. Another alternative is to graze cattle and horses together with a proper stocking rate to keep the forage from getting tall and mature.

V. FORAGE MATURITY AND NUTRITIONAL VALUE

Forages (pasture and hay) make up a large portion of horse diets. Horses by instinct are forage consumers. They should be fed at least 1.0% of their body weight as forage (on a dry matter basis). This means 10 lb of hay, or other forage equivalent, for a 1000 lb horse. Some people feed forage at a higher level and some at a lower level. It is thought that the minimum forage requirement is 0.5% of body weight or 5 lb per 1000 lb horse. But, many prefer to feed forage at a level of at least 1.0% of body weight.

Forages are an important factor in preventing digestive upsets in the horse. Moreover, forages supply fiber and bulk in the diet and also furnish protein, energy, minerals, and vitamins.

Of most importance in determining the nutritional value of a forage is its maturity at the time it is consumed. As the forage becomes more mature, the crude fiber and lignin contents increase and the levels of protein, energy, minerals, and vitamins decrease (Table 13.1). Moreover, as the plant becomes high in fiber and lignin, the body enzymes are less able to digest the nutrients that are in the forage. Therefore, forages become increasingly less valuable for horse feeding as they increase in maturity (12). The information shown in Tables 13.2 and 13.3 illustrates the decreasing level of nutrients in forages as they increase in maturity. A study at the Pennsylvania Agriculture Experiment Station (13) also showed that the level of nutrients decreases as forages mature and that their digestibility also decreases. This information is shown in Tables 13.4 and 13.5. The information in Tables 13.2 and 13.3 can be used as a guide as to when a forage has optimum nutritional value and when it should be grazed or cut for hay. It is apparent that grazing should be done before a forage gets too mature. This can be accomplished by using smaller pastures or controlling horse numbers so that the forage does not become too tall and mature. Frequent rotation of pastures adequately stocked with horses is one way to keep forages short, lush, and green. If pastures are utilized when they have maximum nutritional value, it decreases the amount of nutrients which need to be provided by other feeds. Moreover, horses do better and look better when high-quality pastures are used.

TABLE 13.2

Influence of Stage of Maturity on Nutrient Content of Pasture Forages (On Dry Basis)[a]

Forage	Stage of maturity	Crude fiber (%)	Crude protein (%)	Digestible energy[b] (Mcal/lb)	Calcium (%)	Phosphorus (%)
Bluegrass pasture	Immature	25.1	17.3	1.44	0.56	0.47
	Early bloom	27.8	14.8	1.38	0.46	0.39
	Midbloom	29.3	13.2	1.36	0.38	0.38
	Milk stage	30.3	11.6	1.34	0.19	0.27
Buffalograss pasture	Immature	26.8	12.4	1.28	0.56	0.23
	Full bloom	29.0	9.8	1.24	—	—
	Mature	30.7	5.9	1.04	0.40	0.16
Bermudagrass pasture	Immature	24.9	17.1	1.42	0.89	0.32
	Early bloom	23.9	17.4	1.38	0.58	0.23
	Full bloom	26.1	10.3	1.26	0.54	0.20
	Mature	28.5	5.8	1.24	0.40	0.18
Crested wheatgrass pasture	Immature	22.2	23.6	1.07	0.46	0.35
	Full bloom	30.3	9.8	0.84	0.39	0.28
Western wheatgrass pasture	Immature	26.7	16.4	1.22	0.42	0.31
	Early bloom	31.4	11.8	1.14	0.58	0.34
	Full bloom	32.4	8.6	1.06	0.45	0.29

[a]Data adapted from Tables of Feed Composition, NRC Publ. 1684, by Dr. John P. Baker of the University of Kentucky (12).
[b]Digestible energy values are for cattle; no data for horses were available then.

TABLE 13.3

Influence of Stage of Maturity on Nutrient Content of Selected Hays (On As-Fed Basis)[a]

Forage	Stage of maturity	Crude fiber (%)	Crude protein (%)	Digestible energy[b] (Mcal/lb)	Calcium (%)	Phosphorus (%)
Alfalfa hay	Immature	23.4	19.1	1.02	1.89	0.27
	Prebloom	24.1	16.4	1.07	1.06	0.19
	Early bloom	26.8	16.6	1.03	1.12	0.21
	Midbloom	27.6	15.2	1.04	1.20	0.20
	Full bloom	29.7	14.0	1.00	1.13	0.18
Clover hay	Immature	17.8	18.7	1.03	1.54	0.27
	Early bloom	23.7	13.7	1.07	1.44	0.32
	Midbloom	25.3	13.2	1.15	1.81	0.30
	Full bloom	25.9	12.5	1.02	1.43	0.22
	Mature	30.6	9.3	1.05	0.97	0.19
Brome hay	Immature	25.8	15.0	1.19	0.59	0.33
	Early bloom	28.2	10.9	1.12	0.40	0.23
	Full bloom	31.9	9.2	0.94	—	—
	Mature	31.7	5.4	0.98	0.40	0.20
Orchardgrass hay	Immature	25.1	14.8	0.95	0.32	0.28
	Early bloom	28.2	12.0	0.90	0.22	0.34
	Mature	35.3	5.9	0.76	0.21	0.20
Prairie hay	Immature	28.4	7.8	0.92	0.51	0.17
	Midbloom	29.2	7.4	0.91	0.31	0.19
	Late bloom	29.7	6.0	0.90	0.33	0.12
	Milk stage	30.9	4.4	0.88	0.36	0.12
Timothy hay	Prebloom	29.1	10.9	1.10	0.58	0.30
	Early bloom	29.1	7.6	1.04	0.53	0.23
	Midbloom	29.6	7.5	1.08	0.36	0.17
	Late bloom	28.5	7.3	1.02	0.33	0.16

[a]Data adapted from Tables of Feed Composition, NRC Publ. 1684, by Dr. John P. Baker of the University of Kentucky (12).
[b]Digestible energy values are for cattle: no data for horses were available then.

TABLE 13.4

Apparent Digestion Coefficients of Forages[a]

Stage of maturity	Dry matter			Crude protein			Crude fiber			Ether extract			Nitrogen-free extract		
	First[b]	Second[b]	Third[b]	First	Second	Third	First	Second	Third	First	Second	Third	First	Second	Third
Alfalfa	69.1	61.7	57.3	74.5	72.1	55.2	58.2	54.8	53.2	46.1	6.0	-6.4	75.1	67.2	62.9
Timothy	65.9	60.6	59.3	65.2	62.1	55.3	65.4	59.7	57.9	31.6	17.6	16.0	71.0	64.6	63.9
Orchardgrass	63.1	60.1	54.7	68.3	66.9	52.4	58.5	51.6	49.8	26.3	33.0	17.3	67.8	65.8	61.0

[a]Data from Darlington and Hershberger (13).
[b]First is least mature and maturity increases as one goes to second and third.

TABLE 13.5

Total Digestible Nutrients, Digestible Energy, Voluntary Intake, and Nutritive Value Indices of Forages[a]

Stage of maturity	TDN(%)			Digestible energy (%)			Voluntary intake (g DM/day/W$^{0.7}$ kg)			Nutritive value index		
	First[b]	Second[b]	Third[b]	First	Second	Third	First	Second	Third	First	Second	Third
Alfalfa	63.6	56.0	52.0	64.8	55.9	51.6	95.1	87.4	65.0	77.1	61.0	42.0
Timothy	63.2	57.9	56.9	62.0	56.5	55.1	82.3	87.0	82.2	63.8	61.3	56.6
Orchardgrass	59.2	56.6	52.0	58.0	54.8	49.1	80.9	79.8	82.5	58.6	54.6	50.6

[a]Data from Darlington and Hershberger (13).
[b]First is least mature and maturity increases as one goes to second and third.

VI. PASTURE AND HAY DIGESTIBILITY

It is important to maximize forage utilization. Information on digestibility of forage is important since it provides a guideline on when to graze or when to make hay from excess forage. Most of the present data on digestibility of forages are available with hay. It is reasonable to assume, however, that the information obtained with hay would have considerable application to the pasture from which the hay was made. Therefore, some experimental data with hay will be presented here.

Table 13.6 shows digestibility figures obtained in New Jersey (14) for alfalfa, orchardgrass, timothy, and bromegrass hay, which had 31.3, 34.2, 36.0, and 38% fiber, respectively.

The information in Table 13.6 shows that the steer digests a higher percentage of higher fiber feeds than the horse. This difference is most apparent with bromegrass hay. Accounting for a large portion of this difference is the fact that feed remains in the digestive tract a much shorter period of time in the horse as compared to the steer. The steer is more efficient than the horse in digesting the dry matter, fiber, cellulose, and fat and in utilizing the energy content of the four hays. However, the horse is more efficient in digesting the protein and ash (mineral) content of the hays. Just why the horse digests the protein more efficiently in hay is not known. There are differences in the digestibility figures depending on the hay being fed. For example, the higher the crude fiber in the hay, the lower the dry matter, protein, and energy digestibility obtained for the horse. It is interesting that the horse digests 60.8% of the dry matter in alfalfa hay

TABLE 13.6

Digestion Coefficients for Forages by Geldings and Steers (percentage digested)[a]

	Dry matter	Protein	Fiber	Cellulose	Ether extract (fat)	Ash	Energy
Alfalfa							
Steer	61.70	74.38	44.22	60.06	47.45	61.45	58.98
Gelding	60.80	75.43	39.12	54.35	31.24	67.67	56.36
Orchardgrass							
Steer	53.70	55.57	53.44	64.32	53.31	41.48	53.28
Gelding	49.99	60.33	43.10	52.12	46.87	48.50	46.24
Timothy							
Steer	58.01	49.45	61.49	67.18	61.32	40.37	54.64
Gelding	49.45	54.46	43.85	48.33	48.48	48.19	46.33
Bromegrass							
Steer	55.94	39.18	59.88	63.68	56.60	41.76	51.64
Gelding	33.83	43.55	34.49	37.75	39.59	42.67	36.31

[a]Data from Vander Noot and Gilbreath (14).

but only 33.83% in bromegrass hay. Another study showed that the average digestibilities of three different species of hay were 12% less in horses than in cattle (15). This indicates why it is important to use high-quality forages for horses. It appears that the higher the crude fiber content in the hay, the lower the digestibility of its dry matter. This emphasizes the need to keep horse pastures short, lush, and green since the fiber level increases and the digestibility of nutrients decreases as pastures mature.

In another study in New Jersey (16) a nutritive value index (NVI) for horses was established for hays. The NVI was highest for red clover (68.6), followed by alfalfa (56.8 and 62.5), timothy (52.8), bromegrass (48.1 and 48.8), canarygrass (46.9 and 46.0), orchardgrass (45.2), bermudagrass (44.0), and fescue (36.8 and 44.6). (The two values given in parentheses indicate the hays were studied in two different experiments.) This research indicated that the nutritional value of the two legumes, red clover and alfalfa hay, were superior to that of all the grass hays used in this study.

Only the surface has been scratched on the information needed on forage digestibility for the horse. Present information indicates that there is a large amount of variation between horses in their ability to digest forages (16). Many factors affect the digestibility of pasture or hay, which include: (1) kind of soil; (2) kind and level of fertilization; (3) kind of forage; (4) stage of maturity when grazed or cut for hay; (5) method of processing hay; (6) how well hay was processed; and (7) method of storage and how well and how long the hay is stored before being fed. Moreover, there are seasonal differences in the digestibility of forages (17). Texas A & M studies showed that digestion of hemicellulose was more efficient in mature horses than in yearlings fed on bermudagrass pastures and bermudagrass hay. However, the digestibilities for dry matter and other nutrients did not differ between the two classes of horse (18,19). Until more is known, the following are good recommendations to use: (1.) graze forages which the horses like and when they are short, lush, and green; (2.) make hay from forages which are green and have plenty of leaves and before they become too mature.

VII. PASTURE ACREAGE REQUIRED

The acreage needed to supply adequate nutrition for a horse depends on many factors such as soil type, its level of fertility, forage species, climate, rainfall, pasture management, and other factors. Many feel that 1 acre of legume–grass pasture with good management and growing conditions can provide enough feed for a horse during the grazing season. Less acreage might be needed with a good irrigation program. But, under most conditions, it may require 2 and sometimes 3–5 acres of pasture per horse (2,4,20). Larger acreages of pasture may be needed if native grasses are used under range conditions (3). The acreage required can also be decreased by the amount of supplemental feed used. High-level performance and race horses being pushed for early and maximum development are usually fed a considerable amount of supplemental feed and thus can get by with less pasture acreage. In areas where pasture acreage is limited, many high-level performance horses are rotated out to pasture periodically during the day in order to give all the horses some access to pasture for romping, exercise, and sunlight. Horses are athletes and need to exercise to keep their muscles in good condition.

VIII. IRRIGATION

In many areas irrigation may be necessary for optimum pasture quality and production. One cannot depend on rainfall, which may not occur when needed or occur in inadequate quantity, to produce top quality pastures. Irrigation can be provided by either ditch, sprinkler or other means. Irrigation can greatly improve pasture growth, especially during the dry periods or when rainfall is limited or sporadic. The cost of irrigation needs to be considered, however, since it can be high in some locations. In some areas, an irrigation program may be justified only with breeding farms, or with high-quality horses used for racing, performance, show, or other activities that justify the extra cost involved. In many areas, however, irrigated pastures can be a low-cost item and actually provide the lowest cost feed for the horses since it may enable near year-round grazing. Another alternative is to consider irrigating only a small area to: (1) provide some green forage for the mares during the breeding season; (2) provide part of the grazing for a few hours daily, or on alternate days, for the horses most likely to need green feed (such as the mares and young horses); or (3) use the irrigated pastures in rotation with the other pastures every 2–4 weeks. This procedure would provide all horses with some green feed periodically.

IX. WEED CONTROL

Once pastures are established, it is necessary to control weeds. Pastures should be kept free of weeds, especially poisonous weeds such as are shown in Table 13.7 (5). Even nontoxic weeds need to be controlled since they compete directly with pasture forages for space and nutrients. Weed control can occur by either chemical or mechanical means. Clipping or mowing pastures on a regular schedule does a good job of controlling most weeds on pastures. Where moving cannot be used, cutting the weeds with a hoe may sometimes be the only alternative. If one starts cutting weeds as soon as a few appear, it is relatively

TABLE 13.7

Poisonous Plants in North Carolina that Affect Horses (5)

Plant	Symptoms
Locoweed	Nervous
Bracken fern	Thiamin deficiency; occurs during droughts
Wild cherry trees	Cyanide poisoning, hard breathing, rapid death
Wild tobacco	Paralysis
Milkweed	Dizziness, respiratory paralysis
Poison hemlock	Depression, paralysis, and death
Water hemlock	Muscle stiffness, staggering
Bunch flowers	Depressed appetite, labored breathing
Black Snakeroot	Decreased temperature, staggering, coma
Oak (Acorns)	Must be consumed in large quantities
Pokeweed	Convulsions, lung paralysis
Rattle Box	Drooling saliva, bloody feces, death
Lupines	Frothing at mouth, convulsions
Black Locust	Weak pulse, gastroenteritis
Climbing Bittersweet	Uncertain
Buckeye	Lack of coordination, paralysis
Laurel	Respiratory failure, death
Ivy	Unknown
Carolina Jessamine	Muscular weakness, lower temperature, death
Dogbone	Fever, sweating but cold extremities
Jimson-weed	Frequent urination, partial blindness, death
Nightshade	Paralysis
White Snakeroot	Pungent odor to the breath, death
Bitterweed (Sneezeweed)	Salivation, convulsions, death

easy to control them. If weeds are left for any length of time, however, they become a major problem to control since they can spread rapidly.

If chemical weed control is practiced, one needs to follow the label directions carefully on its use. Horses should be removed from the pasture before spraying and should be kept out for the time interval indicated for the herbicide used. Only products labeled for pasture use should be utilized and proper advice should be sought before using them.

A perusal of Table 13.7 shows the poisonous weeds which affect horses in North Carolina (5). The County Agent or Farm Advisor in each county has a list of the poisonous plants in their area.

X. PRECAUTIONS WITH FORAGES

A. Alfalfa

The high level of protein in alfalfa will cause horses to drink more water which results in increased urination. There is no evidence, however, that this hurts the kidneys of healthy horses although it might aggravate an existing kidney disease. A University of Illinois study (7) detected no metabolic evidence for excess protein (18.5% versus 12.9% protein diet) causing any detrimental effect on horses exercising on a graded treadmill. Horses fed lush alfalfa hay without being gradually accustomed to it may develop looser stools and a few may develop colic. So, horses should be gradually accustomed to high-quality alfalfa hay. Some horse owners prefer to use an alfalfa–grass hay mixture or to feed alfalfa at one feeding and grass hay at the next as a means of preventing a loose stool. Blister beetles can be present in alfalfa. Therefore buying alfalfa from a responsible grower is very important to make sure the hay is free of beetles since they can be harmful and may cause death in horses. Some feel that alfalfa may cause increased sweating but it may be largely due to the higher level of energy in alfalfa (6).

B. Clovers

Clovers are sometimes difficult to bale properly since they may cause mold problems. Clovers, especially red clover, may sometimes cause slobbering (excess salivation) in horses (5,6). Blackpatch disease is present in some red clover pastures and has been implicated as sometimes causing the slobbering syndrome (21).

C. Bermudagrass

Bermudagrass is sometimes blamed for causing impaction colic in horses, but mature varieties of any hay high in lignin content may also cause impaction problems (6). Horses grazing only bermudagrass during the winter may dig up and consume rhizomes along with sand, resulting in impaction or sand colic (4,5).

D. Fescue

Reproductive problems such as prolonged gestations, abortions, thickened placentas, and agalactia (no milk) have been reported in some situations where brood mares are grazed on fescue pastures (infected with endophyte) or are fed fescue hay during the last 90 days of the gestation period. It is recommended that fescue not be fed to mares during the last 90 days of gestation or during lactation (5,6,20). Fescue pastures can be tested for the presence of endophyte through the local County Agricultural Extension Service.

E. Ryegrass

Grazing ryegrass pastures during dry weather when the forage is growing slowly may be harmful. Decreased moisture may cause production of a toxin in ryegrass (causing rhygrass staggers) which results in muscle spasms after exercise (5).

F. Prussic Acid and Cystitis Hazard

There is some hazard in grazing sudangrass or sudangrass hybrids since they may form prussic acid in the new growth that follows a frost, a period of drought, or heavy trampling (5). Prussic acid poisoning is very rapid. Sometimes the first sign of trouble is finding horses dead or dying. Prussic acid poisoning results in abnormal breathing, trembling muscles, spasms or convulsions, nervousness, respiratory failure, and death. Animals that show no evidence of being poisoned and are removed from the sudangrass are not likely to be affected.

The sorghum–sudan hybrids and sudangrass hybrids may also cause a disorder known as cystitis (4–6,22,23). This disease (which is a urinary tract inflammation) causes continuous urination and incoordination in gait, and the mares appear to be constantly in heat. Animals seldom recover after the incoordination or dribbling of urine occurs. Hay made from sudan or sorghum–sudan hybrids will not produce the disease (4–6,23). This discussion indicates that one needs to exercise caution if sudan or sorghum–sudangrass hybrid pastures are used for horses.

XI. CONCLUSIONS

The pastures to use will vary with the geographical area of the country. The County Agent, Farm Advisor, or State University can be of considerable help in advising on the pasture programs that are best for the area. Legumes such as alfalfa, ladino clover, red clover, birdsfoot trefoil, and others are excellent and high in nutritional value. Sometimes legumes are used alone, but usually they are mixed with grasses. Grasses such as timothy, bromegrass, orchardgrass, canarygrass, bermudagrass, bluegrass, buffalograss, crested wheatgrass, and others are used extensively with horses. Temporary pastures such as oats, wheat, rye, barley, and others are also used to supplement the permanent pastures. If properly managed and cared for, many kinds of pastures can be used successfully.

As pastures increase in maturity, they increase in fiber content. The higher the fiber level in pasture or hay, the lower their digestibility. This emphasizes the need to keep horse pastures short, lush, and green for maximum feeding value. Hay should also be made while the forage is green, leafy, and before it becomes too mature.

High-quality pasture is an excellent source of vitamins, minerals, protein, and other nutrients. A lush, green, pasture is very helpful in getting mares settled during the breeding season. It can also provide a significant portion of the feed supply.

Pastures need to be managed properly for best results. Proper fertilization, control of weeds, clipping at timely intervals, pasture renovation, rotational grazing, proper stocking rate, and proper nutrient supplementation are some of the factors involved in a good pasture management program.

REFERENCES

1. Aiken, G. E., G. D. Potter, B. E. Conrad, and G. W. Webb. *Proc. Equine Nutr. Physiol. Symp., 9th,* p. 20 (1985).
2. Wheaton, H. N., and M. Bradley. *Mo., Agric. Ext. Serv., Mimeo* (1980).
3. Freeman, D. W., D. Rollins, and R. W. Treadwell. *Okla., Agric. Ext. Serv., Fact Sheet* No. 3971 (1984).
4. Mueller, J. P., and J. T. Green, Jr. *N. C., Agric. Ext. Serv., Forage Mimeo* No. 12 (1985).
5. Mowrey, R. A. *N. C., Agric. Ext. Serv., Rev.* **9** (1985).
6. Householder, D. D., and G. D. Potter. *Tex. A&M Mimeogr. Horse Nutr. Feed.* (1988).
7. Miller, P. A., and L. M. Lawrence. *J. Anim. Sci.* **66,** 2185 (1988).
8. Hansen, D. K., F. M. Rouguette, Jr., G. W. Webb, G. D. Potter, and M. J. Florence. *Proc. Equine Nutr. Physiol. Symp., 10th,* p. 25 (1987).
9. Hermann, E. D., L. W. Garrett, W. E. Loch, J. S. Morris, and W. H. Pfander. *Proc. Equine Nutr. Physiol. Symp., 7th,* p. 56 (1981).
10. Hermann, E. D., L. W. Garrett, W. E. Loch, J. S. Morris, and W. H. Pfander. *Proc. Equine Nutr. Physiol. Symp., 7th,* p. 62 (1981).
11. Cunha, T. J. *Feedstuffs* **60**(7), 23 (1988).
12. Baker, P., Kentucky Agricultural Experiment Station, Lexington (personal communication), 1977.
13. Darlington, J. M., and T. V. Hershberger. *J. Anim. Sci.* **27,** 1572 (1968).
14. Vander Noot, G. W., and E. B. Gilbreath. *J. Anim. Sci.* **31,** 351 (1970).
15. Johnson, D. E., M. M. Borman, and R. L. Rittenhouse. *Proc., Annu. Meet.—Am. Soc. Anim. Sci., West. Sect.* **33,** 294 (1982).
16. Fonnesbeck, P. V., R. K. Lydman, G. W. Vander Noot, and L. D. Symons. *J. Anim. Sci.* **26,** 1039 (1967).
17. Moffitt, D. L., T. N. Meacham, J. P. Fontenot, and V. G. Allen. *Proc. Equine Nutr. Physiol. Symp., 10th,* p. 79 (1987).
18. Aiken, G. E., G. D. Potter, B. E. Conrad, and R. W. Blake. *Proc. Tex. A&M Agric. Ext. Serv., Horse Short Course* (1986).
19. Aiken, G. E., G. D. Potter, B. E. Conrad, and J. W. Evans. *Proc. Tex. A&M Agric. Ext. Serv., Horse Short Course* (1987).
20. Huff, A. N., T. N. Meacham, and M. L. Wahlberg. *Va., Agric. Ext. Serv. Publ.* pp. 406–472 (1986).
21. Nesmith, B. *Univ. Ky. Equine Data Line* Sept. p. 2 (1985).
22. Evans, J. W., A Borton, H. F. Hintz, and L. D. Van Vleck. "The Horse," p. 750. Freeman, San Francisco, California, 1977.
23. Ensminger, M. E., and C. G. Olentine, Jr. "Feeds and Nutrition—Complete," p. 949. Ensminger Publ. Co., Clovis, California, (1978).

14

Hints on Feeding Horses[1]

I. INTRODUCTION

A good feeding program with horses depends as much on the feeder as it does on the feed. A well-balanced diet is the first requirement. However, competent feeders are indispensable in a good feeding program designed to develop top quality horses for racing or performance purposes. There is no substitute for dependability, regularity, alertness, hard work, and integrity of the person doing the feeding. Therefore, a good feed is only the beginning of a good program.

There is a need for individual feeding and for a constant study of the peculiarities and needs of each individual horse. In a sense, horses are similar to humans and will vary considerably in likes and dislikes for certain feeds, levels of feed intake, and time of feeding. Proper attention to these small details will mean the difference between developing a champion or just another horse. Feeding horses for racing, show, or performance is more complicated than feeding any other farm animal.

II. HINTS ON FEEDING

There are many hints that can be given on feeding horses. They are very useful and can serve as reminders to the beginner who is learning how to feed and handle horses.

1. The level of feed needed for optimum performance varies with different horses. Some horses run or work best when they are in a trim, light condition. Other horses, however, may need more condition to perform at the maximum. Some horses require a near-empty stomach to perform at their best in a race or show. Therefore, feed intake varies considerably. The quality of the diet needs to be high enough, however, with all horses in order to supply them with all the essential nutrients they need to build sound feet and legs and a healthy body.

---

1 Written by T. J. Cunha and reviewed by Norman K. Dunn, Professor of Animal Science and Director of Equine Operations, California State Polytechnic University, Pomona, California.

2. The nutritional needs of the horse will vary according to the demands made on it. The rapid development of a race horse to win as a 2 year old will require a well-balanced diet and an excellent job of feeding. Horses that are developed for show or for pleasure riding will also have different requirements depending on how much they are used. Horses developed for rodeos, working cattle, and other uses will also have requirements based on how extensively they are worked. Horse owners will also vary on how they want their horses to look. Many do not care how their horses look, whereas others take great pride and pleasure in having their horses in excellent condition, with a sleek, beautiful appearance.

3. Feeding the same diet to all classes of horses results in overfeeding some nutrients and underfeeding others. The amount of diet used will vary with the response obtained. It also depends on the condition of the horse, as well as the growth or activity desired. For best results, horses should be kept in a thrifty condition, but should not be allowed to get too thin or too fat. Overfeeding, as well as underfeeding, hurts many horses. A good horse owner will regulate the feeding program according to many factors and the response of the animals to it.

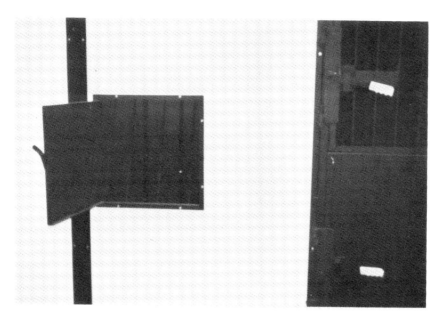

Fig. 14.1 Feeder door located in aisle way of barn. Allows feeding from outside of stall and is desirable as a time saver and safety factor. Person feeding must, however, make effort to observe the condition of the horse. (Courtesy of Professor Norman K. Dunn, Cal Poly University, Pomona.)

4. No two horses can usually be fed exactly alike in developing a candidate for championship honors. Some horses eat slowly, whereas others eat fast. Some will consume more feed than others. One horse may do well on only part of the feed consumed by another. Some horses may like more concentrate or hay than others. These differences need to be considered in the nutrients added or contained in the diet in order to provide an adequate level of energy, protein, minerals, vitamins, and other nutrients for all horses regardless of their eating habits.

5. Frequent inspection of the feed box will help determine the eating habits of each horse. This allows one to learn the eating pattern and then quickly detect when the animal goes off-feed or when something is wrong.

6. Top quality hay, which is as free as possible from weeds, dust, and mold, should be used. Otherwise, heaves, colic, and other digestive disturbances may occur. The hay should also be as green and leafy as possible, which indicates that it was cured properly and has a high nutritional value. The higher the quality, the more digestible the hay is for the horse. The final answer as to how good a hay is, however, will depend on how well the horse likes and responds to it.

7. One of the most important needs on a horse farm is an excellent quality pasture. Pastures not only provide top quality feed, but they serve as an area for

Fig. 14.2 Plastic bucket which is portable and can be placed in the corner of a pen to feed concentrate feed to horses. It can be easily washed and cleaned frequently to prevent feed accumulation and souring. (Courtesy of T. J. Cunha, Cal Poly University, Pomona.)

Fig. 14.3   Hay storage in the loft of a horse barn with a trap door which allows hay to be dropped directly into hay rack. (Courtesy of T. J. Cunha, University of Florida and Cal Poly University, Pomona.)

exercise and a natural environment for developing horses, especially breeding animals.

8. On many large horse farms there will be one person feeding the weanling horses, another feeding the yearlings, and others feeding the different age groups. This lessens the number of horses fed by each person, and allows each animal to be studied, treated, and fed as an individual. It also allows feeders to develop expertise in taking care of horses in each age category where feeding and management requirements differ.

9. There is a definite need for constant study of the peculiarities and needs of each individual horse. Horses will vary greatly in likes and dislikes for certain feeds, as well as in their temperament and working habits. Some will respond well to one caretaker and not another. Choice of caretaker needs to be considered if the horse is a valuable animal.

10. Regularity of feeding is very important. Horses should be fed at the same time every day, including holidays and weekends. This is necessary if they are to be kept at maximum feed intake without going off-feed. Regularity of feeding is critical with horses being developed for racing or performance.

11. Diets should not be changed abruptly. A necessary change should always be done gradually. Abrupt and rapid changes in diet will throw animals off-feed and can result in digestive disturbances and increase the chance of colic and founder occurring.

12. Only top quality feeds, free from mold and dust, should be used. One who is feeding only a few horses and does not have the experience to know quality feeds should consider purchasing a good, commercial horse feed. There are many on the market.

13. Horses need adequate exercise. It helps improve their appetite, digestion, overall condition, and well being. It also keeps them in shape for riding, racing, work, or performance. Horses are athletes and a lack of exercise can ultimately affect their ability or willingness to perform. Horses can become lazy with a lack of exercise.

14. The droppings of the horse should be observed frequently. They should be watched for consistency, unusual odor, or any deviation from normal which may indicate digestive problems. This is a good way to detect, at an early stage, anything that may be developing.

15. Caretakers should be quiet and gentle when working around horses. This

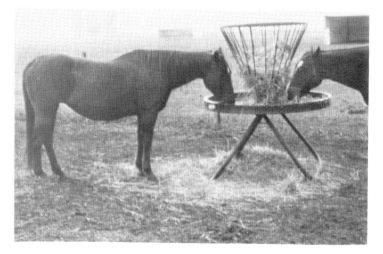

Fig. 14.4 Circular hay rack which allows horses to eat hay from all sides of it. The photo was taken at Sycamore Farms, Waterloo, Nebraska. (Courtesy of Larry Slade, Utah State University.)

helps in gaining the confidence of the animal. People who are irresponsible and reckless should not be allowed to work with valuable horses.

16. There is no substitute for experience in feeding horses properly. The beginner should work with an experienced person and profit by their years of experience and knowledge. The "eye of the master" is still a good guide to use in evaluating a feeding program. Spending a little time observing an animal each day can indicate a great deal about how the horse is doing and how good the feeding and management program is.

17. A clean, fresh water supply should always be available. Stagnant water may cause horses to go off-feed. Horses should not be allowed to drink a heavy fill all at once if they have been deprived of water for any length of time. This same recommendation applies to horses exercised or worked heavily. They should be allowed to rest and cool off for a while before being given a heavy fill of water. Walking out hot horses and allowing them a few sips at a time is a safe practice.

18. One should not let fads, fancies, and trade secrets influence good judgment in feeding horses. There are still too many supposed magical potions and tonics being used. This will decrease as more research information is obtained and horse feeding becomes more scientific.

19. If a horse is to be worked or raced hard soon after feeding, one should limit the amount of feed given, as well as the kind of feed used. The level and kind of feed will depend on the likes and disposition of the horse. Each one needs to be observed individually in order to meet its needs properly.

20. Salt and/or a complete mineral mixture should be available for horses at all times. This gives them an opportunity to consume extra minerals in case their diet or diet intake is inadequate in supplying their needs. Mineral boxes can be placed in the corner of a stall or in the corner of a pasture so the horses will not run into them. Plastic mineral boxes can be used to prevent injury when consuming minerals from the box.

21. Feed boxes should be kept clean at all times. Old, wet, moldy, or spoiled feed should not be allowed to accumulate since it will cause digestive disturbances or even more serious effects.

22. Many prefer to feed horses twice daily. Some prefer to have the concentrate mixture cleaned up in about 30 minutes after feeding. Some horses are fed three or four times daily while at the race track. When a horse does not clean up the feed quickly, the amount used should be reduced until the appetite of the horse justifies feeding more of the concentrate diet. Many feed an equal part of the concentrates at each feeding during the day, although there are variations in this procedure.

23. Feeds that are too heavy (lack bulk) may tend to pack in the stomach. Therefore, some bulk should be provided in the diet by using fibrous feeds. Feeds that are too high in fiber may also tend to cause discomfort or digestive distur-

bances, especially with horses that are exercised, worked, or raced heavily. High-fiber diets do not supply the energy necessary for a heavily worked horse.

24. Horses sometimes eat too rapidly. This can be helped by using a large feed box and spreading the feed thinly in it. This keeps the horse from getting a large mouthful at once. Placing a few smooth rocks or a 5-lb mineral block in the feed box also helps. This forces the horse to eat the feed around the rocks or block carefully and slowly.

25. Sale, show, or racing horses, which are in good body condition, should always be brought down in condition gradually. Harmful effects can occur if they are brought down too quickly. Many horses have been ruined by an abrupt change from heavy concentrate feeding to none whatsoever. Turning a highly conditioned horse out in a large field may also prove to be dangerous.

26. Feeds should be kept in a dry, well-ventilated room with good air circulation. If the feed becomes wet or the room is too humid, the feed can become moldy, musty, or rancid. Fresh feeds should be purchased and not kept too long before feeding. Moldy, musty, or rancid feeds are harmful for horses.

27. Vitamins, trace minerals, or other nutrients that are needed in minute quantities such as a few parts per million (ppm) should not be added to the diet. This requires expertise and sophisticated weighing and mixing facilities. Unless one has these, it is best to buy supplements, or feeds, from reputable feed companies that are experienced in this area. Vitamins, minerals, and other nutrients are very valuable at proper levels but excessive amounts can be harmful.

28. If the feeds used are not giving successful results, then expert advice should be sought to correct the situation. If the horses do not look good, if the hair and coat condition is dull, if feet and leg problems persist, if the foal crop continues low, then something may be wrong with the feeding program. Factors other than feeding may be involved, however, so one needs to evaluate the total management program.

29. There is always something new being developed in feeding and nutrition. Today's recommendations may not be adequate tomorrow. A good feeder keeps up-to-date on new developments and is constantly trying to improve. One who is satisfied with "status quo" has stopped making progress and will be left behind by those who constantly keep trying to do better.

30. Changes in feeding programs should allow time to observe the results obtained. Allowing at least 1 year to elapse before making another significant change provides enough time to evaluate what has happened relative to feet and leg problems, conception rate, services per conception, performance level, endurance, skin and hair coat appearance, and other criteria an owner may wish to use in evaluating their horses.

31. Proper feeding and water facilities should be provided. They should be kept clean to avoid digestive disturbances or even more serious consequences. Good sanitation practices will reduce the possibility of colic or founder occurring.

32. The horse's mouth should be examined regularly. The teeth should be checked and dental care provided since teeth problems limit the horse's ability to chew. Sharp edges should be removed and the teeth floated on a regular basis. One also needs to check for the presence of any irritant such as weed seeds, seed awns, pieces of lodged hay stems, and others. These can cause sore spots or, worse, can affect teeth, mouth tissue, and the ability to eat properly.

33. Horses should be weighed periodically to determine how well they are doing and to catch any significant changes in weight. One should know the horse's age and what its weight should be when in a thrifty condition.

34. Poisonous plants in pastures for horses should be avoided. They are usually not a problem if there is adequate forage of high quality so the horses are not forced to eat them to supply some of their nutrient needs. Exceptions to this occur, however, with yellow star thistle and other plants containing toxic compounds. Therefore, to be on the safe side with valuable horses, it is best to eliminate poisonous plants as soon as they are seen on pastures. Hay should also be inspected for the presence of poisonous plants.

35. Horses on pasture should be fed in groups according to their age, growth or development stage. This will aid in preventing overfeeding or underfeeding certain animals which occurs when horses of varying size and ages are fed together. Feeders spread far enough apart so that one or two horses cannot guard more than one feeder at a time will allow the more timid horses an opportunity to eat.

36. Foals should be allowed access to creep feed. This is especially beneficial if the mares are not good milkers or if the foals rely too heavily on forage. Moreover, foals that are eating feed from a creep are less nutritionally stressed at weaning time. They are also used to consuming feed which is beneficial in getting them used to feeding after weaning. Any creep feed that is wet or moldy should be removed daily. Feeders for mares with foals at side should be placed low enough so that the foal will learn to eat with the mare.

37. Different sources of feed weigh differently per unit of volume. Feeding by weight instead of volume will aid in decreasing overfeeding or underfeeding. It will also aid in determining if one is meeting the horse's nutritional requirements. For example, feeding a bucket of corn in place of oats may result in twice as much energy or conversely almost half as much energy if oats are substituted for the corn since corn contains about 1.8 times the energy level of oats. If pelleted feeds are used, they will also weigh more per unit of volume than when unpelleted. Feeding high-level performance horses should be done on a feed weight basis in order to more adequately provide their total nutrient needs.

38. Regular deworming and health inspections are important in keeping horses healthy. Feeding and other areas should be cleaned frequently to remove manure and thus reduce parasite infestation. Internal parasite prevention is critical since infections decrease feed intake, feed efficiency, and body weight.

39. The horse has a unique digestive tract which is limited in size. Feeding

large amounts of grain or concentrate feeds at one time results in a higher frequency of colic and founder because of the limited size of the digestive system. High-level performance horses, which require higher amounts of energy from concentrate or grain sources, should be fed them three or four times daily instead of twice a day. They may be fed one-fourth of the hay at the morning and noon feedings and the remaining one-half at the evening feeding. The horse's digestive system will better adapt itself to digesting the feed if it is not over-loaded at each feeding. More frequent feedings, with less portions in each one, will reduce founder, colic, and other digestive problems.

40. Wastage and parasite contamination occurs when hay is fed on the ground. It is preferable, therefore, to feed hay in bunks or racks. A trough under the hay rack reduces loss of leaves as the hay is pulled from the rack by the horses.

41. All feeding equipment should be constructed to minimize chances of injuring the horse. Moreover, it should be located in areas where the horses will not run into it and cause injury. Feed and hay mangers should be at a height where the horse can assume a normal eating position.

42. On days when high-level performance horses are idle, about one-half of their grain allotment should be withheld and their hay portion increased. This will prevent azoturia which is commonly known as Monday-morning sickness which causes muscle spasms or tetany to occur. It helps to have them on pasture and not restricted to a stall when not being used. When activity resumes these horses should be warmed up slowly.

43. All diets should contain some long roughage since the horse requires some fiber or bulk. This will decrease digestive disorders and should reduce wood chewing, and tail and mane biting. Other causes of wood chewing are boredom, lack of exercise, and possibly a nutrient deficiency.

15

Feeding the Foal

I. INTRODUCTION

Feeding the foal starts *in utero* with feeding the mare a well-balanced diet which supplies all the nutrients needed by the developing foal during gestation. Proper feeding during gestation also enables the mare to be a better milk producer after foaling.

It is very important that the foal receive a good start in bone development and growth while still suckling the mother. If it does not, it may be too late for improvement after weaning. This is especially the case, if the foal is expected to become a high-level performance horse.

II. MARE'S MILK AND COLOSTRUM

Nature provides mare's milk for the nursing foal. The milk will do a good job in meeting the foal's needs during the first 2 or 3 weeks of life. How adequately it meets these needs depends on how good a milk producer the mare is. It also depends on how much growth and development one expects of the foal. In addition to mare's milk, other feed is also needed by the foal.

One should help the foal suckle, if there is need to, as soon as possible after birth. A majority of normal foals will nurse within 1–2 hours after foaling. Those that are the most vigorous will nurse within 30–45 minutes. The first milk, or colostrum, is very important for the newborn foal. Colostrum is secreted before, during, and for a short time after foaling. It contains antibodies, and for approximately the first 24–36 hours of the foal's life its intestinal tract is permeable to these antibodies so they can be readily absorbed into the body. Without this colostrum, a foal begins life at a disadvantage. It is more susceptible to constipation and infections during its early days of life. Colostrum also has a high vitamin and other nutrient content, as well as a laxative substance that promotes bowel movement and elimination. A lack of colostrum is one of the most important factors causing infection and death in newborn foals. The foal should have about a quart of colostrum within 24–36 hours after birth. After that period the immunoglobulins in colostrum cannot be absorbed by the digestive tract.

Some farms and veterinarians keep a supply of frozen colostrum which may be used if the mare does not milk, the foal does not suckle, or the mare has leaked the colostrum prior to foaling. If this occurs the colostrum should be thawed and fed to the foal via bottle feeding or stomach tube shortly after birth and periodically thereafter. Sometimes it is possible to have the foal suckle another mare which has sufficient colostrum for two foals.

Even though many think that mare's milk is a perfect food, it is not. It is deficient in iron, copper, and possibly other nutrients. If the foal is restricted to milk alone, it soon becomes anemic. The young foal soon needs a source of these minerals plus other nutrients for maximum growth and development. Copper is especially needed for normal bone development in the foal. Therefore, milk alone becomes less sufficient for the foal as it grows older. Peak milk production usually occurs at 6–12 weeks after foaling. Therefore, as the foal grows the milk becomes a smaller percentage of its total dietary intake. Studies at Michigan State (1) showed that the crude protein in milk dropped from 19.1% shortly after birth to 3.8% 12 hours later and to 2.2% 2 months later. Other nutrients in milk such as calcium, phosphorus, fat, sodium, potassium, and others also decline as the lactation period continues; the energy value in milk also decreases. This is the reason why a good creep feeding program is recommended by many scientists, especially for high-level performance horses. Table 15.1 shows why milk production is not adequate for the foal according to Dr. E. A. Ott of the University of Florida. It shows that the digestible energy in milk is inadequate, especially after 1 month of age. The digestible protein is inadequate during lactation,

TABLE 15.1

Milk and Nutrient Production by the Brood Mare[a]

Month of lactation	Milk production (lb/day)	Digestible energy (kcal/day)	Digestible protein (g/day)	Calcium (g/day)	Phosphorus (g/day)
0–1	30.6	7256 (8153)[a]	375 (450)[a]	17.4 (18)[a]	5.9 (14)[a]
1–2	32.3	7276 (10411)	330 (500)	14.7 (18)	4.4 (15)
2–3	37.2	7909 (12747)	319 (546)	13.5 (17)	4.5 (15)
3–4	33.2	6795 (14161)	292 (535)	10.0 (17)	3.5 (15)
4–5	24.0	4905 (15372)	196 (522)	6.5 (16)	2.2 (15)
5–6	16.5	3375 (16351)	135 (510)	4.5 (16)	1.5 (15)

[a]All the figures in parentheses are the requirements for the foal. The others are those contained in the milk. Table prepared in a report by Dr. E. A. Ott, University of Florida, Gainesville.

TABLE 15.2

Milk Production—Average of 14 Mares (3).

Days in lactation	Milk production per day (lb)	Milk (% of body weight)	Solids in milk (%)	Protein in milk (%)	Fat in milk (%)
10	25.0	2.2	11.0	2.7	1.5
30	26.0	2.3	10.8	2.3	1.1
45	24.9	2.2	10.5	2.2	1.6
60	22.0	2.2	10.4	2.1	1.4
90	23.3	2.1	10.3	2.0	1.5
120	22.9	2.0	10.2	1.9	1.0
150	21.6	1.9	10.1	1.8	1.0

especially during the late stages. Both calcium and phosphorus in the milk are inadequate but phosphorus is the most deficient.

A Cornell University study (2) with Thoroughbred and Standardbred mares from 5 to 20 years old showed that milk intake by foals was 35.2, 33.0, and 39.6 lb at 11, 25, and 39 days postpartum, respectively. Milk production was equivalent to 3.1% of the mare's body weight at 11 days postpartum, 2.9% at 25 days, and 3.4% at 39 days. At 11, 25, and 39 days postpartum, respectively, dry matter intake equaled 3.1, 2.1, and 2.0% of the foal's body weight, and daily gross energy intake was 9380, 7590, and 8910 kcal.

A Texas A & M study (3) with 5–18-year-old Quarter Horse mares showed that average daily milk yield ranged from 26 lb in early lactation to 21.6 lb in late lactation (150-day lactation period). Average percentages of total solids, protein, and fat over the 150-day lactation were 10.5, 2.1, and 1.3, respectively. They decreased significantly with advancing lactation. The data are shown in Table 15.2.

A Cornell study (4) with Thoroughbred mare's milk is shown in Table 15.3. These data on minerals are for the first 8 weeks of the foal's life.

A Michigan State study (5) gives data on milk composition on Arabian and Quarter Horse mares that averaged 7.5 years of age (6.5–8.5 years old). The data shown in Table 15.4 indicate that mare's milk is low in iron.

The data shown on milk production are limited but they give some idea as to milk composition and level of production. As more sophisticated methods of measuring milk production, foal milk intake, milk composition, and other variables, which include mare's age, diet adequacy, stage of lactation, breed of mare, and many other factors are devised, it will be easier to evaluate the role of milk and its adequacy for foals being developed at varying rates of growth and for different purposes.

TABLE 15.3

Mineral Composition of Mare's Milk (4)

Weeks postpartum	Dry matter (%)	Ash (%)	Ca[a] (ppm)	P[a] (ppm)	Mg[a] (ppm)	K[a] (ppm)	Na[a] (ppm)	Cu[a] (ppm)	Zn[a] (ppm)
1	12.0	0.61	1345	943	118	664	237	0.85	3.1
2	11.5	0.57	1317	866	108	665	196	0.69	2.7
3	10.8	0.51	1160	742	92	547	184	0.42	2.4
4	10.5	0.45	1070	659	86	469	161	0.55	2.2
5	10.3	0.42	919	615	74	391	184	0.46	1.8
6	10.3	0.41	931	593	69	391	184	0.41	2.1
7	10.2	0.40	896	600	63	430	161	0.34	2.0
8	10.0	0.38	831	574	58	469	138	0.29	1.9

[a]Minerals given as ppm in fluid milk.

III. CREEP FEED

The foal soon starts to nibble on the feed given to its mother. This will occur at about 7–15 days of age. Sometimes what the foal consumes from its mother's feed will result in an imbalanced diet. Because of this possibility, most horse owners prefer to provide foals with a well-balanced creep feed. If the foal

TABLE 15.4

Dry Matter, Ash, Iron, Zinc, and Copper in Mare's Milk[a] (5)

Stage of lactation	Total dry matter (%)	Ash (%)	Iron (ppm)	Zinc (ppm)	Copper (ppm)
Parturition[b]	25.2	0.72	1.31	6.4	0.99
12 hours	11.5	0.50	0.95	2.8	0.83
24 hours	11.4	0.53	1.05	3.6	0.73
48 hours	12.0	0.54	0.86	3.7	0.66
5 days	11.6	0.54	0.88	3.5	0.44
8 days	11.5	0.55	0.88	3.3	0.44
3 weeks	11.3	0.50	0.83	2.8	0.29
5 weeks	11.2	0.43	0.71	2.2	0.25
2 months	10.3	0.37	0.61	2.1	0.23
3 months	10.4	0.32	0.55	2.1	0.25
4 months	10.0	0.27	0.49	2.4	0.20

[a]All values are expressed on a liquid milk basis.
[b]Milk sample obtained 15–30 minutes after birth of foal.

becomes thin, it is an indication that the mother is a poor milker and creep feeding will help. The creep feed plus the milk the foal gets from nursing its mother should be designed to provide a well-balanced diet. It is usually recommended that creep feeding be started within 1–2 weeks of foaling. The creep should be located where the mare goes periodically during the day for water or shade.

The creep feed given to the foal should be kept clean and fresh so that no moldy or sour feed is consumed. If the mother is a good milker and the foals are on excellent pasture, it may be difficult to get them to consume a creep feed. In such cases it may be necessary to get them started by letting them nibble some feed from one's hand (see Fig. 15.1).

It is difficult for most horse owners to formulate and mix a high-quality creep diet. Therefore, it may be best to rely on a good commercial creep feed that is properly fortified with vitamins, minerals, protein, and other nutrients needed by the young foal. The use of a creep feed helps to ensure that the inherited potential for growth and development is realized. At 5–6 weeks of age, a foal should be consuming at least 0.5 lb of creep feed daily per 100 lb body weight. By weaning time, the foal should be consuming 5–8 lb of creep feed per day. the amount fed depends on the milking ability of the mother, the development desired in the foal, the kind of creep feed used, and the economics involved, etc. The creep feed helps avoid set backs that can occur when the foal is weaned from its mother. At weaning the foal is changed from dependence on mother's milk to a man-made diet. This means the foal should be prepared or conditioned to this change beforehand and to have learned to eat a concentrate feed. The best way to prepare

Fig. 15.1 Students at Cal Poly, San Luis Obispo are teaching foals to eat creep feed. A creep is an enclosure with openings for the foals to get in but not the mares. (Courtesy of Richard Johnson, Cal Poly University, San Luis Obispo.)

Fig. 15.2 A safe creep feeder that features strength, openness, and safety. Sufficient room is available for several foals and a smooth fence line is at the edge of the opening. Creep feeder located at Elsinore Arabian Stud, Elsinore, California. (Courtesy of Professor Norman K. Dunn, Cal Poly University, Pomona.)

Fig. 15.3 A type of creep feeder panel that is safe and strong. It should be placed along the same fence line where mares are fed. The creep feeder is located in a pasture at Cal Poly University, Pomona. (Courtesy of Professor Norman K. Dunn, Cal Poly University, Pomona.)

a foal for weaning is to have it consuming a sufficient amount of a creep feed during the later part of the suckling period.

If the foal and mare are doing well during the first 6–12 weeks of age, creep feeding may be deferred until then when the mare's level of milk production starts to decrease and the foal's nutrient requirements are increasing. If the creep feed is fed free choice, and the foal is consuming more than desired, it can be mixed with a chopped high-quality hay, or the high-quality hay may also be made available in a rack in the creep.

IV. EARLY GROWTH RATE

At birth, the weight of the foal will be about 8–10% of the mare's weight. The size and weight of the foal at birth is influenced by the nutrition of the mare and heredity. Usually the colt is heavier than the filly. The newborn foal will have a wither height of about 60% of its mature height (6). The first 3–4 months of the foal's life are one of the most critical periods in its development. Tables 15.5–15.7 show some data presented by Dr. L. H. Breuer (6). The information in Table 15.5 is very interesting. It shows that the foals gained 4 lb per day during the first month of life. They essentially doubled their weight during this period. This indicates why it is so important that the mare be a good milker and why creep feed is helpful, especially if the mare is a poor milker.

The data also show that the foals again about doubled their weight by the time they were 3.5 months of age. They averaged a gain of 3.15 lb daily from birth to 3.5 months of age. After 3.5 months, however, the rate of growth starts to decrease as is shown in Table 15.7.

The wither height increased 10 inches between birth and 3.5 months of age and reached about 80% of its mature height. This rapid increase indicates the high rate of bone growth and development which occurs in the young foal. This indicates the need for proper nutrition, especially with minerals. Mare's milk is low in iron, copper, and possibly other trace minerals. Therefore, it is very

TABLE 15.5

Growth of Suckling Foals from Birth to 3.5 Months of Age[a]

	Birth	1 Month	$3\frac{1}{2}$ Months
Body weight (lb)	113	234	444
Weight gain (lb)		121	331
Wither height (in)	38	42	48
Feed per head per day (lb)			1.7

[a]Average of 21 foals. Data from Breuer (6).

TABLE 15.6

Growth Data on Suckling and Weaned Foals at 3 or 6 Months of Age[a]

	Group A suckling foals weaned at 6 months[b]	Group B weaned foals at 3 months fed on dry feed[b]
Number of foals	14	14
Initial body weight (lb)	448	458
Average daily gain (lb)	2.5	2.3
Wither height increase (in)	3.7	3.9
Supplemental feed per head daily (lb)	4.6	13.2

[a]Data from Breuer (6).
[b]Weights at 6 months of age.

important that trace minerals be supplied by trace mineralized salt or a complete mineral mixture in the creep feed or in the mineral box. Table 15.5 also shows that the suckling foals consumed about 1.7 lb of supplemental feed daily (in addition to the mare's milk). This amount of feed would account for about 1 lb of gain daily in light-weight foals (6). Therefore, if one wishes to produce a rapid gain in foals, it is necessary to provide them with supplemental feed. The return in gain is very high for each pound of supplemental feed used during early growth.

The information in Table 15.6 compares foals of about the same body weight and expected mature weights. Group A was kept with the mare, and Group B was weaned early at 3 months of age. Group B was fed a dry diet in drylot while Group A was kept with their dams and had access to a supplemental dry diet and pasture.

The information in Table 15.6 shows that the foals kept with their dams gained faster but the increase in height at the withers was less than the foals that were weaned at 3 months of age. Foals are usually weaned at an average age of

TABLE 15.7

Growth of Horses From 7 to 18 Months of Age[a]

Months of age	7	12	15	18
Body weight (lb)	630	814	928	1,041
Wither height (in)	52	56	57.25	58.25
Average daily gain (lb)				1.21

[a]Average of 18 horses. Data from Breuer (6).

about 6 months. The weaning age will vary somewhat depending on a number of factors. The data in Table 15.6 indicate that keeping the foal with the mare saved 8.6 lb of supplemental feed daily or 774 lb of total feed during the 90 days from 3 to 6 months of age. Thus, unless there are special circumstances where early weaning is needed or advantageous, then it appears best to keep the foal with the mare.

Table 15.7 shows the weight, height, and gain of horses from 7 to 18 months of age. Body weight increased 65.2% but height at the withers increased only 6.25 inches. Rate of gain decreased considerably, as compared to early growth, and averaged 1.21 lb per day. The data shown in Table 15.7 indicate that the horses used in the study had the potential to approach 100% of their mature weight and wither height at 18–24 months of age (6).

V. CAUTION ON RAPID GROWTH

Horses fed for rapid body and skeletal growth may develop bone abnormalities and lameness (7–9). Too high a level of feeding has been suggested as a possible cause of epiphysitis, osteochondrosis, contracted tendons, and other bone abnormalities. But, other causes and conditions could also be involved (see Chapter 20).

There is still a lack of knowledge concerning the nutritional requirements for different rates of growth in the young horse. Once this information is obtained it may decrease, or eliminate, bone abnormalities in rapidly growing young horses as improved diets are used. A recent University of Kentucky report stated that young horses fed for rapid growth may not maximize bone deposition (7). But, they also showed that a creep feeding program which supplies NRC-recommended nutrient levels can increase the rate of skeletal growth with little decrease in quality of bone. This emphasizes feeding a well-balanced diet in energy, protein, vitamins, and minerals in order to maximize proper bone development.

In many cases where bone problems are occurring, decreasing the level of feed intake, and thus growth rate, is beneficial in decreasing certain bone abnormalities.

VI. MILK REPLACERS AND EARLY WEANING

Milk substitutes are being improved and developed by the feed industry for all classes of animals. There are some situations for the horse where some degree of early weaning may be needed. These could include the following: (a) sickness or death of the mare: (b) mare is a very poor milker; (c) injury to mare's udder; (d) early weaning to help get the mare to breed again; (e) an orphan foal; and (f) early weaning to permit showing or training the mare. If one is interested in early weaning, the age of weaning should be decreased gradually in line with the success being obtained. If decreasing it by 1 month is successful, then an earlier weaning age might be attempted. One should not go into early weaning unless a good diet is used to substitute for the mare's milk and a good management program is followed so the foal is not harmed. Most horse owners wean foals at about 6 months of age, although some horse farms are weaning foals at an earlier age. Early weaning requires superior managerial ability and greater attention to diet quality and nutrient content. Unless one can do as good a job as the mare in feeding her foal, it is best to let her nurse it.

An Ohio study (10) involved removing foals from their dams at 3 days of age and feeding them a reconstituted 26% crude protein milk replacer free choice for 1 month, at which time *ad libitum* solid feeding began. Control foals were weaned from their dams at 2 months of age and fed a 21% crude protein concentrate *ad libitum* until the trial terminated after 26 weeks. Average daily gains for the artificially reared and 2-month-old-weaned foals were 2.09 and 2.16 lb, respectively at 26 weeks. This study demonstrated that artificial rearing systems can be used for an orphan foal or when early separation of the foal and mare is indicated. The foals receiving the milk replacer fed free choice for 1 month consumed on average 26.6 lb per day.

VII. FOAL WEIGHTS AND BIRTH DATES

Cornell University published birth and growth data obtained in cooperation with Winfield Farms of Oshawa, Ontario, Canada on Thoroughbred horses (11). Every foal at Winfield Farms was weighed and measured for height at withers and front cannon bone circumference at the midpoint on about the 15th day of each month. They studied a total of 19,883 records on 1992 foals out of 813 dams and by 365 sires collected from January, 1958 to June, 1976. They found that colts were heavier than fillies at birth and the difference increased with age. Dams under 7 years of age and older than 11 years of age had foals of lighter weight at birth than mares 7–11 years of age, and the difference in colt size continued to 510 days of age. Dams under 7 years of age had shorter foals with smaller cannon bones than mares 7–11 years of age, and these differences were still evident at 510 days of age.

The foals born in January, February, or March were lighter, shorter, and had a smaller cannon bone than foals born in April, May, or June. These differences continued throughout the study period. The percentage of foals born during the various months were as follows: January, 1.5%, February, 7.3%, March, 22.8%, April, 31.4%, May, 30.3%, and June, 6.6%.

VIII. SELF-FEED MINERALS

If the foal depends heavily, or almost entirely, on mare's milk, most horse owners will self-feed trace mineralized salt in a mineral box which is easily accessible to the foal. If the foal has access to a properly balanced creep feed, mineral supplementation may not be needed while the foal is suckling its mother. But, having the trace mineralized salt available is a sound practice in case the foal's needs are greater than its total diet provides. Horses have been shown to have a specific appetite for salt if the diet is deficient in sodium (12, 13). Therefore, they would consume the salt in the mineral box if the diet lacked sodium.

IX. SUGGESTED DIETS

There are many diets which can be fed as a creep feed. Some creep feeds are very complex whereas others are quite simple. Most complex creep diets are used in developing high-level performing horses whereas the simple creep feeds are used with other horses used for pleasure riding and other activities which are not as demanding on the horse.

A. Complex Creep Feed

Table 15.8 is an example of a complex creep feed. Soybean meal is used to provide an adequate lysine level. The grains are rolled or flaked in order to add palatability to the diet. Oat groats are an excellent and very palatable feed. If they cannot be obtained, then oats should be substituted for the oat groats. Either corn, barley, milo, or a combination of these three grains can be used. Dried skim milk is an excellent feed for foals. The molasses is added for palatability and to help control dustiness in the feed. Sometimes it takes 7 or 8% molasses to do the best job in this regard. The trace mineralized salt shown in Salt Premix A in Table 15.10 is suggested. The Vitamin Premix A shown in Table 15.9 is also recommended. Both are designed for use with high-level performance horses.

B. Simple Creep Feed

There are many simple creep feeds used for pleasure horses and their foals. Table 15.11 gives one example. The simple creep feed does not contain oat groats and dried skim milk which are excellent feeds but which are also expensive. It also contains a lower level of trace minerals in the salt and does not contain any B-complex vitamin supplementation. Moreover, the option is given

TABLE 15.8

Complex Creep Diet for Nursing Foal

Feed	Percent in diet[a]
Oat groats, rolled	15.0
Oats (heavy oats), rolled or flaked	20.0
Corn, barley, milo, or combination of them rolled or flaked	35.4
Soybean meal	15.0
Dried skim milk	5.0
Blackstrap molasses	5.0
Dicalcium phosphate (or other calcium and phosphorus source)	2.0
Limestone, ground	0.8
Salt, trace mineralized—Salt premix A shown in Table 15.10	1.0
Vitamin Premix A shown in Table 15.9	0.8
	100.0

[a]The level of feed ingredients in the diet should be adjusted to provide 18% crude protein, 0.90% calcium, and 0.80% phosphorus if the feeds used vary from expected analytical values.

TABLE 15.9

Vitamin Premix for Horse Diets[a]

Vitamin	Level of vitamins in 1 lb of premix with carrier			
	Premix[b] A	Premix B	Premix C	Premix D
A (IU)	400,000	400,000	400,000	400,000
D (IU)	40,000	40,000	40,000	40,000
E (IU)	1600	1600	1600	—
K (mg)	200	200	—	—
Thiamin (mg)	240	—	—	—
Riboflavin (mg)	400	—	—	—
Niacin (mg)	1,200	—	—	—
Pyridoxine (mg)	120	—	—	—
Pantothenic acid (mg)	480	—	—	—
Choline (g)	6	—	—	—
Vitamin B12 (mg)	1.2	—	—	—
Folacin (mg)	120	—	—	—
Carrier (includes all vitamins in it) (lb)	1	1	1	1

[a]These vitamin premixes are added to the concentrate portion of the diet. The remainder of the diet consists of milk (in the case of the foal) or hay and/or pasture. This vitamin premix is usually added at a level of 1.5 lb or less per 100 lb of concentrate mix.

[b]Other vitamins could be included if so desired at the following levels in the above 1 lb premix: vitamin C, 5.0 g; D-biotin, 1 mg; and PABA, 2 g. If myo-inositol is used it can be added at a level of 0.1–0.3% of the diet.

TABLE 15.10

Trace Mineralized Salt for Horses[a]

Trace mineral	Level of mineral element added to		Equivalent in concentrate feed if salt is added at 1.0% level	
	Salt Premix A (%)[b]	Salt Premix B (%)[c]	Concentrate feed with Salt Premix A (ppm)	Concentrate feed with Salt Premix B (ppm)
Iodine	0.010	0.006	0.5	0.3
Iron	1.600	0.800	80.0	40.0
Copper	0.400	0.200	20.0	10.0
Cobalt	0.006	0.004	0.3	0.2
Manganese	0.800	0.600	40.0	30.0
Zinc	1.600	0.800	80.0	40.0
Selenium	0.002	0.002	0.1	0.1

[a]The levels of trace minerals in Salt PreMix A and B will be diluted by the level of forage (hay and/or pasture) which is fed in addition to the concentrate feed containing the trace mineralized salt.

[b]Salt PreMix A with a higher level of trace minerals is suggested for high-level performance horses that require a higher level of nutrition.

[c]Salt PreMix B with a lower level of trace minerals is suggested for horses used for pleasure riding and other activities which are not as demanding on the horses. Salt PreMix B can be self-fed to horses which are fed on pasture and/or hay with little or no concentrate feeding.

TABLE 15.11

Simple Creep Feed[a]

	Percent in diet
Oats, rolled or flaked	20.0
Corn, barley, milo or a combination of them, rolled or flaked	47.0
Soybean meal	25.0
Blackstrap molasses	5.0
Salt, trace mineralized–Salt Premix B shown in Table 15.10	1.0
Vitamin Premix B, C, or D shown in Table 15.9[b]	0.5
Limestone	1.0
Dicalcium phosphate	0.5
	100.0

[a]The level of feed ingredients in the diet should be adjusted to provide 18% crude protein, 0.90% calcium, and 0.80% phosphorus if the feeds used vary from expected analytical values.

[b]This vitamin premix does not contain the B-complex vitamins. It also gives the option as to whether vitamins E and K are to be added to the feed.

TABLE 15.12

Expected Feed Consumption by Horses (14)

Stage of life cycle	Forage as % of:		Concentrate[a] as % of:		Total feed intake as % of body wt.
	Body wt.	Diet	Body wt.	Diet	
Mature horses					
Maintenance	1.5–2.0	80–100	0–0.5	0–20	1.5–2.0
Mares, late gestation	1.0–1.5	65	0.5–1.0	35	1.5–2.0
Mares, early lactation	1.0–2.0	45	1.0–2.0	55	2.0–3.0
Mares, late lactation	1.0–2.0	60	0.5–1.5	40	2.0–2.5
Working horses					
Light work	1.0–2.0	65	0.5–1.0	35	1.5–2.5
Moderate work	1.0–2.0	40	0.75–1.5	60	1.75–2.5
Intense work	0.75–1.5	30	1.0–2.0	70	2.0–2.5
Young horses					
Nursing foal, 3 months[b]	0	20	1.0–2.0	80	2.5–3.5
Weaning foal, 6 months	0.5–1.0	30	1.5–3.0	70	2.0–3.5
Yearling foal, 12 months	1.0–1.5	45	1.0–2.0	55	2.0–3.0
Long yearling, 18 months	1.0–1.5	60	1.0–1.5	40	2.0–2.5
2 year old, 24 months	1.0–1.5	40	1.0–1.5	60	2.0–2.5

[a]Air-dry feed (about 90% dry matter).
[b]Does not include milk consumption.

as to whether vitamin E or vitamin K is to be included or left out in the vitamin pre-mix used.

Table 15.12 gives information on expected feed consumption by horses during various stages of their life cycle as suggested by Dr. E. A. Ott of the University of Florida (14). The figures will vary depending on the mature weight of the horses being used, the quality of the diet being fed, the rate of growth desired, the level of performance required, reproduction rate, level of milk production, and many other factors. However, the data shown in Table 15.12 can be a valuable guide in determining concentrate, forage, and total feed intake to expect during various stages of the life cycle of the horse.

X. CONCLUSIONS

Proper feeding of the foal starts *in utero* by feeding the mare a well-balanced diet. Colostrum is very important since a lack of it is a major cause of infections and death in foals shortly after birth. Creep feeding is very important for foals whose mother's milking level is inadequate and especially for foals being developed for high-level performance. Young horses fed for rapid body and skeletal growth may develop bone abnormalities and lameness. But, this may be mini-mized, or eliminated, if a properly balanced diet is fed which is adequate in all nutrients. Early weaning requires superior managerial ability and excellent quali-ty milk replacer diets. Unless one can do as good a job as the mare in feeding her foal, it is best to let her nurse it. Most foals are born during March, April, and May. Suggested concentrate diets for creep feeding are given and discussed.

Foals given a creep feed learn how to eat concentrates and are better prepared to withstand the shock of weaning. Information on expected feed consumption by horses during various stages of their life cycle is given and discussed.

REFERENCES

1. Ullrey, D. E., R. O. Struthers, D. G. Hendricks, and B. E. Brent. *J. Anim. Sci.* **25,** 217(1966).
2. Oftedal, O. T., H. F. Hintz, and H. F. Schryver, *J. Nutr.* **113,** 2196(1983).
3. Gibbs, P. G., G. D. Potter, R. W. Blake, and W. D. McMullan. *J. Anim. Sci.* **54,** 496(1982).
4. Schryver, H. F., O. T. Oftedal, J. Williams, L. V. Soderholm, and H. F. Hintz. *J. Nutr.* **116,** 2142(1986).
5. Ullrey, D. E., E. T. Ely, and R. L. Covert. *J. Anim. Sci.* **38,** 1276(1974).
6. Breuer, L. H., *Proc. Md. Nutr. Conf.* p. 102(1974).
7. Thompson, K. N., J. P. Baker, and S. G. Jackson. *J. Anim. Sci.* **66,** 1692(1988).
8. Glade, M. J., and T. H. Belling, Jr. *Growth* **48,** 473(1984).
9. Donoghue, S. *Proc. 26th Annu. Conv. Am. Assoc. Equine Pract.* p. 65(1980).
10. Knight, D. A., and W. J. Tyznik. *J. Anim. Sci.* **60,** 1(1985).
11. Hintz, H. F., R. L. Hintz, and L. D. Van Vleck. *J. Anim. Sci.* **48,** 480(1979).
12. Ralston, S. L. *J. Anim. Sci.* **59**(5), 1354(1984).
13. Meyer, H., A. Linder, M. Schmidt, and E. Teleb. *Proc. Equine Nutr. Physiol. Symp., 8th,* p. 11(1983).
14. Ott, E. A. *Feedstuffs* **60**(31), 78(1988).

16

Feeding the Weanling, Yearling, and Long Yearling Horse

One of the most critical times in the life of a growing horse occurs between weaning and about 1 year of age. But, the second year is also very important for the maturation of bone, muscle, and other parts of the body. Foals which have been given a creep feed and are used to consuming concentrates are better prepared for the shock of weaning. They are also heavier, because of the creep feed they have consumed, and are better able to handle forages (hay and/or pasture). Foals that are not given a creep feed may look good while still nursing; however, they may need special care in order to cope with the shock of weaning, otherwise, they may go backward for awhile and not develop to their full potential in size and productivity (or not develop it as rapidly).

I. FEEDING THE WEANLING

At weaning time, which is usually at 6 months of age, the foal is changed from considerable dependence on mother's milk to a man-made diet. Many people who are feeding high-level performance horses feel the diet fed to the foal immediately after weaning should be as good as the mother's milk. Some horse owners wean at 7–8 months of age in order to take advantage of the nutritional value of the mare's milk. This is especially the case when the feed supply available for the weaned foal is limited in quantity and quality. Other owners prefer to wean earlier than 6 months because they are able to provide creep and starter diets which are excellent substitutes for or supplements to mare's milk. Early weaning, however, should not be attempted unless an excellent diet, properly fortified with minerals, vitamins, amino acids, and protein is used. Early weaning also requires superior managerial ability. Unless one can do as well as the mare in feeding the foal, it is best to let her nurse the foal and not wean too early.

A. Rate of Gain

Weanlings grow rapidly and develope considerable bone and muscle. Therefore, it is very important to feed them a well-balanced diet in ample quantity to supply the energy, protein, vitamins, minerals, and other nutrients needed (1).

Fig. 16.1 Special bucket holders are attached to fence posts to reduce wear and tear on fencing. Feeders allow for feeding of grain for horses on pasture and being above ground level reduces damage to buckets. Feeders located at Cal Poly University, Pomona. (Courtesy of Professor Norman K. Dunn, Cal Poly University, Pomona.)

Table 16.1 gives information on the weight and the rate of gain of growing horses at various ages. The information was prepared by the 1978 NRC Committee on nutrient requirements of the horse (2). The data show that horses of heavier mature weights gain faster than horses of lighter mature weights. It is important to note that the fastest gains are made during the first year. Diets fed to the young growing horse need to reflect the higher level of energy, protein, minerals, vitamins, and other nutrients needed during the stages of most rapid growth.

B. Weight and Height from Birth to 18 Months of Age

Assuming that the average mature weight of Thoroughbred stallions is 1200 lb and mares is 1100 lb, the Thoroughbred foals used in a Cornell study (3) attained about 46, 67, and 80% of their mature weight at 6, 12, and 18 months of age, respectively. This is similar to a Quarter Horse study at Louisiana State University (4), where the horses attained about 44, 63, and 79% of their mature weight at 6, 12, and 18 months, respectively. A Cal Poly University, Pomona, study (5)

TABLE 16.1

Daily Gain by Growing Horses[a]

	Age (months)	Weight of growing horse (lb) for mature weight of horse (lb)			Daily gain of growing horse (lb) for mature weight of horse (b)		
		880	1100	1320	880	1100	1320
Nursing foal	3	275	341	374	2.2	2.64	3.08
Weanling	6	407	506	583	1.43	1.76	1.87
Yearling	12	583	715	847	0.88	1.21	1.32
Long yearling	18	726	880	1045	0.55	0.77	0.77
Two year old	24	803	990	1188	0.22	0.33	0.44

[a]Data from Hintz et al. (2).

showed that Arabian horses attained 44, 66, and 80% of their mature weight at 6, 12, and 18 months of age. A European study (6) showed that half Arabs and Anglo-Arabs had 45, 67, and 81% of their mature weight at 6, 12, and 18 months of age. The similarity of these results with four different breeds of horses would indicate that values of 45, 66, and 80% of mature weight at 6, 12, and 18 months of age is an approximate guide for rate of development for most light horse breeds.

The University of Minnesota (7) reported that ponies with a mature weight of 400 lb might be expected to reach 55, 75, and 84% of their mature weight at 6, 12, and 18 months of age. Therefore, ponies reach their mature weight at an earlier age. Large draft horses appear to attain their mature weight at a later age. Results from the University of Missouri (8) showed that Percherons attained 38, 55, and 73% of their mature weight at 6, 12, and 18 months of age. Studies at Cornell (9) showed that draft horses attained 34, 52, and 69% of their mature weights at 6, 12, and 18 months of age.

The Cornell study (3) showed that Thoroughbred foals attained 83, 90, and 95% of their mature height at the withers at 6, 12, and 18 months of age. The Cal Poly University, Pomona, study (5) with Arabian horses showed they attained 84, 91, and 95% of their mature height at 6, 12, and 18 months of age. With half Arabs and Anglo-Arabs the European study (6) showed they reached 83, 92, and 95% of their mature height at 6, 12, and 18 months of age. The results from a Quarter horse study (10) were somewhat different from those indicated above since the values reported were 90, 93, and 99% of their mature heights at 6, 12, and 18 months of age.

The data obtained in the Cornell study (3) with Thoroughbreds show that early growth is very rapid. The foals gained 242 lb during the first 90 days after birth or 2.69 lb per day, 165 lb during the second 90 days or 1.83 lb per day, 132 lb during the third 90 days or 1.46 lb per day, and only 99 lb during the fourth 90 days or 1.1 lb per day. Therefore, during the first few months after birth the fastest growth and most elongation of bones occurs. It is very important that the mare be fed a well-balanced diet in order to produce plenty of milk for the foal. Moreover, the foals should be fed a well-balanced creep feed to supplement the mother's milk if they are being developed for high-level performance at an early age. This should especially be the case if the mother is a poor milker.

C. Amount of Feed to Use

Shortly before weaning the foal should be eating from 0.5 to 0.75 lb (or more) of a creep diet per 100 lb body weight. After weaning, the concentrate feed should be increased to 1–1.5 lb (or more) of concentrates per 100 lb body weight. In addition, the foal should be given at least 1 lb of forage per 100 lb body weight. Some owners will feed 1.5–2.0 lb of forage per 100 lb body weight. The amount of concentrates and forage to feed will vary and depend on a

Fig. 16.2 Hay feeder suitable for two or three horses. A safe and efficient method of feeding hay. Feeder located at Old English Rancho, Ontario, California. (Courtesy of Professor Norman K. Dunn, Cal Poly University, Pomona.)

Fig. 16.3 A partially filled 55-gal drum makes a safe, sturdy, and practical feeder, especially for pellets and grain. Bottom half of barrel is filled with sand which is covered with a 2 in layer of concrete. Feeder located at Old English Rancho, Ontario, California. (Courtesy of Professor Norman K. Dunn, Cal Poly University, Pomona.)

number of factors. These include the individuality of the horse and its likes, dislikes, and eating habits. It also depends on the quality of the concentrate feed as well as the quality of the hay or pasture provided. Of importance too will be whether the weanling is being developed for racing, performance, or a sale. Weanlings being pushed for early racing or performance should be fed concentrates more liberally. They should also be fed a more sophisticated diet, which supplies all the energy, vitamins, minerals, protein, and other nutrients that are needed for fast growth, proper bone development, and maximum performance.

D. Suggested Concentrate Diets

Table 16.2 gives an example of a concentrate diet which might be fed to weanling foals being developed for high-level performance. Many other diets can also be used. Oats are an excellent feed for young, growing horses. Corn, milo, or barley are added primarily as a source of energy. Soybean meal is the best plant protein supplement to use in young, growing horse diets. It is high in

TABLE 16.2

**Sample Concentrate Diet for Weanling Foal Fed
for High-Level Performance**

Feed	Percentage in diet[c]
Oats (heavy oats) rolled or flaked	25.0
Corn, barley, or a combination of them, rolled or flaked	30.85
Milo (or corn or barley), rolled or flaked	7.0
Soybean meal	23.2
Dehydrated alfalfa meal (20% protein)	5.0
Blackstrap molasses	5.0
Vitamin supplement[a]	0.7
Dicalcium phosphate (or other calcium and phosphorus source)	2.0
Limestone, ground	0.25
Salt, trace mineralized[b]	1.0

[a]Vitamin Pre-Mix A in Table 15.9.
[b]Same as Salt Pre-Mix A in Table 15.10.
[c]These levels should be adjusted to provide 18% protein, 0.85% calcium, and 0.75% phosphorus if the feeds used vary from expected analytical values.

lysine, which is very important for the weanling (11). This indispensable amino acid is low in the other plant protein supplements such as linseed meal, peanut meal, sunflower seed meal, and others. The weanling can utilize 5% of dehydrated alfalfa meal, which is an excellent source of many nutrients. The molasses is added for palatability and to help control dust. The vitamin supplement is used to reinforce the diet and make sure it is adequate in all the vitamins needed. The trace mineralized salt and other minerals are added to make sure all the essential mineral elements required by the horse are supplied. In addition, it is recommended that minerals be self-fed in case the weanling needs more than is contained in the diet used.

The sample diet shown in Table 16.2 would constitute about 70% of the total diet fed the weanlings (Table 15.12). The remainder of the diet would be a high-quality hay and/or pasture, which contains at least 12% protein which would result in about 16% protein in the total diet. The vitamin premix added would be the same as shown in Table 15.9. The trace mineralized salt would contain iodine, iron, copper, cobalt, manganese, zinc, and selenium at levels shown in Table 15.10.

Table 16.3 gives an example of a more simple concentrate diet which might be used for weanling foals used for pleasure riding and other activities which are not as demanding on these horses as those horses being developed for high-level performance. Many other diets may also be used. The diet provides a lower level

TABLE 16.3

Sample Concentrate Diet for Weanling Foals Not Fed
for High-Level Performance

Feed	Percentage in diet[c]
Oats, rolled or flaked	18.0
Corn, barley, milo, or a combination of them, rolled or flaked	48.0
Soybean meal	25.0
Blackstrap molasses	5.0
Vitamin supplement, premix B, C or D shown in Table 15.9[a]	0.7
Dicalcium phosphate (or other calcium and phosphorus source)	1.5
Limestone, ground (or other calcium source)	0.8
Salt, trace mineralized, salt premix B shown in Table 15.10[b]	1.0

[a]This gives the option as to whether vitamin E or vitamin K is to be included or left out in the vitamin premix used.

[b]This provides a trace mineralized salt with a lower level of the trace minerals.

[c]These levels should be adjusted to provide 18% protein, 0.85% calcium, and 0.75% phosphorus if the feeds used vary from expected analytical values.

of oats, and leaves out dehydrated alfalfa meal. The vitamin premix used does not contain the B-complex vitamins. Moreover, it gives the option as to whether all four fat-soluble vitamins (A, D, E, and K) are to be used and also gives the option as to whether vitamin E or vitamin K are to be included or left out in the vitamin premix used. The trace mineralized salt used contains a lower level of the trace minerals. This concentrate diet should be fed with a high-quality hay and/or pasture which contains at least 12% protein.

E. Feeding Management

Aggressiveness begins to appear in weanlings as they try to establish dominance via the pecking order system (12). The more aggressive weanlings start to fight off the more timid weanling from the feeders. If possible, the weanlings should be individually fed in order to ensure an even consumption of concentrate feeds. If individual feeding is not possible, then separate feeders or long troughs should be used which provide enough room for all horses to eat without fighting. It also helps to group weanlings and other classes of horses into groups with a similar disposition. Concentrate feeds should be fed at least twice daily (12). Top quality pasture and/or hay should be provided on a free-choice basis (12).

These recommendations apply to all classes of growing and developing horses. It is very important that they get their fair share of concentrate feeds and that they receive an adequate level of exercise.

F. Exercise and Skeletal Abnormalities

Exercise is needed to stimulate body systems and to help develop sound bone and fitness. A few hours of free exercise daily is excellent for all young growing horses. They will usually quit exercising and then fatigue sets in. But, forced exercise should be well regulated and not forced to extremes. Excessive forced exercise may be harmful to the young horse's skeletal system which is still quite immature (12)., Excessive forced exercise causes some horses to develop joint inflammation, soreness, lameness, pulling up on their pasterns, bending over on the knees, etc. (12). These external disorders represent a variety of internal skeletal problems and are generally and inaccurately all lumped together under the term "epiphysitis" (12). Considerable research is still needed on the optimal skeletal development rate in horses. Presently, three factors known to cause skeletal disorders in horses are: (1) genetic predisposition associated with large size at maturity; (2) nutrient imbalances or deficiencies in the total diet; and (3) confinement coupled with forced exercise (12). Other conditions are also involved.

G. Anabolic Steroids

Texas A & M scientists (12) state that "due to a lack of demonstrated benefits on growth rates and a clear cut depression on fertility levels it is strongly recommended that horse owners refrain from using anabolic steroids on healthy, young, growing horses."

Studies at Colorado (13) reported that 2–4-year-old stallions receiving anabolic steroids had severely reduced testicular size, sperm production, number of sperm per ejaculate, and sperm motility. Another Colorado study (14) reported that anabolic steroids in fillies suppressed estrous and ovulation, induced male-like behavior, enlarged the clitoris, and upset the normal hormonal balance and fertility.

II. FEEDING THE YEARLING

If the horse reaches a year of age and is well grown, thrifty, in excellent shape, and has sound feet and legs, it has successfully passed a critical period in its life cycle. Weight gains will decrease during the second year of the foal's life as is shown in Table 16.1, but the foal is still growing and should continue to be fed a high-quality diet.

A. Amount to Feed

The yearling should be placed on a feeding program of 1–1.5 lb of forage and 1–1.5 lb of concentrates per 100 lb body weight to obtain the growth rate desired (Table 15.12). The ratio of concentrates to forage will depend on how rapidly one expects the yearling to develop and how well it is responding to the diet fed. If the yearling is being developed for racing or high-level performance, then a higher level of concentrates relative to forage should be fed. This will help produce bone and muscle and assure that a well-developed, sound horse is ready for racing or performance toward the end of the second year. Yearlings that are not being developed for racing or performance can be fed a higher proportion of forage relative to concentrates. In many cases, where the yearlings are being developed for pleasure riding, the forage part of their diet may be approximately 75% or more of the total feed intake. The level of feeding, therefore, can vary considerably depending on how the horses are to be used, the kind and quality of diet used, and the response of the horses to the feeding program being followed.

B. Ratio of Concentrates to Forage

Table 15.12 presents information on the ratio of concentrates to forages. The levels suggested in Table 15.12 can be used as a guide. Deviations can be made from the ratio of concentrates to forage depending on the objectives of the feeding program.

C. Suggested Concentrate Diets

Table 16.4 gives a sample concentrate diet that can be used as a guide for yearlings being developed for high-level performance. The total diet would be fed in a ratio of about one-half concentrates to one-half forage (hay and/or pasture). Since the concentrate portion of the diet would contain 16% protein, it should be fed with forages that have at least 10% protein. The total diet would then provide about 13–14% protein.

The total diet for yearlings is lower in protein, calcium, and phosphorus than the weanling diet. It also contains less oats, soybean meal, and dehydrated alfalfa meal, and a higher level of grains other than oats. A group in Florida reported that corn can replace oats in yearling diets with good results (15). Linseed meal also replaces some of the soybean meal in the diet. The diet is less costly than the weanling diet shown in Table 16.2.

Table 16.5 gives an example of a concentrate diet for yearlings not fed for high-level performance. The diet is lower in oats and higher in other grains. It also replaces some of the soybean meal with linseed meal. Dehydrated alfalfa meal is omitted from the diet. The vitamin premix used does not contain the B-

TABLE 16.4

**Sample Concentrate Diet for Yearlings Being Developed
for High-Level Performance**

Feed	Percentage in diet[a]
Oats, rolled or flaked	15.0
Corn, barley or combination, rolled or flaked	38.1
Milo (or other grain), rolled, flaked or micronized	14.0
Soybean meal	13.0
Linseed meal	6.0
Dehydrated alfalfa meal	5.0
Blackstrap molasses	5.0
Dicalcium phosphate (or other calcium and phosphorus source)	2.2
Salt, trace mineralized[b]	1.0
Vitamin supplement[c]	0.7

[a]These levels should be adjusted to provide 16% protein, 0.8% calcium, and 0.65% phosphorus if the feeds used vary from expected analytical values. This concentrate feed would be fed with hay and/or pasture which has at least 10% protein to provide 13–14% protein in the total diet for yearlings.

[b]Same as salt premix A in Table 15.10.

[c]Same as vitamin premix A in Table 15.9.

complex vitamins. It also gives the option of using the four fat-soluble vitamins (A, D, E, and K) or the option of whether vitamin E or K are to be included or left out in the vitamin premix used. The trace mineralized salt used contains a lower level of the trace minerals. This concentrate diet should be fed with hay or pasture which has at least 10% protein. Ordinarily this concentrate portion of the diet should constitute 50% of the total feed intake. If the concentrate diet shown in Table 16.5 makes up less than 50% of the total feed intake, then a forage with more than 10% protein should be used.

III. FEEDING THE LONG YEARLING

The long yearling requires a little less protein, calcium, and phosphorus than the yearling. It also consumes more forage than the yearling. The forage intake is about 60% of the total feed intake (Table 15.12) as compared to 40% concentrates.

The sample concentrate diet used for yearlings being developed for high-level performance (Table 16.4) can also be used for long yearlings being developed for

TABLE 16.5

Sample Concentrate Diet for Yearlings Not Fed
for High-Level Performance

Feed	Percentage in diet[c]
Oats, rolled or flaked	10.0
Corn, barley, milo, or a combination of them, rolled or flaked	60.2
Soybean meal	11.0
Linseed meal	10.0
Blackstrap molasses	5.0
Dicalcium phosphate (or other calcium and phosphorus source)	1.1
Limestone, ground (or other calcium source)	1.0
Salt, trace mineralized, salt premix B shown in Table 15.10[a]	1.0
Vitamin supplement, premix B, C, or D shown in Table 15.9[b]	0.7

[a]This provides a trace mineralized salt with a lower level of the trace minerals.

[b]This gives the option as to whether vitamin E or vitamin K is to be included or left out in the vitamin premix used.

[c]These levels should be adjusted to provide 16% protein, 0.8% calcium, and 0.65% phosphorus if the feeds used vary from expected analytical values.

high-level performance. The only difference is that about 15% less concentrate feed is fed to the long yearling. This automatically lowers the level of protein, calcium, phosphorus, and other nutrients in the total diet.

The long yearlings not being developed for high-level performance can be fed the concentrate diet shown in Table 16.5. In some cases, when excellent quality forage is available, some long yearlings not being developed for high-level performance are fed largely on forage (pasture and/or hay). This is especially the case when the horses are used for pleasure riding and other activities which are not very demanding on the animal. The Texas Station (16) fed yearling horses an exclusive forage diet with satisfactory results, but indicated that the activity and training schedules of the yearlings should be considered before doing so. If one depends heavily on forage (hay and/or pasture) it should be of good quality. Moreover, whether some concentrate feeding is used will depend on how well the horses look and how well they respond to the forage program. Horses fed entirely on forage should be self-fed a mineral supplement adequate in all the essential minerals required.

IV. USE OF FAT AND CORN IN DIETS

In all the diets shown for weanling, yearling, and long yearling horses there may be situations, with high-level performance horses, where a diet of higher energy density may be desirable.

The addition of 5–10% fat to the concentrate diet may be considered. The added fat increases the caloric density of the diet and may allow a reduction in total feed intake. A lowered feed intake avoids overloading the digestive tract which minimizes digestive disturbances which might otherwise occur. In some cases, horses involved in high-intensity work (anaerobic) may need a higher level of grain, such as corn, to supply more starch and simpler sugars. How much grain will satisfy this need is not known. Corn might replace 5–10% or more of oats in the concentrate feed in the diet. But, excess grain is to be avoided, however, since it may cause digestive disturbances due to carbohydrate overloading.

The level of added fat and corn to use will depend on (1) it having a beneficial effect and (2) the success derived from its use in the diet.

See Chapter 22 on energy physiology for more detail on the use of added fat and higher levels of corn in the diet, and the diet adjustments required when this is done.

V. CONCLUSIONS

Early weaning of foals requires an excellent quality diet and superior managerial ability. Foals which have been given a creep feed and are already used to consuming concentrates are better prepared for the shock of weaning. Preliminary studies indicate that light horses have about 45, 66, and 88% of mature weight at 6, 12, and 18 months of age. They also have about 83, 91, and 95% of their mature height at the withers at 6, 12, and 18 months of age. Example concentrate diets are given for weanlings, yearlings, and long yearling horses being developed for (1) high-level performance or (2) pleasure riding and other activities which are not very demanding on the animal. These diets are only a guide and many other concentrate diets can be used. The need for proper exercise is stressed for developing young horses. But, excessive forced exercise may be harmful to the young horse's skeletal system which is still quite immature.

REFERENCES

1. Ott, E. A., and R. L. Asquith. *J. Anim. Sci.* **62,** 290 (1986).
2. Hintz, H. F., J. P. Baker, R. M. Jordon, E. A. Ott, G. D. Potter, and L. M. Slade. "Nutrient Requirements of the Horse." NAS-NRC, Washington, D.C., 1978.

3. Hintz, H. F., R. L. Hintz, and L. D. Van Vleck. *J. Anim. Sci.* **48,** 480 (1979).

4. Cunningham, K., and S. H. Fowler. *La., Agric. Exp. Stn., Bull.* **546** (1961).

5. Reed, R. R., and N. K. Dunn. *Proc. Equine Nutr. Physiol. Symp., 5th,* p. 99 (1977).

6. Budzynski, M., E. Sasimowski, and R. Tyszkowski, *Rocz. Nauk Roln., Ser. B* **93,** 21 (1971).

7. Jordon, R. M., *Proc. Equine Nutr. Physiol. Symp., 5th,* p. 63 (1977).

8. Trowbridge, E. A., and D. W. Chittenden, *Mo., Agric. Exp. Stn., Bull.* **316** (1932).

9. Harper, M. W., *Bull.—N. Y., Agric. Exp. Stn. (Ithaca)* **403** (1921).

10. Heird, J. C., *Proc. Equine Nutr. Physiol. Symp., 3rd,* p. 81 (1973).

11. Ott, E. A., R. L. Asquith, and J. P. Feaster. *J. Anim. Sci.* **53,** 1496 (1981).

12. Householder, D. D., and G. D. Potter. *Tex. A & M Univ. Horse Feed. Nutr. Leaf.* (1988).

13. Squires, E. L., G. E. Todter, W. E. Berndtson, and B. W. Pickett. *J. Anim. Sci.* **54,** 576 (1982).

14. Squires, E. L., J. M. Maher, and J. L. Voss. *Proc. Equine Nutr. Physiol. Symp., 8th,* p. 279 (1983).

15. Ott, E. A., R. L. Asquith, and J. P. Feaster. *Fla., Agric. Exp. Stn.* **AL-1980-12** (1980).

16. Rouquette, F. M., Jr., G. W. Webb, and G. D. Potter. *Proc. Equine Nutr. Physiol. Symp., 9th,* p. 14 (1985).

17

Feeding the High-Level Performance Horse

Unfortunately, the difference in nutrient needs for the horse used for pleasure riding and other less demanding activities versus the horse being developed for high-level performance at an early age is not known. The person developing a racehorse wants to make sure that all vitamins and minerals are provided in the diet, even though there is a lack of research information to indicate whether they are all needed. They also use a higher level and a higher quality of protein to make sure that the horse obtains the amount necessary. Unfortunately, not much can be done to alter this until research information is obtained to guide the development properly of both the high-level performance horse and the horse used for pleasure riding and less demanding activities. Horse owners should be cautioned, however, that there is a limit to the level of vitamins, minerals, protein, and other nutrients that can be added to horse diets. Too high a level of nutrients can cause harmful effects or imbalances in the diet. Therefore, one can add nutrients, even though they may not be needed, but this must be done in moderation and with expert advice to prevent problems that could occur.

I. NUTRITION AND THE HIGH-LEVEL PERFORMANCE HORSE

There is much to learn about the nutrition of the racing and high-level performance horse. Most of the present nutritional requirement levels are based on studies with horses kept in stalls or under experimental conditions, which are not similar to those of the performance horse. This is not meant as a criticism of the studies. Rather, it is mentioned to point out that until nutritional requirement studies also involve the horse in training and performance the nutritional needs of horses will not be thoroughly known. The stress of heavy exercise and the weight of a rider on the horse undoubtedly increase the need for certain nutrients over that of a horse kept idle in a stall. Oxygen need and tissue repair are increased by muscular activity. This, in turn, increases certain nutrient needs. Therefore, it is hoped that studies in the future will involve the stress of exercise and the weight of a rider in determining nutritional needs. These nutritional studies should involve the horse in training and performance.

The heavy breakdown of feet and legs that occurs in horses indicates the need

Fig. 17.1 An excellent program of feeding and management is needed to develop a distinguished horse such as Secretariat shown here at Claiborne Farms, Paris, Kentucky. (Courtesy of John P. Baker, University of Kentucky.)

for research information to decrease these losses. It is estimated that only 50% of the Thoroughbred horses being trained reach the track for racing and only 20% of those remain sound through their first year of racing. This fact should indicate that there is a great deal to learn on how to feed a horse in order to minimize and, if possible, eliminate these costly losses. Many people have seen on national television a top horse break a leg while racing. In some cases, the break is so bad the horse is destroyed shortly afterward. This is an example of a need for considerably more research information to prevent such occurrences in the future. By contrast, the human athlete does not have the kind of breakdowns that occur with the performing or racing horse. Occasionally, an injury does occur to a human athlete, but it is a rare occurrence when compared to the horse. Much of the difference is caused by the horse being developed to race at a much earlier age than the human athlete. However, the practice is going to continue, and research is needed to determine the effect of training and racing at a young age (both in years of age and early physiological development) on nutritional requirements. Until this is known, the horse will be performing against great odds.

Professional horseowners find that in order to be successful, they must base many of their decisions on personal observation or practices that others have found to be successful—there is little scientific information available to guide them. Fortunately, more research on the horse is underway. More universities are developing horse research centers, and more horse scientists are being trained. However, it will be many years before the nutrition of the horse will be thoroughly understood. For example, research with other farm animals has been underway in the United States for over 70 years, but, even today, there are many areas in which very little nutritional knowledge is available for beef, dairy, swine, and sheep. During the period when the greatest advances were being made in the nutrition of other farm animals, there was an unfortunate widespread belief that horses had lost their usefulness. This resulted, in part, from the continuous replacement of horses by the mechanization of armies, transportation, and farms. The horse then increased in popularity as a sport and pleasure animal. The light horse population has been increasing rapidly and now there are 6–9 million horses in the United States depending on whose estimate one uses. The light horse industry is thus a big and important phase of the livestock industry. They can profit a great deal from studies which would enable them to develop a racing or performance horse without incurring the tremendous losses they have each year.

Fig. 17.2 One of the great race horses of all time, Secretariat. Secretariat is shown in action. (Courtesy of Professor Norman K. Dunn, Cal Poly University, Pomona.)

II. TREATMENT OF A HORSE LIKE AN ATHLETE

A horse being trained to race and perform should essentially be treated like a human athlete. Both need to be in top condition in order to compete effectively and win. The human has the advantage of training and competing at an older age whereas the horse is competing at a comparatively young age. If a human has a life span of 77 years and the horse a life span of 22 years, a 2-year-old horse is comparable to a 7-year-old human. Most horses start training at about 1 year of age, which is comparable to a boy starting to train at about $3\frac{1}{2}$ years of age. Therefore the horse is in training and competition at a very young age compared to the human. This means a great deal of stress is being placed on the legs, body, and skeletal system of the young horse which is still quite immature. This accounts for some of the breakdowns that occur in the feet and legs of many horses. They are being trained at such a young age physiologically that their legs and bodies are not developed enough to take the stress of training and performance. Studies in Florida (1) showed that maximum bone strength does not occur in horses until they are 4–7 years of age. Another report (2) showed that maturity in the metacarpus in the racehorse is reached at approximately 59 months of age. By contrast, human athletes, who start training at an older age, compete effectively year after year until they reach an age where a younger person is more effective.

The United States races more horses at 2 years of age than any other country in the world. Moreover, all horses have a birthday on January 1. Therefore, many horses are classified as 2 year olds before they are actually 24 months of age. So we are asking horses to become top athletes at a very young age. To accomplish this requires a very well-balanced diet as well as excellent management. The feeding and nutrition of the young horse being trained for racing is more complicated than feeding other classes of animals. This is due to the stress and strain to which the horses are put and the requirement for complete soundness of the animal. Unfortunately, there is still very little known about the optimum nutritional needs of the horse. One needs to use information obtained with other animals and with the human athlete, as well as the experience of top horsemen, to help formulate feeding and nutritional programs for the horse. These can be used until actual research information is obtained which may improve the program followed. In a sense, the racehorse is an athlete whose nutritional requirements are very high and exacting, but the requirements are poorly known. It is, therefore, not surprising that only a small percentage of Thoroughbred horses that start training are still sound and racing after 1 year of competition.

To make matters more complicated, most racehorses are individual prima donnas when it comes to eating. They vary in their likes and dislikes for certain feeds, time of eating, level of feed intake, proportion of forage to concentrates, and response to different caretakers. They need to be catered to and fed as individuals, which is true, to a similar extent, with human athletes. The big

Fig. 17.3   A statue of Man O' War at Kentucky State Horse Park, Lexington, Kentucky. This distinguished horse required the utmost in care, feeding, and management to win and contribute as he did. (Courtesy of John P. Baker, University of Kentucky.)

difference is that humans can talk and make their needs known, whereas it takes much longer to discover the needs, likes, and dislikes of the horse. This requires a patient and knowledgeable trainer to develop the horse to its maximum capability. Competent trainers are scarce, and, hence, many temperamental horses never develop their full potential in performance.

It is interesting that human athletes have continually reduced the amount of time required to run various races. The time required to run the Kentucky Derby has been decreased slightly but to a much smaller extent. Therefore, not as much progress has been made with the horse as compared to the human athlete. How much this might be due to racing at an early age and how much to nutrition or other factors is not known, but it is apparent that more progress has been made in improving the performance of human athletes than the horse athlete.

III. INCREASE ENERGY AVAILABILITY

The information shown in Chapter 9 (Table 9.1) shows that the energy needed for a strenuous effort, such as racing at full speed, is 70 times greater than that required for walking. This tremendous difference in energy requirement indicates the need for understanding how to provide energy for maximum racing speed and endurance.

The racehorse is trained to compete at its maximum capability during an approximate 1–3-minute racing period. This means that maximum availability and utilization of energy must occur during this short time. Diet must be manipulated to maximize the amount of energy available in the muscles during exercise or racing. More needs to be known on how dietary factors can be used to increase oxygen-carrying capacity and availability in the body during a race. A lack of available energy and oxygen may be the cause for many horses giving out toward the end of a race. There is no doubt that proper nutrition is needed for a race horse to perform at its maximum potential (3).

New Jersey scientists (4,5) showed that muscle glycogen and free fatty acids play a dominant role in supplying energy to the muscles for work in both unconditioned and conditioned horses. The conditioned horse adapts itself to utilize fat, in addition to glycogen, to meet the increased energy requirements of exercise and training. The unconditioned horse, however, is not able to oxidize or use fat as efficiently as the conditioned horse. It uses, instead, more muscle glycogen for exercise. This is similar to the human. The untrained human uses one-half as much free fatty acids for energy as the human who is trained and in good condition. The New Jersey study indicates that fat is made available to the muscle as free fatty acids, which can be used as a form of energy for exercise or running. It may be that the horse preferentially uses this source of energy during heavy exercise (6). It is important to know whether adding fat to the diet will increase energy availability and maximize endurance for the racing and heavily exercised horse. If it does, how much fat should be used, when should it be fed, what kind of fat is best, and many other questions related to this problem need to be answered. A good start has been made in this area by a number of scientists (4–11). A Texas A & M report (12) showed that adding 5 and 10% fat to the diet of the exercising horse had a sparing effect on muscle glycogen reserves. This glycogen reserve could later be mobilized to help defer fatigue in the exercising horse. See Chapter 22 for more detail on the value of adding fat to the diet.

The effect of fat on energy availability needs much more study. Studies with other species showed that the addition of fat improved the performance of pigeons flying more than 200 miles (13). It also benefited sled dogs racing for 12 hours (14).

As one adds fat to diets, the level of protein used needs to be increased since amino acid needs are directly related to the calorie level in the diet. Vitamin and mineral levels may also need to be altered as the fat level in the diet is increased. Studies in Utah (10) showed that a combination of 12% vegetable oil plus 1000 IU of vitamin E gave the most desirable response in blood packed cell volume, hemoglobin, and glucose after exercise in endurance horses. Vitamin E appears to be beneficial with the unsaturated vegetable oil. Dr. R. F. Sewell at the

Fig. 17.4 Walkers are becoming more widely utilized to "cool out" or "warm up" horses as labor availability and costs continue to be major problems. This four-horse walker at Cal Poly University, Pomona, California features safety factors such as fenced area, safety releases and high and strong cross arms. (Courtesy of Professor Norman K. Dunn, Cal Poly University, Pomona.)

University of Georgia showed that adding 10% fat to the diet of the growing pig increased the requirement for pantothenic acid by 50%.

IV. EFFECT OF EXERCISE ON CALCIUM NEEDS

Cornell scientists (3) studied the effect of exercise on calcium balance and the turnover of calcium in four yearling Standardbred horses in a series of experiments in which the horses were: (1) at rest; (2) worked 10 miles in 2 hours each day at a trot; (3) worked 6 miles in 1 hour each day at a trot; and (4) at rest. Each study period was 1 week preceded by 3 weeks of equilibration to each activity. Two of the horses were fed a diet containing 0.4% calcium and two were fed 0.6% calcium. The calcium to phosphorus ratio in each diet was 1 : 1.

The results showed that exercise did not appear to change the proportion of calcium intake that was absorbed or retained. However, urinary calcium decreased markedly during the exercise periods. Retention of calcium by the body increased during the exercise periods. The increase was due largely to decreased urinary excretion of calcium. The excretion of calcium in the feces was unaffected by exercise. The rate of calcium deposition in bone increased 15–20% above the level of the horses resting when the horses were exercised 10 miles per day. The rate of calcium deposition was not increased, however, when the horses were exercised only 6 miles per day. The results of this study indicate that bone tissue is more active in terms of calcium turnover during exercise than during rest. The Cornell scientists state that the experiments were of short duration and left unanswered the question of the effect of exercise on the calcium requirement of the horse. See Chapter 22 for more detail on the effect of exercise on the high-level performance horse.

More studies are needed to determine the effect of exercise, degree of exercise, frequency of exercise, age of the horse when exercise is initiated, effect of the riders (and their weight), effect of temperature and humidity, and many other factors on calcium as well as other mineral and nutrient requirements of the horse. Bone density, breaking strength of the bone, and other criteria of bone

Fig. 17.5 Swimming pools are becoming more widely utilized on horse farms as a method of providing extensive exercise with a minimum amount of concussion. Swimming is also an excellent form of physical therapy for horses with leg, muscle, and joint complications. This pool is located at the J Bar D Training Center, Chino, California. (Courtesy of Professor Norman K. Dunn, Cal Poly University, Pomona.)

integrity and strength are also needed. Until this, and other information is obtained, there will continue to be a lack of information on bone problems with the horse. Since strong bone is so important to the value and usefulness of a horse, there is no doubt that proper formation and maintenance is one of the highest priority items in horse research. A Kentucky study (14a) showed that exercise training improved the stress bearing characteristics of the third metacarpal without affecting the quantity of body growth.

V. EFFECT OF HARD AND PROLONGED EXERCISE

As a horse runs or exercises heavily, excessive sweating occurs. Any minerals that are lost in the sweat will increase their need in the diet. Sweat losses are influenced by temperature and humidity. With high temperatures and high humidity, there is more sweat loss than with low temperatures and low humidity. The loss of sodium and chlorine appears to be of great concern since sweat contains 0.7% salt. Horses that sweat considerably become exhausted and fatigued. They react similarly to men who perspire excessively during strenuous activity in hot climates. They need to take extra salt tablets to avoid exhaustion, fatigue, and collapse. This is why salt should always be self-fed to horses even

Fig. 17.6 Horses being led and started in training in Ocala, Florida on the farm of the late Carl Rose, a pioneer in the development of the Thoroughbred industry in Florida. (Courtesy of T. J. Cunha, University of Florida and Cal Poly University, Pomona.)

though it is added to the diet. There may be a loss of 50–60 g of salt in the sweat and 35 g in the urine of horses doing moderate work. This amounts to about 0.2 lb of salt. A horse working really hard would lose even more salt.

Dr. Gary P. Carlson of the University of California reported on the effect of exercise and heat stress on endurance horses in trail rides, which may cover distances of 25–100 miles in a given period of time (14b). If the temperature is high, many of these horses develop problems that include excessive fatigue, muscle spasms and cramps, dehydration, and exhaustion. Body fluid and electrolyte losses through sweating play a role in the development of many of these conditions. The body attempts to maintain a balance against fluid and electrolyte loss so that its organs can continue to function. Excess heat is generated by the horse during prolonged exercise. This heat must be dissipated by sweating. If the exercise or running is excessive and prolonged, there may be a significant electrolyte loss (salts dissolved in the body fluids). An imbalance of electrolytes can adversely affect the thirst mechanism and failure to replace water loss adequately may occur. There may also be a failure to eat. Many riders taking part in long-distance endurance trail rides give their horses electrolytes in the feed or water.

In trail endurance rides, with high environmental temperatures, Dr. Carlson found there was a major decrease in chloride and calcium despite dehydration. Once these horses had rehydrated themselves, there was only a slight but significant decrease in sodium and a rather large decrease in potassium. This reflects the relative losses from the body through sweat and epinephrine mechanisms, as well as the body's ability to come up with stores from body reservoirs. Some of the horses that become stressed during endurance rides develop a condition known as "thumps," spasm of the diaphragm related to the heart beat and electrolyte alterations, particularly decreases in calcium and potassium.

Following endurance rides, Dr. Carlson examined the horses' blood, which revealed an elevated pH, or tendency to develop alkalosis, despite the fact that major physical problems were not evident. Respiration rates were also quite high. Following the 30-minute resting period, the pH declined to a slight degree, and respiratory rates soon returned to normal levels. After the 30-minute rest, a mild alkalosis was still present; the levels of carbon dioxide and bicarbonate were also higher than normal. Almost all horses developed alkalosis during the ride even under mild climate conditions (70–80°F range).

There is still much to learn about horses that become distressed during endurance rides and require fluid and electrolyte supplementation. Horse owners should determine whether electrolyte supplementation is necessary on the basis of the heat stress and fluid losses. Some riders have reported problems with excess electrolyte intake. Therefore, electrolyte supplementation should be undertaken with the advice of someone who is knowledgeable in this area.

It is hoped that more research will be conducted on endurance trail rides. They provide an excellent opportunity to study the physiology and metabolism of the

Fig. 17.7   Horses being ridden and exercised in the training process in Ocala, Florida area. (Courtesy of T. J. Cunha, University of Florida and Cal Poly University, Pomona.)

horse under considerable stress. From these studies may come information that will unlock secrets of stamina, endurance, and speed in racing. It's interesting to note that the winning time in the Kentucky Derby in 1896 was 2.07 $\frac{3}{5}$ minutes, whereas in 1968 it was still over 2 minutes (2.02 $\frac{1}{5}$ minutes). On the other hand, the human athlete has broken the 4-minute mile, the 10-second 100-yard dash, and many other records. The horse athlete has obviously not made the progress the human athlete has. It is apparent, therefore, that much needs to be learned before the horse can perform to its full capability.

VI. AVOIDANCE OF EXCESS FAT

Horses being developed for racing or high-level performance must not be allowed to become fat. They need to be kept in a trim, thrifty condition. Some call them "racehorse thin" or "hard and lean" by the time they are ready for racing. They are beautifully muscled and show no fat. Some show people may think the horses look starved.

On the other hand, one needs to make sure the horse does not get too thin. There is a difference between a thrifty, lean horse without excess fat and a horse that is too thin because of a lack of enough to eat. In developing a race or high-level performance horse, it is important that they receive the protein, minerals, vitamins, energy, and other nutrients needed to develop the body of a well-trained athlete. They must also obtain adequate nutrition to perform to the maximum of their inherited potential. An Irish study (15) indicates that heritability of performance in Thoroughbred horses is 35%. Other studies on the effect of

Fig. 17.8 Horses in training at a track in Ocala, Florida area. Elmer Huebeck (near dog) of Ocala is supervising the training. (Courtesy of T. J. Cunha, University of Florida and Cal Poly University, Pomona.)

heritability on performance have been made with Thoroughbreds (16), Standardbreds (16, 17), and Quarter Horses (18). An Oklahoma study (19) showed that age and sex of the animal and age of the dam affect Thoroughbred racing performance; the age of sire and season of birth may also affect racing performance. Thus, one should place some emphasis on breeding in selecting for potential winners in racing.

VII. SUGGESTED CONCENTRATE DIETS

There are many concentrate diets that can be fed to develop high-level performance or race horses. The diet shown in Table 17.1 is one example. It can be modified by using other feeds or other levels of nutrients.

The diet shown in Table 17.1 should be fed to horses once they start training. It should be continued during training and can also be used during racing or performance. The amount to feed will vary with the condition of the horse, how it responds to the diet, and the amount and quality of the hay and/or pasture used. The concentrate diet (in Table 17.1) should be fed at a level of 40–50% of the total diet intake. If this seems too much, it can be decreased during periods of

TABLE 17.1

**Sample Concentrate Diet for Race or High-Level
Performance Horses**

Feed	Percentage in diet[a]
Oats (heavy), steam rolled	30.00
Corn, ground or rolled	10.75
Barley, steam rolled	9.50
Wheat bran	7.00
Alfalfa meal, dehydrated (20% protein)	8.00
Soybean meal (expeller, with 4–5% fat)	23.00
Blackstrap molasses	7.00
Dicalcium phosphate (or other calcium and phosphorus source)	2.00
Limestone, ground	0.75
Salt, trace mineralized, same as salt premix A (Table 15.10)	1.00
Vitamin supplement, same as vitamin premix A (Table 15.9)	1.00

[a]The level of feeds in the diet should be adjusted to provide 18% protein, 0.95% calcium, and 0.85% phosphorus if the feeds used vary from expected analytical values.

light training, racing, or performance. The level of concentrate feeding can likewise be increased during heavy training, racing, or performance. The condition of the horse can serve as a guide as to whether too much or too little concentrates are being fed in relation to forage intake (hay and/or pasture).

Only high-quality hay and/or pasture should be used. The hay should be green, leafy, and free from dust, molds, and weeds. The pasture should be short, green, and free of weeds, holes, and other obstacles that might injure the horse.

The horse should always have access to a mineral box in case they need more minerals than are contained in the diet. Clean, fresh water should always be available to horses. The feeds used should be fresh and stored in a cool, dry room with good air circulation around the feed sacks. Feed that gets old, musty, or moldy, or that develops rancidity should be avoided.

A. Fat Addition to Concentrate Diet

Many horse trainers prefer to add 5–10% fat to the concentrate diet for high-level performance horses. Whether higher levels of fat might give additional benefits to the high-level performing horse is not definitely known. These horses require extra energy and the addition of fat increases the energy density of the diet. In turn, this may allow a reduction in total feed intake required to meet total

Fig. 17.9 Horses being trained to enter and come out of the starting gate in Ocala, Florida area. (Courtesy of T. J. Cunha, University of Florida and Cal Poly University, Pomona.)

energy requirements. This is important since it is difficult to get some horses to consume enough total feed during intense training or performance. Moreover, it minimizes the possibility of colic, founder, and other digestive disturbances occurring from too heavy concentrate consumption during intense activity.

A Texas A & M study (12) showed that the addition of 5–10/% fat to the concentrate diet increased the muscle glycogen of exercising horses. Muscle glycogen is a stored form of energy which can be mobilized for muscular work and thus may delay fatigue in the exercising horse. The possibility exists, therefore, that adding fat to the diet is one means of helping defer fatigue in many high-level performance horses which fade out in the latter stages of performance. This is an area requiring further research since sustained high-level performance in the horse has not improved to the same level as it has in the human athlete. More detail on fat and carbohydrates and their role in the exercising horse is given in Chapters 8, 9, and 22.

If fat is used it needs to be a high-quality product and protected against rancidity by a proper antioxidant. Rancid fat can destroy many nutrients as well as cause digestive disturbances and decrease the palatability of the diet.

As the fat level is increased in the diet, the level of protein should be increased since the protein level required is related to the energy or calories in the diet. The calcium and phosphorus level in the concentrate diet also needs to be adjusted. Therefore, the concentrate diet shown in Table 17.1 could have 5 or 10% fat added to it in place of grain. The protein, calcium, and phosphorus levels should be adjusted to the same levels shown in Table 17.1. The vitamins and trace mineral levels shown in Table 17.1 should still be adequate for a 5 or 10% fat addition to the diet.

Several surveys have indicated that horses at the track need 35–50 Mcal of digestible energy daily whereas the estimates based on 1978 NRC values were 25–30 Mcal (20). Therefore, the addition of fat to the diet of high-level performance horses is one method of increasing their energy level intake.

B. Hay Protein Level with Concentrate Diet

The protein level in the concentrate diet shown in Table 17.1 is 18%. This would allow the use of grass hay which may vary from 7 to 10% protein. This would provide a protein level of about 12–14% in the total diet. Many horse trainers prefer to use a grass hay during periods of intense training or high-level performance. The 18% protein level in the concentrate diet would ensure an adequate level of total protein intake when grass hays are fed.

If the trainer is using some alfalfa–grass hay or feeds alfalfa at one feeding and grass hay at the next, then the protein level in the concentrate diet shown in Table 17.1 could be decreased to about 16% depending on the level of alfalfa hay fed. The objective is to have a protein level in the total diet of 12–14%. One report from the University of Kentucky (21) states that the protein needs of the 2 year old in training can best be met by feeding a 13–14% protein concentrate plus good-quality hay.

It is best to ensure an adequate level of protein rather than run the risk of a lack of protein. A Cornell study (22) involved feeding young foals a level of 9, 14, and 20% protein in the diet. Growth, feed intake, and feed utilization were significantly depressed at the 9% protein level. The foals fed the 20% protein diet, which is considerably higher than the 14% NRC-recommended protein level, were neither helped nor harmed by the higher protein level. The high level of protein caused no skeletal problems. Other scientists have fed high levels of protein with no harmful effects (see Chapter 22).

C. Grain in Concentrate Diet

The energy level in the concentrate diet shown in Table 17.1 could be increased by replacing some of the oats with corn. The wheat bran and alfalfa meal could also be decreased a little as a means of increasing the energy level of the diet. It is important, however, to keep some bulk in the diet. These changes can be considered if the level of energy required for intense high-level performance is greater than that supplied by the concentrate diet shown in Table 17.1 and the forage being fed.

An Illinois study (23) with Quarter Horses showed that replacing oats with corn resulted in a faster racing time. The corn diet occupied less volume per pound and decreased the gastrointestinal fill which may have affected the horse's racing performance. Excess high-energy grain is to be avoided, however, since it may cause digestive disturbances because of carbohydrate overloading.

The level of corn and fat to use should be guided by the success being obtained. This means that close observation of the horses is important as one increases fat and corn levels in the diet. No more should be used than the high-level performance can utilize properly.

Tables 12.2 and 12.3 give information on the digestible energy, crude protein, crude fiber, calcium, and phosphorus of some commonly used feeds from a University of Kentucky report (21). It can be used as a guide to evaluate fat and grain changes in the concentrate diet shown in Table 17.1.

VIII. CONCLUSIONS

Very little is known about the nutritional requirements of the high-level performance horse. During training and performance considerable stress is placed on the feet, legs, body, and skeletal system of young horses whose bones are still quite immature. This accounts for the many breakdowns which occur during this period. Two research reports indicate that maximum bone strength does not occur in horses until they are 4–7 years old, but they are in training and performance much before this time.

The high-level performance horse needs to be treated like an athlete and its likes and dislikes catered to in feeding, training, housing, and management.

Feeding and training programs are very important in maximizing energy storage and availability for optimum muscular activity and performance. The use of 5–10% fat and a small increase in corn level in the diet to increase energy density, reduce total feed intake, and decrease intestinal tract volume may be beneficial for the high-level performance horse. Recent reports indicate that these horses need a higher energy level than previously thought. They also indicate that some excess protein in the diet is not harmful as previously thought. This makes it possible to ensure adequate protein levels without fear of harmful effects.

REFERENCES

1. El Shorafa, W. M., J. P. Feaster, and E. A. Ott. *J. Anim. Sci.* **49,** 979 (1979).
2. Tomioka, Y., M. Kaneko, M.-A. Ockawa, T. Kanemaru, T. Yoshihara, and R. Wada. *Bull. Equine Res. Inst.* p. 22 (1985).

3. Schryver, H. F., H. F. Hintz, and J. E. Lowe. *Am. J. Vet. Res.* **39,** 245 (1978).
4. Goodman, H. M., and G. W. Vander Noot. *J. Anim. Sci.* **33,** 319 (1971).
5. Goodman, H. M., G. W. Vander Noot, J. R. Trout, and R. L. Squibb. *J. Anim. Sci.* **37,** 1 (1973).
6. Slade, L. M., Utah State University, Logan (personal communication), 1979.
7. Hambleton, P. L., L. M. Slade, D. W. Hamar, E. W. Kienholz, and L. D. Lewis, *Proc. 71st Annu. Meet. Am. Soc. Anim. Sci.* p. 244 (1979).
8. Bowman, V. A., J. P. Fontenot, K. E. Webb, and T. N. Meacham, *Va., Agric. Exp. Stn., Livest. Res. Rep.* **172,** 72 (1977).
9. Hintz, H. F., M. Ross, F. R. Lesser, P. F. Leids, K. K. White, J. E. Lowe, C. E. Short, and H. F. Schryver, *Proc. Cornell Nutr. Conf.* p. 87 (1977).
10. Slade, L. M., Utah Agricultural Experiment Station, Logan (unreported data), 1979.
11. Kane, E., J. P. Baker, and L. S. Bull. *J. Anim. Sci.* **48,** 1379 (1979).
12. Meyers, M. C., G. D. Potter, L. W. Greene, S. F. Crouse, and J. W. Evans. *Proc. Equine Nutr. Physiol. Symp. 10th,* p. 107 (1987).
13. Goodman, H. M., and P. Greminger. *Poult. Sci.* **98,** 2058 (1969).
14. Kronfeld, D. S., E. P. Hammel, C. F. Ramberg, and H. L. Dunlap. *Am. J. Clin. Nutr.* **30,** 419 (1977).
14a. Raub, R. H., S. G. Jackson and J. P. Baker. *J. Anim. Sci.* **67,** 2508 (1989).
14b. Carlson, G. P. Presentation Am. Meet. Morris Anim. Foundation. Denver, Colo. Oct (1974).
15. More O'Ferrall, G. J., and E. P. Cunningham. *Livest. Prod. Sci.* **1,** 87 (1974).
16. Hintz, H. F. *J. Anim. Sci.* **51,** 582 (1980).
17. Tolley, E. A., D. R. Notter, and T. J. Marlowe. *J. Anim. Sci.* **56,** 1294 (1983).
18. Buttram, S. T., D. E. Wilson, and R. L. Willham. *J. Anim. Sci.* **66,** 2808 (1988).
19. Hintz, R. L. *J. Anim. Sci.* **51,** 582 (1980).
20. Hintz, H. F. *Feed Manage.* **37**(2), 15 (1986).
21. Jackson, S. G. *Univ. Ky., Agr. Ext. Serv., Racehorse Symp.* p. 40 (1987).
22. Schryver, H. F., D. W. Meakim, J. E. Lowe, J. Williams, L. V. Soderholm, and H. F. Hintz. *Equine Vet. J.* **19,** 280 (1987).
23. Burke, D. J., W. W. Albert, and P. C. Harrison. *Proc. Equine Nutr. Physiol. Symp., 7th,* p. 197 (1981).

mmm pg 346-347, tp 172

18

Feeding the Mare and Stallion

The breeding mare and stallion should be kept in a thrifty condition. Mares and stallions that become obese may decrease in reproductive efficiency. Recent studies at Texas A & M (1,2) indicate, however, that having mares in a fat condition benefited reproduction.

If the mares or stallions get too thin, from a lack of enough to eat or from nutritional deficiencies, it interferes with reproduction. More studies with other animals are available than with the horse. Presumably, the research findings with other animals may have some application to the horse. It appears that level of feeding should be gauged to the condition of the horse and how it responds to it. It is best to avoid any extreme conditions of fatness or thinness.

I. EFFECT OF CONDITION ON REPRODUCTION

Studies have shown that a deficiency of energy delays puberty in the heifer. If the energy level is particularly low, heifers will stop expressing estrus and ovulation. Severe energy restriction in pony mares results in embryo death (3). Other reports indicate embryo deaths in mares greatly restricted in energy intake. In bulls, a low-energy intake results in a decreased desire for mating and low-quality semen. This information was obtained with beef cattle, but there is some possibility that the same might occur with the horse. Similar findings have been observed on horse farms, but experimental evidence to back it up is not available.

Calving difficulties occur with excessively fat beef heifers (3a). This is due to an extremely large fetus and the accumulation of internal fat. Difficulty can also occur in breeding extremely fat cows. Studies have shown that fat bulls become inactive, are more prone to injuries during mating, are reluctant to mate, and are not in condition to be turned out to pasture without first being gradually let down in condition. Some may become temporarily sterile. Studies have shown that overfeeding dairy heifers during their early growth period has turned potentially good milkers into poor ones. This may be due to structural damage, which occurs in the mammary gland during the early growth period. There are indications that the same may also occur with beef heifers. Not many years ago, sows were fed 6–8 lb of feed daily during the gestation period. Studies showed this was too high a level of feed intake (3b). Sows are now fed an average of 4–4.5 lb of feed

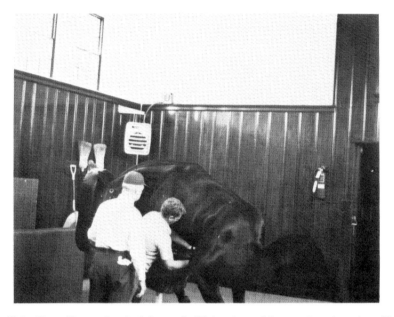

Fig. 18.1 The stallion needs to be fed properly if it is to be used for natural service or by artificial insemination as is shown here. (Courtesy of John P. Baker, University of Kentucky.)

during the gestation period. This has actually increased reproduction rate. More-over, the sows fed less feed farrowed their pigs in a shorter time than the overfat sow. Overfat sows can take 24 hours to farrow their pigs, whereas some sows in a trim condition can do it in 2–4 hours. This information indicates that an overfat condition in cattle and swine is detrimental. Texas A & M research (1) has demonstrated that mares should be fed to a fatter condition than previously recommended to enhance reproductive performance. Another Texas A & M study (2) demonstrated that, in the multiparous mare, foaling was not impaired by an extremely high degree of fatness produced by overfeeding during gesta-tion. Evidently horses can tolerate a higher level of fatness than can cattle and swine. It appears prudent, however, to avoid too much obesity since one report (4) indicated that fat mares had a lower conception rate than mares in good condition.

It should be stressed that if one lowers total feed intake to keep horses in a thrifty condition, they still need their daily requirements of protein, minerals, and vitamins. Therefore, if the total feed intake is decreased, one also needs to increase the level of protein, minerals, and vitamins in the feed so that it will supply the animal's daily needs for them. This is very important and has been done with sow feeds. At first, swine producers decreased the total feed intake but

Fig. 18.2 This horse is not getting enough to eat. Note poor body condition and rough, hair coat. Horses in this poor condition are not productive. (Courtesy of Larry Slade, Utah State University.)

did not increase the level of minerals, vitamins, and protein in the diet. As a result they encountered certain nutrient deficiencies (3b). Therefore, horse producers need to make sure the feed they use is adequate in all nutrients when fed at an energy level to keep the horses in a trim, thrifty condition but not too fat or too thin.

A Texas A & M study (5) indicated that mares in poor body condition apparently have impaired reproductive efficiency even when energy requirements for lactation are met. Increasing the dietary energy fed to these mares increased reproductive performance. They also stated that mares foaling in a high level of condition are not impaired reproductively and can utilize body energy for reproduction and efficient foal growth when lactation energy requirements are not met. Workers at Texas A & M (6) developed a condition score system (see Table 18.11) which can be easily utilized by the horse producer to monitor energy stores within the animal.

A Minnesota study (7) showed that no reduction occurred in conception among pony mares fed to lose 20% of body weight during gestation but fed to gain weight during lactation. Another study (8) showed that energy restriction during the last 90 days of gestation had no effect on foal birth weight and did not affect reproductive efficiency.

II. FACTORS AFFECTING REPRODUCTION RATE

The practice of having January 1 as the birthday for horses used for racing and showing places a handicap in getting mares to foal near this date. A recent study (9) at Virginia Polytechnic Institute (VPI) showed that January and February were the poorest months for breeding mares. Conception rate was only 28.6 and 27.8%, respectively, for these 2 months. It increased to 45.4% in March. The following percentage conception rates occurred in subsequent months: April, 51.8%; May, 57.2%; June, 51.9%; July, 59.3%; August, 45.0%; and September, 57.1%. The best breeding conception rate occurred from March onward. For those farms where breeding continued through the summer months there was no decline in fertility. However, the number of mares bred in August and September were small. The VPI study is similar to a Cornell study (10) at Winfield Farm in Ontario, Canada, with Thoroughbreds where the percentage of foals born during the various months was as follows: January, 1.5%; February, 7.3%; March, 22.8%; April, 31.4%; May, 30.3%; and June, 6.6%.

If the birth date for horses could be changed to the first of March, April, or May, there would be a greater increase in conception rate of mares. This would more nearly coincide with the VPI survey data and would allow the breeding of mares in the spring when pasture becomes available. Most horse people observe that their mares start conceiving better when they are placed on a lush, green pasture. Whether something in the pasture causes the increase in conception rate or whether some other factors are involved is not known. It is possible that some environmental factor (length of day, sun's rays, temperature, humidity, etc.) could be involved. The pasture could be supplying some nutrients that increase conception rate. Two of the Quarter Horse farms in the VPI study reported a conception rate of 94% with one stallion at one farm and 100% with three stallions on another farm when pasture breeding was used. This means that above 90% conception can be occasionally obtained.

It is difficult to get mares bred to foal near the January 1 deadline. Since mares have an average gestation period of about 11 months. It is apparent that only a small percentage will foal in January. Changing the birth date to March, April, or May 1 would eliminate many of the problems breeders encounter in attempting to breed mares for January foaling. However, getting the birth date changed would be a difficult job. It would cause problems in racing and in registration. It would also involve breaking tradition, which is not easy to do. Most people are in agreement with the change in date but cite the difficult task involved even if it had a chance for success. If this is true, then the best solution is research to find methods of increasing conception rate during the early part of the year. The survey by VPI was made in five Thoroughbred, five Quarter Horse, and one each of Standardbred, Appoloosa, Arabian, and Pony farms. It involved records on 1876 mare years. It is interesting to note that conception, foaling, and weaning percentages were 80.1, 73.8, and 70.8. This means that out of 80.1 mares that conceived, 6.3 mares did not foal and 3 mares lost their foals between birth and weaning. This is an 11.6% loss of foals between conception and weaning. The VPI scientists also found that there was an average of 1.71 estrus periods per mare per year, 2.75 services per mare yearly, and 1.61 services per period in this study. The conception rate of young mares fluctuated until age 7, reached a peak of 89.6% at age 9, remained on a plateau above 70% until age 15, and declined thereafter. Percentage conception rate from 16 to 19 years of age was 63.6% and decreased to 50.0% for mares between 20 and 24 years of age. Only 0.6% of the mares did not show heat or estrus in this study. The incidence of twinning was also low. Only 11 of the 462 Thoroughbred mare conceptions (2.4%) were twins and only one twin conception (0.16%) was noted among the 624 non-Thoroughbred mare conceptions. It is interesting that only 2 of the 12 mares that conceived twins carried them to term or foaling time. In this study the abortion rate was 4.8, 7.2, 0.0, 4.2, and 6.2%, respectively, for lactating mares, barren mares, maiden mares, mares that had previously aborted, and mares that had lost foals. It is interesting that no abortions occurred with the maiden mares in this study.

This study provides some very interesting data and additional surveys are needed to get background data on what is occurring on horse farms with good management conditions. It is evident that the foal crop will average from 65 to 80% on the better managed farms. Previous surveys indicate that the national average foal crop is somewhere between 50 and 65%, which is much too low. Essentially it means that the majority of the mares foal one year and skip the next. It is also apparent that a few select breeders can sometimes average over a 90% foal crop. These figures indicate the great variation occurring in the foal crop percentage in the United States. It stresses the need for research, which is long overdue, to start finding the answers to this very serious and costly problem confronting the horse producer.

Early pregnancy loss which occurs between fertilization and 150 days postovulation has been reported to be approximately 11% (11) and 13% (12). A Texas A & M study (13) showed that pregnant mares in an energy-deprived situation had a higher incidence of abortion in the first 90 days of gestation than mares that were kept in energy equilibrium at constant body weight.

Mares bred at the first postpartum estrus usually have a lower conception rate and a higher incidence of abortion. Studies are needed to determine more effective methods of breeding mares early after parturition since early foaling dates are quite important in developing high-level performance horses.

A Texas A & M study (14) with Quarter Horse and Thoroughbred mares showed that the average time from foaling to first ovulation, length of estrus, and time from ovulation to end of estrus were 12, 5.5, and 1.65 days, respectively. Conception rates on "foal heat" and in the first three cycles were 85.7% and 92.5%, respectively.

A Texas A & M report (15) indicated that twin conceptions occur approximately 5% of the time in Thoroughbred mares and about half that in other breeds of horses. Since less than 2% of births (or stillbirths) are twins, Dr. G. W. Webb calculated that approximately 60% of twin conceptions result in abortion.

III. FEEDING THE MARE DURING GESTATION

The most important period during gestation is the last 90 days. This is the period when growth rate of the embryo is the greatest. About 60–65% of the weight of the fetus is deposited during these last 90 days. The products of conception (foal, fluid, afterbirth, etc.) total 12% of the body weight of mares that weigh less than 990 lb and 10% for mares weighing over 990 lb (16). An 1100-lb mare, for example, will increase in weight by about 110 lb from the products of conception, which include the foal. The foal will probably weigh from 90 to 100 lb or more at birth.

Table 18.1 gives information on the energy needs of the mare during the last 90 days of the gestation period. The 1100-lb mare will consume 16.2 lb of feed (on a 100% dry matter basis) daily, which is 1.47% of its body weight. During the last 90 days of gestation, the mare will gain at the rate of 1.21 lb per day for a total gain of 108.9 lb, which is a large part of the weight of the products of conception.

Table 18.2 gives a suggested concentrate diet that can be used during the gestation period. During early gestation only a few pounds of this concentrate

TABLE 18.1

Feed Intake and Digestible Energy in Feed during last 90 Days of Gestation (16).

Mature mares (lb)	Daily feed[a] per animal (lb)	Daily feed[a] (% of live weight)	Digestible energy per lb of feed (Mcal)	Daily gain (lb)
440	8.1	1.84	1.14	0.59
880	13.7	1.56	1.14	1.17
1100	16.2	1.47	1.14	1.21
1320	18.5	1.40	1.14	1.47

[a]All values are on a 100% dry matter basis.

TABLE 18.2

Sample Concentrate Diet for Gestation

Feed	Percentage in diet[a]
Oats, steam rolled or crimped	30.0
Corn or milo, rolled or crimped	10.0
Barley, steam rolled	13.0
Wheat bran	10.0
Soybean meal, 44% protein	11.0
Linseed meal (expeller, with about 4% fat)	4.0
Alfalfa meal, dehydrated, 17% protein	10.0
Blackstrap molasses	7.0
Dicalcium phosphate (or other calcium and phosphorus source)	2.0[b]
Limestone, ground	0.5[b]
Salt, trace mineralized, premix A or B, (Table 15.10)[c]	1.0[b]
Vitamin supplement, premix A, B, C, or D, (Table 15.9)[d]	1.5

[a]The levels in the concentrate diet should be adjusted to contain 16% protein, 1.0% calcium, and 0.9% phosphorus if the feeds used vary from expected analytical values. The diet has a higher level of calcium and phosphorus since this concentrate mixture will be fed at a lower level compared to forages.

[b]In addition, minerals should be self-fed in case the horse needs more than provided in the total diet and in case concentrate level feeding is low.

[c]Use either trace mineralized salt premix A or B shown in Table 15.10. It would be preferable to use premix A with high-level performance horses.

[d]Use vitamin premix A, B, C, or D shown in Table 15.9. It would be preferable to use vitamin premix A with high-level performance horses. The other vitamin premixes do not contain the B-complex vitamins and give the option as to whether vitamin E or K are to be included or left out.

diet need to be fed with a high-quality hay and/or pasture. During the last 90 days of gestation 25–35% of the diet (see Table 18.5) should consist of the concentrate mixture shown in Table 18.2. This means that 4–6 lb of the concentrate feed should be fed per mare daily. These levels can be increased or decreased depending on the condition of the mare, its response to the feed, and the quality of the hay and/or pasture used. It is recommended that the total diet fed during gestation contains at least 12% protein. This means that the hay and/or pasture used during gestation should have at least 10% and preferably 11–12% protein. The higher level of protein should supply a safety factor for hay and/or pasture, which may have protein of low digestibility depending on the kind of pasture or hay used and its stage of maturity when consumed or harvested.

Research at Texas A & M (17) indicates that the 1978 NRC protein requirements for the pregnant mare are adequate to excessive, but that the NRC protein

Fig. 18.3   One of many types of hay and pellet feeders currently on the market. Feeder is located on outside of paddock fence for safety reasons which should also add to its longevity. Photo taken at Cal Poly University, Pomona. (Courtesy of Professor Norman K. Dunn, Cal Poly University, Pomona.)

requirements for the lactating mare may be marginal to low. The protein level was marginally effective in maintaining milk production in the mares and thereby growth in foals.

An Oregon State study (18) with pony mares (from 2 to 20 years of age) indicated that dietary crude protein differences of 8.6, 11.4, and 17.2% did not affect general reproduction parameters (estrus and ovulation) but did affect progesterone concentrations which in turn may be responsible for the differences in conception rate, which was lowest at the 8.6% protein level and highest at the 11.4% protein level diet.

A VPI study (19) involved feeding Saddlebred mares 100 and 120% of 1978 NRC energy levels during the last 3 months of gestation. The 120% NRC energy level resulted in an advantage in average daily weight gain in the mares during pregnancy and in the foals and mares during 30 days postpartum.

A Minnesota study (20) with pony mares showed that foal weight is related to mare weight and averages 7.7% of the dam's weight. Total foaling weight loss averages 11.6% of the mare's prefoaling weight.

Pregnant and lactating mares require high-quality protein for proper fetal development and high milk production.

During the last 90 days of gestation the mares should be separated from barren mares and other horses. If they are fed a concentrate mixture, the feeding system should allow all mares, especially the timid ones, to eat their fair share of feed. Similar procedures should be followed to ensure that lactating mares are able to consume their feed needs.

IV. FEEDING THE MARE DURING LACTATION

Proper nutrition is very critical since the mare needs to recover from parturition, produce enough milk, and rebreed successfully. It is very important that the mare be fed enough to provide for milk production as well as to maintain her body. If her feed intake is inadequate in energy, vitamins, minerals, protein, and other nutrients, milk production will decrease and so will her ability to breed back while suckling the foal. An inadequate diet may account for much of the alternate year foaling that occurs. The mare fails to conceive while nursing the foal. If mares get too thin because of an inadequate feed intake, chances are good their rebreeding performance will be poor, with delayed postpartum intervals, low conception rates, and increased embryo mortality. It has been suggested that having mares in a gaining state during the breeding season may improve chances of conception (19). To get mares to conceive each year requires both an excellent feeding program and excellent management.

A Cornell study (21) showed that feeding pony mares 80% of the 1978 NRC energy requirements was inadequate for maintenance of body weight during lactation. The mares produced less milk energy as evidenced by the fact that their foals did not grow as fast as other foals fed a higher energy level.

A. Feeding during First 3 Months of Lactation

Table 18.3 gives information on the feed and energy intake required for mares of various mature weights. Table 18.4 compares the feed and energy intake of

TABLE 18.3

Feed Intake and Digestible Energy in Feed during First 3 Months of Lactation (16).

Mature mares (lb)	Daily feed[a] per mare (lb)	Daily feed[a] (% of live weight)	Digestible energy per lb of feed (Mcal)	Milk production per day (lb)
440	11.5	2.61	1.27	17.6
880	18.4	2.09	1.27	26.4
1100	22.2	2.02	1.27	33.0
1320	26.0	1.97	1.27	39.6

[a]All values are on a 100% dry matter basis.

1100-lb mares during gestation and lactation. During the first 3 months of lactation feed intake increases 6 lb daily or 37% over feed intake during the last 90 days of the gestation period. This is an increase from about 1.5 to 2.0% of body weight as feed intake (Table 18.4).

During the first 3 months of lactation, the concentrate diet should be 45–55% of the total feed intake of the mare (Table 18.5). The amount of concentrate intake should be varied, however, depending on the quality of the hay and/or pasture used, the milk production level of the mare, her condition, and other factors. There is no exact proportion of concentrates to feed. It depends on many factors, and the feeder needs to consider these in deciding on the level of feed intake to use. The concentrate mixture will also influence how much of it to feed.

Table 18.6 gives a sample diet which can be used during lactation. Other diets

TABLE 18.4

Feed Intake and Digestible Energy in Feed for 1100-lb Mares (16).

Stage	Daily feed per mare[a] (lb)	Daily feed[a] (% of live weight)	Digestible energy per lb of feed (Mcal)	Milk production per day (lb)
During last 90 days of gestation	16.2	1.47	1.14	—
During first 3 months of lactation	22.2	2.02	1.27	33.0
During 3 months of lactation to weaning	20.6	1.87	1.18	22.0

[a]All values are on a 100% dry matter basis.

TABLE 18.5

Ratio of Concentrates to Roughage during Gestation and Lactation[a]

| | Level of energy in the hay[b] | | | |
| | 1.0 Mcal/lb | | 0.9 Mcal/lb | |
Stage	Concentrate (%)	Roughage (%)	Concentrate (%)	Roughage (%)
Last 90 days of gestation	25	75	35	65
Lactation, first 3 months	45	55	55	45
Lactation, 3 months to weaning	30	70	40	60

[a]Data from Hintz et al. (16).
[b]More concentrate feeding is needed with lower energy level roughages. This is necessary to supply additional energy as well as the vitamins, minerals, protein, and other nutrients in the concentrate diet.

and other feeds can be used instead. The concentrate diet contains nutrients that are used to balance those supplied by the hays fed. During the first 3 months of lactation, the total diet of the mare should contain at least 12.5% protein (16). Many prefer to use 14% protein. Cornell University recommends 12–14% protein in the diet of lactating mares (22). The higher protein level would be safer to use with higher milk-producing mares. It would also provide a safety factor to compensate for hays and/or pastures with lower digestibility because of stage of maturity at which they are consumed or harvested. If the concentrate diet shown in Table 18.6 is used at one-half the total feed intake, it means the hay and/or pasture fed needs to contain at least 10% protein to give a total diet level of 13% protein. Since hay and/or pasture can vary considerably in protein level and digestibility, it stresses the importance of making sure the forage part of the diet is of high quality and balances out the nutrients supplied in the concentrate mixture. Minerals should always be self-fed to the mares in case they need more than is supplied in the concentrate feed. They will most likely need additional minerals depending on their level of milk production, the level of concentrate feed consumed, the quality and digestibility of the pasture and/or hay used, and many other factors.

Glade and Luba (23) reported that the substitution of soybean meal for the lower quality protein of a complete pelleted horse diet resulted in an increase in the protein content of early lactation equine milk and was accompanied by faster growth and improved lysine and methionine status in nursing foals. They indicate that their study suggests that the improvement in mare nutrition provided by the soybean meal was translated into higher concentrations of these two important amino acids in the blood of their foals.

A Florida study (24) involved supplementation of foaling mares with levels of vitamins A, D, and E (above 1978 NRC recommended levels) and supplemental levels of thiamin, riboflavin, niacin, pantothenic acid, pyridoxine, folacin, choline, vitamin B12, and trace minerals. The supplementation failed to produce significant responses in the mares or their suckling foals. They concluded that the control diet provided adequate levels of these nutrients for the foaling mare.

An English study (25) with Thoroughbred mares and ponies with access to green forages had blood serum folacin levels of 10.6 and 10.9 μ/l, respectively, while Thoroughbred racehorses confined to stalls had average folacin levels of 3.3 μ/l. He suggested that the high erythropoietic activity of horses in training and reduced availability of folacin in the diet may make folacin supplementation of horses desirable.

TABLE 18.6

Sample Concentrate Diet for Lactation

Feed	Percentage in diet[a]
Oats (heavy), steam rolled or crimped	15.0
Corn, milo, or combination, rolled or crimped	23.0
Barley (or corn), steam rolled	11.0
Wheat bran	7.0
Soybean meal, 44% protein	15.0
Linseed meal, (expeller, with about 4% fat)	4.0
Alfalfa meal, dehydrated, 17% protein	7.0
Blackstrap molasses	7.0
Dicalcium phosphate (or other calcium and phosphorus source)	1.25[b]
Limestone, ground	0.75[b]
Salt, trace mineralized, premix A or B, (Table 15.10)[c]	1.0[b]
Vitamin supplement, premix A, B, C, or D, (Table 15.9)[d]	1.0

[a]The levels in the concentrate diet should be adjusted to 16% protein, 0.80% calcium, and 0.7% phosphorus if the feeds used vary from expected analytical values. The diet has a lower level of calcium and phosphorus than the gestation diet since it will be fed at a higher level compared to forages.

[b]In addition, minerals should be self-fed in case the horse needs more than in the concentrate diet and in case concentrate level feeding is low.

[c]Use either trace mineralized salt premix A or B shown in Table 15.10. It would be preferable to use premix A with high-level performance horses.

[d]Use vitamin premix A, B, C, or D shown in Table 15.9. It would be preferable to use vitamin premix A with high-level performance horses. The other vitamin premixes do not contain the B-complex vitamins and give the option as to whether vitamin E or K are to be included or left out.

A Cornell study (21) with pony mares indicated that the 1978 NRC energy requirements for lactation is 10–15% too high for them.

In a study with Quarter Horse mares it was shown that the mares should receive sufficient energy to gain 0.33–0.77 lb/day for the first 3 months post partum (26). This requires a daily caloric allowance of 16 Mcal of digestible energy in excess of the maintenance requirement for mares weighing 1100–1320 lb.

B. Feeding from 3 Months to Weaning

During this period, milk production decreases to about two-thirds the level of milk produced during the first 3 months after foaling (Table 18.4). As a result, the level of feed intake decreases somewhat (Tables 18.4 and 18.7). During this period, the foal starts to consume more of its total feed intake from a creep feed as well as from hay and/or pasture. Therefore, the dependence on mother's milk decreases and at the same time the mare decreases its milk output.

The intake of concentrates by the mare also decreases to 30–40% of the total feed intake (Table 18.5). This level of concentrate intake is a guide that can be modified depending on the condition of the mare, the quality of the pasture and/or hay being used, her level of milk production, etc.

The diet shown in Table 18.6 can also be used during the second 3 months of lactation, although other feeds and other diets can also be fed. During this period the mare should receive at least 11% protein in the total diet (16). A level of 12.0–12.5% protein is preferable, however. It would be safer to use with high milk-producing mares. It would also provide a safety factor to compensate for hays and/or pastures with a lower digestibility because of stage of maturity when harvested or consumed. If the concentrate diet shown in Table 18.6 is used at about one-third of the total feed intake, the hay and/or pasture used would need to have at least 10% protein to give a total diet intake of 12.0% protein. The hay

TABLE 18.7

Feed Intake and Digestible Energy in Feed for Mares from 3 Months to Weaning (16).

Mature mares (lb)	Daily feed[a] per mare (lb)	Daily feed[a] (% of live weight)	Digestible energy per lb of feed (Mcal)	Milk production per day (lb)
440	11.0	2.50	1.18	13.2
880	17.1	1.94	1.18	17.6
1100	20.6	1.87	1.18	22.0
1320	23.9	1.81	1.18	26.4

[a]All values are on a 100% dry matter basis.

Fig. 18.4 An example of well-fed mares and their foals on pasture. (Courtesy of T. J. Cunha, Cal Poly University, Pomona.)

and/or pasture used can vary considerably in protein level and digestibility. Therefore, it is important to make sure that the hay and/or pasture used contains enough protein and is of high quality. Minerals should also be self-fed in case the mares need more than is contained in the concentrate mixture.

V. MARE'S MILK

The mare converts the digestible energy in the feed into milk energy with about 60% efficiency (16). Mares of light horse breeds may produce as much as 52.8 lb of milk per day at their peak of lactation, but the average production is probably within the range of 26.4–39.6 lb per day (16). Horses are estimated to produce milk daily equivalent to 3 and 2% of body weight during early lactation (1–12 weeks) and late lactation (13–24 weeks), respectively. A Texas A & M study with Quarter Horse mares in a 150-day lactation period showed milk production represented 2.1% of their body weight (27). Ponies have an average daily milk production of 4 and 3% of body weight during early and late lactation, respectively.

A. Need for Supplementation of Mare's Milk

The amount of milk a mare produces daily will vary considerably. This is one reason why some foals will do better than others if they rely largely on their mother's milk. The young foal cannot meet all its nutrient needs from mare's milk alone. This is why young foals are usually given a creep feed to help balance out their mother's milk.

A study of Table 15.1 (Chapter 15) shows that the digestible energy in milk is inadequate, especially after 1 month of age. The digestible protein is inadequate during all of lactation, especially during the later part. Both calcium and phosphorus in the milk are inadequate but phosphorus is the most deficient. Milk is also lacking in iron and copper; anemia would result if the foal was to depend solely on mare's milk. Other nutrients in milk such as fat, sodium, potassium, and others also decline as the lactation period continues, which is the reason that a good creep feeding program is recommended. The foal will start consuming creep feed between 1 and 3 weeks of age. If the mare is a good milker, it will delay the time the foal will nibble at extra feed. It is recommended that horse owners help to start the foal eating a creep ration by letting it nibble a little feed from the hand. At 5–6 weeks of age, a foal should be consuming at least 0.5 lb of creep feed daily per 100 lb body weight. By weaning time the foal should be consuming 5–8 lb of creep feed per day. While mare's milk is a good feed, it needs to be supplemented for best results in developing a foal at a maximum rate.

B. Changes in Mare's Milk during Lactation'

Table 18.8 shows information on changes in mare's milk in a study by Dr. D. E. Ullrey at Michigan State University. The milk composition was determined on Arabian and Quarter Horse mares that averaged 7.5 years of age (they varied from 6.5 to 8.5 years in age).

TABLE 18.8

Dry Matter, Ash, Iron, Zinc, and Copper in Mare's Milk[a]

Stage of lactation	Total dry matter (%)	Ash (%)	Iron (ppm)	Zinc (ppm)	Copper (ppm)
Parturition[b]	25.2	0.72	1.31	6.4	0.99
12 Hours	11.5	0.50	0.95	2.8	0.83
24 Hours	11.4	0.53	1.05	3.6	0.73
48 Horus	12.0	0.54	0.86	3.7	0.66
5 Days	11.6	0.54	0.88	3.5	0.44
8 Days	11.5	0.55	0.88	3.3	0.44
3 Weeks	11.3	0.50	0.83	2.8	0.29
5 Weeks	11.2	0.43	0.71	2.2	0.25
2 Months	10.3	0.37	0.61	2.1	0.23
3 Months	10.4	0.32	0.55	2.1	0.25
4 Months	10.0	0.27	0.49	2.4	0.20

[a]All values are expressed on a liquid milk basis. Data from Ullrey et al. (28).
[b]Milk sample obtained 15–30 min after birth of foal.

The highest concentration of dry matter in mare's milk occurs very shortly after the foal is born. By 12 hours after birth, the dry matter content of the milk decreases to about 45% what it was at parturition. The ash, iron, zinc, and copper levels from 12 hours after foaling to 4 months of age gradually decrease. This is the period when the dry matter in the milk is about the same, although there is a small decrease in the dry matter level of milk after 5 weeks of age.

The data in Table 18.8 are on a liquid milk basis. If one assumes that the dry matter in milk (after 12 hours) is about 10% (for ease of figuring), then multiplying all the figures on ash, iron, zinc, and copper by 10 gives the approximate level in the milk on a dry matter basis. The ash, or minerals in milk, begin at about 5% and end up at 2.7% in milk by 4 months of age. The iron level in milk starts at 9.5 ppm and decreases to 4.9 ppm by 4 months after parturition. This level of iron is too low and would cause anemia if the horse consumed only mare's milk. The 1978 NRC committee recommends 50 ppm of iron for the rapidly growing foal (16). Most likely, however, a higher level of iron is needed by the horse. If 50 ppm is the actual requirement, mare's milk supplies about one-fifth the requirement the first day after birth and decreases to one-tenth the requirement by 4 months after parturition. The copper level in mare's milk starts at 8.3 ppm and decreases to 2.0 ppm by 4 months after foaling. The 1978 NRC committee estimates the copper requirement of the growing horse at 9 ppm (16). Therefore, the copper level in milk is close to adequate for the foal only during the first day after birth. Subsequently, the young foal obtains a decreasing amount of copper in the milk until by 4 months of age it is receiving less than one-quarter of its copper needs. A lack of iron and copper results in anemia. Therefore, it is very important that the young foal have access to a high-quality, palatable creep feed and/or a mineral mixture that contains an adequate level of copper and iron. Otherwise, the foal will suffer from anemia and other problems associated with an iron and copper deficiency. The zinc level in mare's milk starts at 28 ppm and continues at a level of 30+ ppm and after 3 weeks decreases to a level of 21 to 24 ppm. A level of 40 ppm of zinc in the diet is recommended for the horse by the 1978 NRC committee (16). This may be increased if excess calcium is fed. To be on the safe side, extra zinc should be supplied in the creep feed and/or mineral mixture used. In fact, the other trace minerals should also be supplied, including manganese, iodine, cobalt, and selenium. All seven trace minerals can be added at very low cost and is good insurance against a possible need.

The Michigan State scientists (28) also showed that the crude protein in milk dropped from 19.1% shortly after birth to 3.8% 12 hours later and to 2.2% 2 months later. Other nutrients in milk such as energy, calcium, phosphorus, fat, sodium, and potassium also decline as the lactation period continues. These changes are shown in Tables 18.9 and 18.10.

Peak milk production usually occurs at 6–12 weeks after foaling. It then gradually declines until the foals are weaned at about 6 months of age.

TABLE 18.9

Changes in Constituents in Mare's Milk[a]

Stage of lactation	Total solids (%)	Crude protein (%)	Lipids (%)	Lactose (%P	Gross energy (kcal/100g)
Parturition	25.2	19.1	0.7	4.6	135
12 Hours	11.5	3.8	2.4	4.8	64
24 Hours	11.4	3.3	2.5	5.2	62
48 Hours	12.0	3.3	2.5	5.8	62
5 Days	11.6	3.1	2.1	5.9	59
8 Days	11.5	3.1	2.0	5.9	59
3 Weeks	11.3	2.7	2.0	6.1	56
5 Weeks	11.2	2.7	2.3	5.7	59
2 Months	10.3	2.2	1.6	6.1	52
3 Months	10.4	2.0	1.4	6.6	52
4 Months	10.0	2.0	1.3	6.5	49

[a]Data from Ullrey et al. (28).

Some early studies were made on the vitamins, minerals, and other nutrients in Percheron and Palomino mare's milk (29–31). A recent study at South Dakota (32) showed that the constituents in milk were not a factor in "foal heat" scours that often develops in young foals during the mare's first estrus, which usually commences about 9 days after foaling. Studies at Illinois (33) showed that at

TABLE 18.10

Changes in Minerals in Mare's Milk[a]

Stage of lactation	Ash (%)	Calcium (μg/g)	Phosphorus (μg/g)	Magnesium (μg/g)	Sodium (μg/g)	Potassium (μg/g)
Parturition	0.72	847	389	473	524	1143
12 Hours	0.50	782	399	138	364	965
24 Hours	0.53	973	442	110	337	841
48 Hours	0.54	1,110	457	92	296	861
5 Days	0.54	1,199	444	101	265	846
8 Days	0.55	1,278	441	94	238	780
3 Weeks	0.50	1,261	391	68	185	606
5 Weeks	0.43	1,110	325	63	188	555
2 Months	0.47	905	285	49	203	456
3 Months	0.32	708	243	40	174	405
4 Months	0.27	614	216	43	161	370

[a]Data from Ullrey et al. (28).

peak lactation the energy requirement was approximately 1.8 times maintenance for Shetland mares.

Zimmerman (34) reported that mare's milk composition was not altered by fiber or fat intake or by feed volume. These dietary variables had no effect on mare or foal performance. However, Elrod and co-workers (35) found that feeding 5% feed-grade rendered fat to mares increased the fat content of milk, foal weight gain, and body fat content of foals.

VI. FEEDING THE STALLION

Unfortunately, little research attention has been given to the feeding and nutrition of the stallion. This is probably related to the difficulty of having a sufficient number of stallions for research studies. Of utmost importance is to keep the stallion in a thrifty, vigorous condition. He should not be allowed to get too fat or too thin. The stallion should be exercised by riding or other means. Access to a good pasture would allow the stallion ample room to romp and exercise. A Connecticut study showed that libido in nonexercised 2-year-old Morgan stallions was higher than in exercised stallions after 12 weeks of forced exercise. Whether the decrease in libido was simply a function of fatigue or a decline in sex drive was not clear. This would indicate that excessive exercise should be avoided (34a).

A. Feeding during Nonbreeding Season

During the year when the stallion is not used for breeding, a high-quality pasture will supply a large part of the feed needed. It should also have access to minerals, which are self-fed in a mineral feeder. This provides an opportunity to obtain minerals not supplied in adequate amounts in the diet. The stallion should also have access to plenty of clean, fresh water. If the pasture is not sufficient, some high-quality, leafy green hay should also be fed. Concentrate feeds should be fed in small amounts to supplement the forage used and to keep the stallion in a trim, thrifty condition. The sample concentrate diet shown in Table 18.2 can be used for stallions during the nonbreeding period of the year. It is a high-quality concentrate mixture that supplements the forage with protein, minerals, and vitamins.

B. Feeding during the Breeding Season

Two or three weeks before the breeding season begins, the concentrate feed given to the stallion should be increased so that it will gain a little weight. During the breeding season the stallion will need more energy, protein, minerals, and vitamins. This can be accomplished by feeding a higher level of concentrates in the diet. It is suggested that the sample diet shown in Table 18.6 be fed during the breeding season. The concentrate feed (in Table 18.6) should be used at about the same level as the roughage (hay and/or pasture) in the diet. The concentrate feed should be fed at the rate of about 1 lb per 100 lb body weight during the breeding season. The remainder of the diet should be forage (hay and/or pasture). The amount of concentrate diet to feed, however, can vary depending on the quality of the forage used, the condition of the stallion, the number of services required

TABLE 18.11

Condition Score for Horses (6,36)

Score	
1	*Poor.* Animal extremely emaciated. Spinous processes, ribs, tailhead, hooks, and pins projecting prominently. Bone structure of withers, shoulders, and neck easily noticeable. No fatty tissues can be felt.
2	*Very thin.* Animal emaciated. Slight fat covering over base of spinous processes, transverse processes of lumbar vertebrae feel rounder. Spinous processes, ribs, tailhead, hooks, and pins prominent. Withers, shoulders, and neck structures faintly discernible.
3	*Thin.* Fat built up about halfway on spinous processes; transverse processes cannot be felt. Slight fat cover over ribs. Spinous processes and ribs easily discernible. Tailhead prominent, but individual vertebrae cannot be visually identified. Hook bones appear rounded, but easily discernible. Pin bones not distinguishable. Withers, shoulders, and neck accentuated.
4	*Moderately thin.* Negative crease along back. Faint outline of ribs discernible. Tailhead prominence depends on conformation; fat can be felt around it. Hook bones not discernible. Withers, shoulders, and neck not obviously thin.
5	*Moderate.* Back level. Ribs cannot be visually distinguished but can be easily felt. Fat around tailhead beginning to feel spongy. Withers appear rounded over spinous processes. Shoulders and neck blend smoothly into body.
6	*Moderate to fleshy.* Slight crease down back. Fat over ribs feels spongy. Fat around tailhead feels soft. Fat beginning to be deposited along the sides of the withers, behind the shoulders, and along the sides of the neck.
7	*Fleshy.* Crease down back. Individual ribs can be felt, but noticeable filling between ribs with fat. Fat around tailhead is soft. Fat deposited along withers, behind shoulders, and along the neck.
8	*Fat.* Prominent crease down back. Difficult to feel ribs. Fat around tailhead very soft. Area along withers filled with fat. Area behind shoulder filled in flush. Noticeable thickening of neck. Fat deposited along inner buttocks.
9	*Extremely fat.* Extremely obvious crease down back. Patchy fat appearing over ribs. Bulging fat around tailhead, along withers, behind shoulders, and along neck. Fat along inner buttocks may rub together. Flank filled in flush.

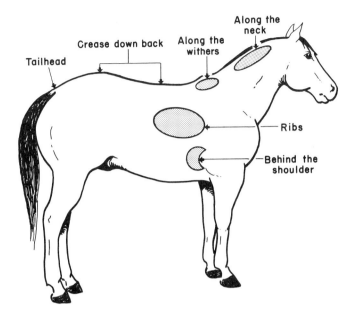

Fig. 18.5   Areas of fat deposition. (Courtesy of D. D. Householder and G. D. Potter, Texas A & M University.)

weekly, and other factors. The forage used should contain at least 10% protein in order to provide about 13% protein in the total diet fed the stallion. The stallion should be fed a quality forage that is green and leafy. Poor-quality forage should not be used for stallions during the breeding season if one expects them to have good libido and fertile semen.

VII. CONDITION SCORING SYSTEM

Workers at Texas A & M (6, 36) have developed a condition scoring system for horses (Table 18.11). The numerical scoring system provides a consistent measure of the degree of body fat scores in brood mares of various breeds and sizes. Fig. 18.5 shows the areas of fat deposition.

Texas A & M scientists recommend for best results that mares should be maintained in condition score 6 or higher. Where potential short-term stress is anticipated (long hauling, foaling, weather changes, etc.) they recommend that mares be brought up to at least condition score 7 prior to the anticipated stress.

The brood mare's energy requirements are about doubled when lactation begins. Colic, founder, or other digestive disturbances may occur if sudden large increases in feed are given to meet the increased energy needs. If the mare is in a fleshy to fat condition, she has the advantage of being fed smaller quantities of feed for 7–10 days and thus utilize stored body fat to make up for some energy lack until she can be fed the level of concentrates required. This allows a gradual increase in feed intake for the first 7–10 days after foaling.

VIII. CONCLUSIONS

Mares restricted in energy intake have impaired reproductive efficiency. Texas A & M studies indicate that mares foaling in a high level of condition are not impaired reproductively as had been previously thought. The highest percentage of foals are born during March, April, and May. There is an approximate 11–13% loss of foals between conception and weaning. The U.S. foal crop is estimated to be 50–65% which means the majority of mares foal one year and skip the next. One study showed that the average time from foaling to first ovulation, length of estrus, and time from ovulation to end of estrus were 12, 5.5, and 1.65 days, respectively. Less that 2 % of births are twins and 60% of twin conceptions result in abortion.

About 60–65% of the weight of the fetus is deposited during the last 90 days of gestation which indicates a need for increased nutrition then. The gestation diet should have 11–12% protein and lactating mares should be fed 12–14% protein. The higher protein levels can be fed with a lower quality hay and to the higher milk producing mares. The young foal cannot meet all its nutrient needs from mare's milk alone. Creep feeding is a good method of supplementing mother's milk, especially for poor milkers and high-level performance animals.

Very few research data are available on feeding the stallion. During the breeding season the stallion should be fed 1 lb of concentrate feed per 100 lb body weight. The total diet should contain about 13% protein and the stallion should not be allowed to get too fat or too thin. The stallion should receive some exercise but not to an excessive level since this might reduce libido.

REFERENCES

1. Henneke, D. R., G. D. Potter, and J. L. Kreider. *Theriogenology* **21,** 897 (1984).
2. Kubiak, J. R., J. W. Evans, G. D. Potter, P. G. Harms, and W. L. Jenkins. *Proc. Equine Nutr. Physiol. Symp., 10th,* p. 233 (1987).
3. Hintz, H. F. *Proc. 33rd Annu. Pfizer Res. Conf.* p. 117 (1985).
3a. Cunha, T. J., A. C. Warnick, and M. Koger. "Factors Affecting Calf Crop." Univ. of Florida Press, Gainsville (1969).
3b. Cunha, T. J. Swine Feeding and Nutrition. Academic Press, Inc. (1977).
4. Zimmerman, R. A., and D. E. Green. *J. Anim. Sci.* **47,** Suppl. 1, 326 (1978).
5. Henneke, D. R., G. D. Potter, and J. L. Kreider. *Proc. Equine Nutr. Physiol. Symp., 7th,* p. 101 (1981).

6. Henneke, D. R., G. D. Potter, and J. L. Kreider. *Proc. Equine Nutr. Physiol. Symp., 7th,* p. 105 (1981).
7. Jordon, R. M. *J. Anim. Sci.* **55,** Suppl. 1, 208 (1982).
8. Banach, M., and J. W. Evans. *Proc. Equine Nutr. Physiol. Symp., 7th,* p. 97 (1981).
9. Hutton, C. A., and T. N. Meacham. *J. Anim. Sci.* **27,** 434 (1968).
10. Hintz, H. F., R. L. Hintz, and L. D. Van Vleck. *J. Anim. Sci.* **48,** 480 (1979).
11. Chevalier, F., and E. Palmer. *J. Reprod. Fertil. Suppl.* **32,** 423 (1982).
12. Bain, A. M. *N. Z. Vet. J.* **17,** 155 (1969).
13. Potter, J. T., J. L. Kreider, G. D. Potter, D. W. Forrest, and W. L. Jenkins. *Proc Equine Nutr. Physiol. Symp., 9th,* p. 392 (1985).
14. Webb, G. W., G. D. Potter, and D. G. Meadows. *Proc. Equine Nutr. Physiol. Symp. 6th,* p. 20 (1979).
15. Webb, G. W. *Proc. Tex. A & M Agric. Ext. Serv., Horse Short Course* (1987).
16. Hintz, H. F., J. P. Baker, R. M. Jordon, E. A. Ott, G. D. Potter, and L. M. Slade. "Nutrient Requirement of Horses." NAS-NRC, Washington, D.C., 1978.
17. Gill, R. J., G. D. Potter, J. L. Kreider, G. T. Schelling, W. L. Jenkins, and K. K. Hines. *Proc. Equine Nutr. Physiol. Symp., 9th,* p. 84 (1985).
18. Holtan, D. W., and L. D. Hunt. *Proc. Equine Nutr. Physiol. Symp., 8th,* p. 107 (1983).
19. Ott, E. A. *Thoroughbred Rec.* p. 14 (1972).
20. Jordon, R. M. *Proc. Equine Nutr. Physiol. Symp., 6th,* p. 27 (1979).
21. Pagan, J. D., and H. F. Hintz. *Proc. Equine Nutr. Physiol. Symp., 7th,* p. 121 (1981).
22. Schryver, H. F., H. F. Hintz, and J. E. Lowe. *Cornell Univ. Coop. Ext. Serv., Inf. Bull.* **94,** 4 (1982).
23. Glade, M. J., and N. K. Luba. *Proc. Equine Nutr. Physiol. Symp., 10th,* p. 593 (1987).
24. Ott, E. A., and R. L. Asquith. *Proc. Equine Nutr. Physiol. Symp., 7th,* p. 44 (1981).
25. Allen, B. V. *Engl. Thoroughbred Vet. Rec.* **103,** 257 (1978).
26. Zimmerman, R. A. *Proc. Equine Nutr. Physiol. Symp., 7th,* p. 127 (1981).
27. Gibbs, P. G., G. D. Potter, R. W. Blake, and D. G. Meadows. *Proc. Equine Nutr. Physiol. Symp., 6th,* p. 24 (1979).
28. Ullrey, D. E., R. D. Struthers, D. G. Hendricks, and B. E. Brent. *J. Anim. Sci.* **25,** 217 (1966).
29. Holmes, A. D., A. F. Spelman, C. T. Smith, and J. W. Kuzmeski. *J. Dairy Sci.* **30,** 385 (1947).
30. Holmes, A. D., A. F. Spelman, and R. T. Wetherbee. *J. Nutr.* **37,** 385 (1949).
31. Holmes, A. D., B. V. McKey, A. W. Wertz, H. G. Lindquist, and L. R. Parkinson, *J. Dairy Sci.* **29,** 163 (1946).
32. Johnston, R. H., L. D. Kamstra, and P. H. Kohler. *J. Anim. Sci.* **31,** 549 (1970).
33. Lucas, V. A., W. W. Albert, F. N. Owens, and A. Peters. *J. Anim. Sci.* **34,** 350 (1972).
34. Zimmerman, R. A. *Proc. Equine Nutr. Physiol. Symp., 9th,* p. 96 (1985).
34a. Dinger, J. E., and E. E. Norles. *J. Anim. Sci.* **62,** 1220 (1986).
35. Elrod, K. A., G. D. Potter, L. W. Greene, W. C. McMullan, and J. W. Evans. *Proc. Tex. A & M Agric. Ext. Serv., Horse Short Course* (1987).
36. Henneke, D. R., G. D. Potter, J. L. Kreider, and B. F. Yeates. *Equine Vet. J.* **15**(4), 371 (1983).
mmm pg 368-369, tp 182

19

Purified Diets for Horses

I. INTRODUCTION

Many horse owners ask what a purified or synthetic diet is, since they have read about studies with these types of diets. The term purified or synthetic diet is used interchangeably. They both refer to a diet that consists of purified ingredients instead of grain, protein supplements, and other feeds. Instead of grain, as a source of carbohydrates, sucrose (sugar), starch, or glucose is used. These carbohydrate feeds are sources of energy and do not contain protein, minerals, and vitamins. Instead of soybean meal, cottonseed meal, linseed meal, or some similar protein source (which also contain vitamins and minerals), casein is used. Casein is protein isolated from milk and is purified so that it contains primarily protein. Instead of alfalfa or some other roughage source (which also contains vitamins, minerals, and protein) cellulose is used. It is primarily cellulose and serves as a source of roughage or bulk in the diet. Corn oil or some other fat is used to supply fat instead of relying on the fat supplied naturally in the various feeds. The purified diet ingredients are used because the scientist wants to know exactly what the diet contains. This makes it possible to subtract a vitamin, mineral, fat, carbohydrate, protein, or amino acid (if synthetic amino acids are used instead of casein in the diet) or to vary the level of protein, energy, minerals, and vitamins and to determine their effect on the horse. This cannot be done with a natural or practical diet consisting of corn, barley, oats, soybean meal, linseed meal, alfalfa meal, bone meal, and salt. However, it can be done with a purified diet such as is shown in Table 19.1. It contains all the vitamins and minerals that are needed by the horse fed a purified diet. Each vitamin or mineral, therefore, can be removed from the diet individually to determine what effect it has on the horse. Most of these vitamins or minerals cannot be studied with a natural or practical diet (both of these words are used interchangeably to refer to the kind of a diet fed on the farm) because the feeds used may naturally contain them at a level adequate for the horse. Moreover, some of the vitamins or minerals that are low or borderline in practical diets cannot be adequately studied to determine what occurs with a deficiency because there may be enough in the diet to prevent observable deficiency symptoms from developing. Therefore, to obtain certain basic knowledge on the function of various nutrients in the horse, the scientist resorts to studies with purified diets. These produce data which can be used to solve some of the feeding and nutrition problems encountered with

TABLE 19.1

Example of a Purified Diet

Ingredients	Supplies
Casein (vitamin free)	Protein
Glucose	Carbohydrates
Cornstarch	Carbohydrates
α-Cellulose	Fiber or roughage
Corn oil	Fat
Dicalcium phosphate	Calcium and phosphorus
Salt	Sodium and chlorine
Individual minerals as mineral salts	Magnesium, iron, copper, cobalt, manganese, zinc, potassium, sulfur, iodine, and selenium
Individual vitamins as synthetic vitamins	A, D, E, K, C, B_1, B_6, B_{12}, pantothenic acid, niacin, riboflavin, choline, inositol, p-aminobenzoic acid, biotin

practical diets on the farm. In addition, the information is used to establish how one nutrient may effect the utilization or need for another one. These nutrient interrelationships are very important in solving some of the nutrition problems in the horse that are not yet understood.

One needs to be very careful, however, in applying nutrient requirement data obtained with purified diets directly to practical diets. This is because the feeds in the purified diets are much more digestible than those in the practical diet. In practical diets, the nutrients are in their natural state. In most cases, they are in different forms from those fed in purified diets. This results in a difference in availability and in requirements between purified and practical diets. Moreover, the two types of diet may have a different effect on the synthesis and requirements of certain nutrients by the intestinal tract microorganisms present in the horse. This means some degree of reservation should be used in applying data from purified diets to practical diets. It also means the nutrients need to be studied with practical diets as well.

The data shown in the tabulation below indicate the big differences that can occur in the nutrient requirements obtained with a purified diet as compared to a practical diet.

	Nutrient requirements with:	
Nutrients	Purified diet	Practical diet
Zinc (ppm in diet)	18	50
Vitamin D (IU/lb feed)	45	227

The data were obtained with the pig and not the horse, but it illustrates what might also occur with the horse. There is a big difference in the need for zinc and vitamin D between the two kinds of diet. This discussion does not mean that purified diet studies are of no value. They are valuable and need to be used in certain research studies. However, one needs to use a certain degree of reservation in their application to practical diets fed on the farm, especially those dealing with nutrient requirements.

II. PURIFIED DIET COMPOSITION

Table 19.2 shows the makeup of a purified diet published by the Kentucky Station in 1969 (1). It does not contain selenium, which should be added, based on current research showing the need for this mineral by the horse. This diet was readily pelleted. It was also consumed by suckling and weanling foals at a level in excess of 2.5% of body weight and resulted in 2.2 lb gain per day. The blood values were within normal ranges and resulted in the passage of formed stools.

This purified diet was compiled by Dr. Howard D. Stowe (1) and required 3

TABLE 19.2

Composition of Purified Equine Diet[a]

Component		Component	
Casein, vitamin free (kg)	16.00	Inositol (g)	40.0
Glucose (kg)	40.00	Choline chloride (g)	24.0
Cornstarch (kg)	25.00	Niacin (g)	4.476
Cellulose (kg)[b]	13.00	p-Aminobenzoic acid (g)	2.200
Cottonseed oil (kg)	1.00	Thiamin·HCl (g)	1.544
Dicalcium phosphate (kg)	2.40	Riboflavin·HCl (g)	0.455
Sodium chloride (kg)	1.00	Calcium pantothenate (g)	0.455
Potassium carbonate (kg)	1.415	Pyridoxine (g)	0.356
MgO (g)	65.0	Folic acid (g)	0.207
$ZnSO_4·7 H_2O$ (g)	8.8	Menadione (g)	0.037
$MnSO_4·H_2O$ (g)	8.0	Vitamin A (g)[c]	1.760
$FeSO_4·7 H_2O$ (g)	4.5	Vitamin D_2 (g)[d]	0.110
$CuSO_4·5 H_2O$ (g)	2.0	Vitamin E (g)[e]	20.000
$CoCl_4·6 H_2O$ (g)	0.6		
KI	0.5	Total ration	100 kg

[a]Data from Stowe (1).
[b]Solka Floc, Brown Company, Boston, Massachusetts.
[c]250 IU/mg.
[d]500 IU/mg.
[e]222 IU/g.

years to develop. He encountered many problems, which are discussed in his paper, and which should be consulted if use of purified diets is contemplated. It is not easy to mix and use a purified diet such as the one shown in Table 19.2. Therefore, care should be exercised by those who plan to use purified diets in horse research studies.

REFERENCES

1. Stowe, H. D. *J. Nutr.* **98,** 330 (1969).

20

Feeding and Health-Related Problems[1]

There are a number of problems which involve the feeding and health of the horse. Some scientists refer to some of them as metabolic disorders since they may be related directly or indirectly to nutrition or feeding methods. Many problems occur because the young horse is being pushed for a level of performance that is too great and too early, which can result in damage to immature, partially ossified bones, which leads to bone abnormalities. Dr. Ben Norman, University of California Extension Veterinarian who, as a DVM and a Ph.D. in nutrition, in an address to veterinarians, stated that, "in 85% of the cases, nutrition will influence the outcome of the disease entity that you work with" (1). Therefore, a number of feeding and health-related problems are discussed in this chapter. They are discussed only briefly to indicate the interaction of feeding methods and programs with some health problems encountered in horses. More details on these problems may be obtained in publications and books dealing with veterinary medicine, feeding, nutrition, and physiology.

Stall confinement also causes many problems. The horse was developed as an animal accustomed to life in the wild. Confinement rearing, if practiced to extremes, causes boredom, frustration, wood chewing, stall walking, wall kicking, and sometimes a bad temper. Moreover, blood circulation, as well as bone and body system development, may be affected by a lack of exercise if confinement is carried too far.

I. LAMINITIS

It is occurring more frequently. It affects the feet, causes extreme pain, a high fever (103–106°F), and the horse has a difficult time moving or walking. It can vary from a mild case to a very severe one. It results in continual pain which causes the horse to avoid putting weight on its feet or subject them to the concussion of walking. If all four feet are affected the horse will lie down to minimize the pain (2).

Symptoms of acute laminitis include lameness (usually in both front feet but it

---

1 This chapter was reviewed by Gerald E. Hackett, Jr., DVM, M. S., Director of Equine Research, California State Polytechnic University, Pomona, California.

can affect all four feet), stiffness, settling back on the hind feet to lessen weight on the forefeet, heat in the hooves, and bouncing pulse in the arteries at the back of the pastern. The wall of the hoof separates from the sensitive laminae (inner portion), allowing the coffin bone to rotate within the hoof wall. If allowed to progress, the rotation of the coffin bone within the hoof encasement is called chronic laminitis, and occurs with varying degrees of severity (2). Chronic laminitis is preceded by acute laminitis and if stopped early, it may be prevented (2). Until more is known about laminitis the following suggestions are offered as a means of minimizing its effects.

1. Avoid overfeeding or irregular feeding of concentrate grain mixture to horses. One should feed the same amount of concentrate feed daily and not vary it from one day to the next. Increasing the amount of concentrates fed to horses should be done gradually. Avoid allowing a horse to get into the grain bin (if the door is left open) where it can gorge itself on grain or other concentrates.

2. If a horse shows the first signs of laminitis on a lush pasture, it should be removed right away. Allow the horse to recover, feed it hay and gradually return it to the pasture after the lush growth period is over.

3. Make sure mares completely expel their afterbirth soon after foaling. If they do not, it may cause uterine infection followed by laminitis.

4. Avoid feeding programs which result in indigestion (irregular feeding, quick changes in kinds of feeds used, the use of moldy, rancid, and wet feeds when the horse is used to dry feeds, etc.). Indigestion may result in laminitis.

5. Act quickly when the horse first starts to show signs of laminitis (walks easy on its feet, acts like it is walking on eggs, walks heel to toe, appears in pain when walking, has elevated temperature, etc.). If one is not experienced with laminitis the assistance of one who is and especially a veterinarian is needed. Quick action may save the horse.

6. It is always safe to change the horse to a diet of good-quality hay. The horse should be allowed to recover, if it is not too late for it to do so. After recovery, the horse should gradually be returned to a safe feeding program.

7. Therapeutic trimming and shoeing can greatly improve chronic cases where laminitis has caused separation of the hoof wall and curling of the toes.

8. Avoid hard work and exercise on hard surfaces since this may bruise the laminae and cause laminitis (2,3).

9. The stress of transport or disease can also cause laminitis. Sometimes an overheated or exhausted horse may suffer metabolic changes which alter normal blood flow and disrupt the laminar integrity (2).

II. NAVICULAR DISEASE

This disease is common to Western stock horses with small feet (3). Navicular disease is an inflammation of the small navicular bone and bursa inside the hoof between the coffin bone and the tendon which flexes or bends back the coffin joint. Symptoms include: pointing; a short, stubby, painful stride; and lameness which is barely perceptible in the early stages. Most navicular disease is caused by small, narrow heels that do not expand and absorb shock while the horse is working (3).

This disease is incurable once bone changes develop, but may be helped by therapeutic shoeing and drug therapy. Small-footed horses may be helped by keeping the walls short so they have frog pressure that will increase circulation through the foot and spread the heels with each step, dissipating some of the concussion. They can be shod with plates or light-weight shoes to keep the frog as nearly on the ground as possible (3).

The risk of navicular disease may be reduced by selecting horses with larger feet, maintaining frog pressure, and working them with caution on hard surfaces (3).

III. COLIC

Colic refers to abdominal pain with some or all of the following signs: restlessness, pawing, repeatedly lying down, rolling and getting up, distress, uneasiness, abdominal straining, mild to profuse sweating, abnormal stance and gait, depression, and loss of interest in feed and water (4).

The horse has a small stomach and if fed too much it cannot relieve the distended stomach by vomiting and thus colic may develop. If the distention is too great, the stomach may rupture and cause death.

The small intestines are long and twisting and herniation through a body opening may occur and cause colic. The cecum and large and small colons are large in relation to the stomach. Impaction may occur in all three and cause colic. Twisting may occur in the large and small colon, and some cases of colic may require surgery. Most cases of colic are mild and respond to simple treatment. The more serious cases can sometimes be confusing to the owner and veterinarian. Even an experienced veterinarian may have some difficulty in deciding whether the condition causing the colic will respond to medical treatment or if surgery is required. Internal parasites play an important role in intestinal disturbances which lead to colic and so an excellent parasite prevention program should be followed. Many cases of colic, and even those which cause death, occur in horses that appear to be free of internal parasites and are on an excellent dietary program. So, colic can be complex and requires competent attention, prevention, and treatment.

It appears that colic may be caused by: (1) obstruction in the digestive tract; (2) digestive tract disturbance which usually produces gas; and (3) parasite infestations which can be great enough to block the intestinal tract. Sand colic can result when horses ingest large amounts of sand while grazing dormant pastures which do not supply enough forage intake.

Following are some feeding and management suggestions which may minimize colic:

1. Each horse should be managed as an individual and fed accordingly.

2. Feed at regular times daily (including weekends and holidays) and avoid sudden diet changes. A necessary diet change should always be done gradually.

3. Horses that are greedy eaters should have large, smooth rocks or a 5-lb mineral block added to the feed box to slow down their eating.

4. The hot horse should be cooled out slowly and allowed only small quantities of water until its thirst is satisfied.

5. A clean fresh water supply should always be available. Many cases of colic occur when the weather is very hot or cold, and are due to an inadequate water intake.

6. Only top quality feeds should be used.

7. The horse's droppings should be observed frequently and watched for consistency, unusual odor, and any deviation from normal as a means of early detection of something wrong.

8. Frequent inspection of the feed box will help detect the horse's eating habits and make it easier to detect quickly when the animal goes off feed.

9. Feed boxes and hay racks should be kept clean at all times. Old, wet, moldy, rancid, musty, or spoiled feed should not be fed since it may cause digestive disturbances, colic, or even more serious problems.

10. Check teeth frequently since dental problems may result in improper chewing of feed which may cause colic.

11. Horses need adequate exercise. Problems with colic are much less in horses running on pastures.

12. Feeds that are too heavy, and which lack bulk, may tend to pack in the stomach. Therefore, bulk should be provided in the concentrate feed used. But, feeds that are too high in fiber may tend to cause discomfort, digestive disturbances, or colic, especially with horses that are going to be exercised, worked, or raced heavily. To avoid this, many feed most of their hay at night and limit total feed intake before working their horses. Some prefer to feed 2 hours before and 1 hour after hard work.

13. Horses should be fed according to the amount of work they perform and their body condition. This prevents overfeeding which may result in colic and overfat horses.

14. Salt should always be available in the stall or pasture. This gives horses an opportunity to consume extra salt in case the diet consumed does not meet their total need. Horses sweat a great deal and it contains 0.7% salt as well as other minerals. Thus the amount of exercise and sweating can greatly affect salt and other mineral needs.

15. Caretakers should be quiet and gentle when working around horses. This minimizes the possibility of the occurrence of digestive disturbances and colic.

This brief summary on colic in the horse drew heavily from an excellent review on this subject by Dr. Leon Scrutchfield of Texas A & M University (4). It is apparent that one should take seriously any problems with colic since it can be devastating to a horse program.

IV. AZOTURIA AND TYING-UP

Monday morning disease originated with hardworking draft horses that were full-fed grain on their day off from work and the condition showed up shortly after starting to work on Monday. Tying-up has occurred in recent years with high-level performance horses and occurs after an idle period and shortly after they resume hard work. The first symptoms may be stiffness, a reluctance to move, lameness, profuse sweating, and a stiffness of gait. The muscles in the loin area are tense and hard. When the horses urinate, it is coffee-colored, indicating muscle damage and the loss of myoglobin.

Combinations of vitamin E and selenium have been used in the treatment of tying-up in horses. Some reports have indicated that either selenium, or vitamin E or both are beneficial for this condition (5-8). Dr. Hill (5) reported that tying-up appears in horses of all ages and affects 2–5% of the horses at racetracks. Two-thirds of the cases occur after 1–2 days rest from a rigid training schedule. Dr. Hill stated that the horses could be divided into three groups: (1) those that tied up regularly and frequently; (2) those that tied up occasionally; and (3) those that had muscle soreness or stiffness as the predominant symptom. Dr. Hill used an injectable preparation which contained 2.5 mg of selenium as sodium selenite and 25 mg of d-alpha-tocopherol acetate per milliliter. Sometimes one injection was sufficient, but a second was needed in some cases. Some horses are cured quickly, whereas others may respond partially or take a longer time to respond. A small percentage of horses may not respond.

Some scientists believe that tying-up is due to excessive accumulation of lactic acid in the muscle. They feel that the rapid oxidation, or breakdown, of the glycogen stored in the muscle can cause the production of large amounts of lactic acid in a relatively short time.

Animals affected with tying-up should be rested until they recover. If they are forced to exercise or move too much they may go down and die. If properly treated, they may recover; but, some may develop tying-up again.

Heavy concentrate feeding during idle periods is a major cause of tying-up. Concentrates should be cut in half (or less) and hay increased during idle periods. Some horses, especially highly strung, nervous types, appear to be more susceptible to tying-up. It also appears that females are more susceptible than males. Cold weather may result in a higher incidence of tying-up.

Tying-up has also been observed in horses following fights with farriers, fearful flights from thunderstorms, long nervous trailer rides, and others (8). It appears, therefore, that muscle exertion beyond the horse's accustomed activity has the potential to occasionally cause tying-up.

V. PERIODIC OPHTHALMIA

It is also referred to as moon blindness or equine uveitis. It results in period of cloudy vision which lasts for 1–14 days and then clears up. It recurs periodically and eventually may cause severe eye damage, cataracts, and blindness. Early work (9) indicated that periodic ophthalmia responded to 40 mg of riboflavin daily even though the horses were on green pasture or hay, which are good sources of riboflavin. Many scientists feel, however, that other factors may be involved. Some think it is probably due to an immunological response to other disease conditions. Others feel that an allergic reaction may be involved or that a genetic predisposition to the disease may be a factor. Periodic ophthalmia is still not well understood and much remains to be learned about it.

VI. CONTRACTED TENDONS

A Cornell University study (10) with 4-month-old weaned foals fed a restricted feed intake diet for 4 months followed by a 4-month period of *ad libitum* feeding developed straight limbs similar to contracted tendons within 1–3 months after being fed all the feed they would consume. The angle of the fetlock returned to normal after 8–9 months. The correction of the contraction coincided with the increase in exercise time. It appears that the period of restricted feeding level had something to do with the contracted tendons although additional research is needed to verify this. More detail on contracted tendons is given in Chapter 11, Section XI.

VII. OSTEOCHONDROSIS

A Texas A & M study (11,12) indicated that osteochondrosis in Thoroughbred foals is clinically similar to the lameness that has been associated with a simple copper deficiency. Their data implicated copper as a significant factor in the pathogenesis of osteochondrosis of suckling foals. They stated that zinc may have played a secondary role by reducing the absorption of copper in the foals.

Three other studies with horses indicate that high zinc levels may cause lower levels of copper in the horse (13–15). The Texas scientists also stated that horses with the genetic disposition for fast growth and those that are fed heavily for faster growth are frequently the ones in which osteochondrosis develops (3). A Maryland study showed that high-energy diets induced osteochondrosis changes in the cartilage of weanling horses (16).

An Ohio study (17) reported results of a subjective survey of horse farms in Ohio and Kentucky which revealed a high incidence of osteochondrosis, epiphysitis, and other skeletal disorders in grazing horses. They observed indications of low copper in the forages and feeds, suggesting that the skeletal disorders may be related to deficiencies of copper, zinc, and possibly other minerals. However, many of their survey sites had adequate amounts of copper and zinc in their feeds and forages and still had a high incidence of skeletal disorders in the foals.

Fig. 20.1 A 3-month-old Thoroughbred foal with osteochondrosis and other defects of collagen metabolism associated with experimental copper deficiency. (Courtesy of Charles H. Bridges, Texas A & M University and the *J. Am. Vet. Med. Assoc.* **193**, 215, 1988.)

A Maryland study (18) reported that similarities between the lesions accompanying chronic overfeeding and those of equine hypothyroidism suggest endocrine involvement linking dietary excess to developmental osteochondrosis.

Osteochondrosis is characterized by cartilaginous hypertrophy, biochemical alterations in the chondroid matrix, chondronecrosis, fracturing, separation of cartilage from its supporting tissues and, in the extreme, physical collapse of joints and growth plates (19).

A Texas A & M study (12) showed that foals deficient in copper developed intermittent, but nondebilitating diarrhea. All the foals developed stilted gaits and ultimately walked on the front of their hooves. Figure 20.1 shows a 3-month-old Thoroughbred foal with osteochondrosis and other defects of collagen metabolism associated with experimental copper deficiency (12). Dr. Charles H. Bridges feels that foals with the residual cartilaginous breaks so characteristic of osteochondrosis may or may not still have the tendon contractions so characteristic of the early phase of copper deficiency. The deficient copper metabolism and the associated collagenous defects can be resolved by the addition of copper to the diet but the broken cartilage will still be broken and the foal and even the eventual horse may still be lame. Dr. Bridges states that this is confusing to many diagnosticians who encounter these cases of osteochondrosis but find normal copper values in the afflicted horse (20).

VIII. EPIPHYSITIS

It occurs most often with horses fed for rapid growth (17, 21). Young horses with epiphysitis usually exhibit some degree of lameness and the affected joints are usually swollen. An Ohio study (17) indicated that copper, zinc, and possibly other minerals might be involved. But, many farms in their survey sites had adequate amounts of copper and zinc in the diet and still had cases of epiphysitis in the foals. Heredity may play a role (22). Horses fed inadequate levels of calcium have developed epiphysitis. But, epiphysitis has also occurred in horses supposedly fed well-balanced diets. Evidently, many factors may contribute to this disease. It appears, however, that feeding for a rapid growth rate is a major contributing factor. One report suggested that feeding for 80–90% of the horse's maximal growth rate may avoid epiphysitis (22).

IX. HEAVES

Heaves, or pulmonary emphysema, results in a loss of elasticity in the lungs, and an accumulation of air in the lung tissue since it cannot be expired properly. Symptoms may occur when horses consume dusty, musty, moldy feeds or when they are exposed to dusty bedding or a dusty atmosphere. The horse coughs, has difficulty breathing, and may show a nasal discharge. The condition may disappear or improve when the horse is turned out to pasture. Feeding pellets, excellent quality grain or hay, or sprinkling the hay with water may help the condition. Sometimes a complete pelleted feed containing fiber sources such as beet pulp or citrus pulp or others is used if it is felt that the horse cannot properly utilize the hay available. Some scientists feel that allergies to the hay might be involved in heaves. Some horses are more susceptible than others.

X. ULCERS

As sophisticated diagnostic procedures and facilities are developed, more cases of ulcers are being detected in horses. The University of Florida recently reviewed the records of 511 foals, which were less than a year old, and found that ulcers were evident in more than one foal out of every four (23). Aided by a video endoscope these veterinary scientists have been able to diagnose ulcers and other gastrointestinal diseases in foals that are just a few days old. Ulcers, which may be caused by diet, disease, stress, or other factors, can kill a horse by destroying protective tissue inside the esophagus, stomach, and intestines (23). With the availability of these diagnostic tools it is important to determine the role which nutrition, disease, management, and other actors may play in preventing ulcers.

XI. ENTEROLITHS

They are also known as calculi or stones. The number of cases of detected enteroliths in horses is increasing at the horse clinic at the University of California, Davis (24) and elsewhere. The presence of a nidus (nails, pins, needles, coins, pebbles, metal fragments, horse hair, etc.) and adequate concentrations of ammonia, magnesium, and phosphorus are needed for calculi formation. Calculi found in the intestines of horses are primarily composed of magnesium, ammonium, and phosphate (24). At the University of California, Davis the average calculi removed during surgery weighed between 200 and 1500 g (24). Sometimes they weigh considerably more than this. The use of wheat bran and alfalfa and the level of water intake have been blamed for calculi in some instances. But, the evidence is only circumstantial and it does not seem reasonable to eliminate alfalfa and wheat bran from the diet because of this possibility. In instances where calculi is a serious problem one could evaluate whether decreasing the level of alfalfa or wheat bran might be of help. Many horses that are not fed either feed develop calculi. The elimination of a nidus from the feed by making sure clean feeds are used is a very important way of preventing calculi formation. Some horses pass small calculi in the feces and never develop any problems but other horses do not pass the calculi. In some cases, enteroliths can grow large enough to block the intestines and cause serious damage.

XII. HYPERLIPEMIA

It is a condition in which there is an unusually high level of fat, especially triglycerides, in the blood. There may also be an impaired utilization of the low-density lipoproteins (25). The normal level of total lipids ranges from 100 to 500 mg per 100 ml of blood. One study indicated an average level of total lipids of 1760 mg per 100 ml of blood in horses with hyperlipemia (26). The condition results when the horse has to draw heavily on body fat stores for energy. It is not able to metabolize all of the increased load of mobilized fat and hence it accumulates in the blood. The use of glucose and heparin or glucose and insulin treatments helps to decrease the lipid level in the blood. But, it is imperative that the horse resumes eating at a normal level. The basic cause of hyperlipemia is when feed is withheld for weight reduction or the horse goes off feed due to disease or a nutritionally inadequate diet. It could also occur if a proper feeding level is not maintained, especially during peak lactation when feed needs are at their highest level.

XIII. WOOD CHEWING

It occurs frequently in horses and results in them eating wood in their stalls, fences, and gates. In some cases they will eat dirt, feces, or tree bark, or chew the manes or tails of other horses. There are a number of causes for wood chewing. A lack of roughage in the diet is one. It also may occur when horses are fed a completely pelleted diet. Increasing the fiber length of hay which is ground for pellets decreases the problem. If horses that are fed a pelleted diet have some access to pasture or a little hay, it will minimize or eliminate wood chewing. Boredom may also cause wood chewing. Keeping horses in a small stall or area could cause wood chewing. The wood chewing could be a manifestation of dissatisfaction with the close surroundings and the lack of a natural area, such as a pasture or paddock for exercise, running, and something to do, see, and explore. Some feel that a shortage of minerals could be involved. Proper mineral supplementation should be followed, of course, but it is not the only cause of the problem. In some cases, wood chewing may be a bad habit similar to nail biting in humans. Even under the best of nutrition, health care, and management conditions wood chewing may occur with some horses. Not all the causes of wood chewing are known yet.

XIV. TEETH FLOATING

Horses should be checked at least annually to determine if the teeth need floating or filing to make it easier for them to chew grains and other hard feeds. Wear on teeth causes sharp points on the outside of the upper teeth and on the inside of the lower ones. Floating is a relatively simple job and does not require much restraint on most horses (3). Any other dental problems should also be taken care of at the same time as teeth floating.

XV. BLISTER BEETLE

The toxic substance in blister beetles is cantharidin. Horses appear to be the most severely affected by the toxin in beetles, although most domestic animals are also susceptible to it. The cantharidin causes severe irritation and hemorrhage in the stomach and intestinal tract. The toxin is absorbed and excreted by the kidneys, causing nephrotoxicity and urinary damage (27). Large doses may cause shock and death within a few hours. Small doses produce signs of abdominal distress and coliclike symptoms. The most frequent clinical signs are: abdominal pain, fever, severe depression, increase in respiratory and pulse rate, and muscle tremors. Horses that survive more than 24 hours may exhibit signs of frequent urinary straining while voiding only small amounts of urine. The urine may be blood tinged or contain blood clots. High fever and severe depression is frequently followed by shock and then death within 48 hours in most of the cases reported. The exact number of beetles required to produce poisoning is quite variable but as few as 3–10 beetles can cause illness and death in horses (27, 28). The cantharidin is deadly whether the horse eats alfalfa with live or dead blister beetles or just the remains of dead ones in last year's hay (28). Alfalfa hay is the only feedstuff where poisoning from dead blister beetles has been reported so far. Crimping of hay usually prevents the escape of beetles at time of cutting. Previously, when the hay was mowed, then left to be raked and baled later on, the beetles would leave the hay before baling (27, 28). Blister beetles travel in swarms, so one part of a field may have beetles while neighboring fields may be free of them.

The best way to avoid beetle poisoning is to use a dependable hay supplier who makes sure the hay is free of beetles. The hay should also be inspected before it is purchased and before it is fed whenever possible. Horses should be watched carefully, and if symptoms appear, the veterinarian should be called immediately.

XVI. FESCUE TOXICITY

Fescue grass may occasionally be toxic if it has been infected with an endo-phyte which is a fungus (29). Symptoms of fescue toxicity include the following: rough hair coat, emaciation, listlessness, increased respiration and heart rates, elevated rectal temperature, cloudy corneas, diarrhea, prolonged gestations, abortions, thickened placentas, and a lack of milk production. Pregnant mares grazing fescue pasture infected with endophyte should be removed from the pasture during the last 90 days of gestation. Since the fungus is primarily trans-mitted via seed, one should make sure the seed used for establishing fescue pastures is free of the fungus. The local county agricultural extension service can be of assistance in determining the presence of endophyte in fescue forage.

XVII. PRUSSIC ACID POISONING AND CYSTITIS

Sorghums, sudangrass, and sudan–sorghum hybrids contain a glycoside which may breakdown to prussic acid (hydrocyanic acid) in the digestive tract. Prussic acid may occur in young immature stands or in the second growth following periods of drought, frost, or after growth is stopped by mechanical harvesting. Proper management and removing horses from these forages during periods when prussic acid deaths may occur is necessary.

Cystitis (urinary tract inflammation) may also occur when grazing the same forages listed above. However, hay made from sudan or sorghum–sudan hybrids is reported not to produce the disease.

More information on both cystitis and prussic acid poisoning is given in Chapter 13, Section X.

XVIII. CONCLUSIONS

There are many health problems that are related directly or indirectly to nutrition or feeding methods. Dr. Ben Norman, a veterinarian and a Ph.D. in animal nutrition, has stated that in 85% of the cases feeding and nutrition will influence the outcome of most health problems. The horse was developed as an animal accustomed to life in the wild. But stall confinement and intensified production methods with high-level performance horses has brought on many problems. Proper feeding and management methods can minimize and eliminate some of these problems.

REFERENCES

1. Norman, B. *Calif., Agric. Ext. Serv. Leafl. AAPB* No. 51 (1977).
2. Loving, N. S. *Horse Care* p. 52 (1988).
3. Coffman, J. R., and M. Bradley. *Univ. MO., Agric. Ext. Leafl.* **2851** (1979).
4. Scrutchfield, L. *Proc. Tex. A & M Horse Short Course* p. 68 (1982).
5. Hill, H. E. *Mod. Vet. Pract.* **43,** 66 (1962).
6. Hartly, W. J., and A. B. Grant. *Fed. Proc., Fed. Am. Sur. Exp. Biol.* **20,** 679 (1961).
7. Stewart, J. M. *N. Z. Vet. J.* **8,** 82 (1960).
8. Smith, H. *Horse Rider* **25**(6), 62 (1986).
9. Jones, T. C. *Am. J. Vet. Res.* **3,** 45 (1942).
10. Hintz, H. F., H. F. Schryver, and J. E. Lowe. *Proc. Cornell Nutr. Conf.* (1976).
11. Bridges, C. H., J. E. Womack, E. D. Harris, and W. L. Scrutchfield. *J. Am. Vet. Med. Assoc.* **185,** 173 (1984).
12. Bridges, C. H. and E. D. Harris. *J. Am. Vet. Med. Assoc.* **193,** 215 (1988).
13. Gunson, D. E., D. F. Kowalczyk, C. R. Shoop, and C. F. Ramberg. *J. Am. Vet. Med. Assoc.* **180,** 295 (1982).
14. Eamens, G. J., J. F. Macadam, and E. A. Laing. *Aust. Vet. J.* **61,** 205 (1984).
15. Kowalczyk, D. F., D. E. Gunson, C. R. Shoop, and G. C. F. Ramber, Jr. *Environ. Res.* **40,** 285 (1986).
16. Glade, M. J., and T. H. Belling, Jr. *Equine Vet. Sci.* **6**(3), 151 (1986).
17. Knight, D. A., A. A. Gabel, S. M. Reed, R. M. Embertson, W. J. Tyznik, and L. R. Bramlage. *Proc. Am. Assoc. Equine Pract.* **31,** 455 (1985).
18. Glade, M. J., and T. H. Belling, Jr. *Growth* **48,** 473 (1984).
19. Glade, M. J. L. Krook, H. F. Schryver, and H. F. Hintz. *Cornell Vet.* **73,** 170 (1983).
20. Bridges, C. H. Texas A & M University, College Station (personal communication, Nov. 29), 1988.
21. Kronfield, D. S. *Proc. Am. Assoc. Equine Pract.* p. 461 (1978).
22. Smith, H. *Horse Rider* **25**(4), 62 (1986).
23. Wilson, J., M. C. Thompson, and A. Merritt. *Horse Rider* **26**(10), 62 (1987).
24. Lloyd, K., H. F. Hintz, J. D. Wheat, and H. F. Schryver. *Cornell Vet.* **77,** 172 (1987).
25. Morris, M. D., D. B. Zilversmit, and H. F. Hintz. *J. Lipid Res.* **13,** 383 (1972).
26. Naylor, J. F., D. S. Kronfeld, and H. Acland. *Am. J. Vet. Res.* **41,** 899 (1980).
27. Lane, T. J. *Fla. Cattleman* **48**(4), 94 (1984).
28. Thomas, H. S. *Horse Rider* **26**(11), 43 (1987).
29. Baker, J. P. *Univ. KY., Equine Data Line* Sept., p. 3 (1988).

21

The Complexity of Proper Bone Formation

I. INTRODUCTION

It is estimated that only one in five Thoroughbred horses which start training get to the race track. Of those that get there, only one in five is still racing after one year. Feet and leg problems account for a large percentage of the horses that drop out along the way. Periodically a horse breaks a leg during an important race shown on national T.V. This highlights the lack of knowledge available to prevent the tremendous toll resulting from unsound feet and leg problems in the horse. The expression "no foot, no horse" is still true and reflects the importance of finding answers to this critical problem area.

II. BONE REQUIRES MANY NUTRIENTS

Unfortunately many horse owners feel that calcium and phosphorus are the main factors in bone formation, but bone is more complex and requires not only calcium and phosphorus, but other minerals, vitamins, protein, hormones, and possibly other factors. Essentially, proper bone formation and maintenance involves a well-balanced diet throughout the life cycle of the horse.

III. BONE IS NOT STATIC

There is the mistaken impression that once bone is formed, it will stay that way. However, bone is not static and is continually being re-formed throughout life. There is a continuous interchange of calcium and phosphorus between the bone, the blood supply, and other body systems. If more calcium and phosphorus leaves the bone than is being replaced, the bone will eventually become porous, weak, and may be pulled out of shape, deformed, or broken down by the weight of the horse and the pull of the body muscles. Moreover, the added stress of walking, running, training, racing, or performance will speed up the process. This accounts for feet or legs being injured or broken during races, where considerable stress occurs.

IV. BONE SERVES AS A BANK

Fortunately, the body has a unique mechanism for using bone as a bank for borrowing calcium and phosphorus during emergencies when the need is greater than the diet provides. This may occur during periods of rapid growth, during the latter part of gestation, or during lactation.

The higher the level of milk production, the more calcium and phosphorus the mare may need to borrow. The level in the diet and that borrowed from bone must provide the mare with enough calcium and phosphorus to allow her to produce the maximum amount of milk she is genetically capable of. A lack of these two mineral elements gives the mare no alternative but to decrease milk production. This occurs because the level of calcium and phosphorus in milk is always about the same. When the calcium and phosphorus supply is not adequate, the mare compensates for it by decreasing milk production in line with the mineral supply available.

During rapid growth there may be periods when the exact level of calcium and phosphorus required is not adequate in the diet and some borrowing from bone is necessary. Rapid growth requires extra phosphorus for muscle formation and extra calcium and phosphorus for bone development. This extra need may also occur during the latter part of the gestation period when the developing foal makes most of its growth. Inadequate calcium and phosphorus during *in utero* development may have carry-over effects later on when the foal is born and is then pushed for early training and performance. Some of the bone abnormalities in the young foal may be traced back to inadequate nutrition *in utero*.

V. BORROWING MUST BE REPAID

It is of extreme importance that borrowing from bone be paid back. The depletion of calcium and phosphorus from bone causes no apparent harm if it is not too heavy and if it is restored by an adequate diet supply as soon as the peak of borrowing is over. However, if the borrowed minerals are not replaced before the next emergency demand period occurs, then serious bone problems may occur and other body functions may be adversely affected. This means proper nutrition is needed throughout the life cycle of the horse in order to produce high-level performance horses. A failure to do so may result in foals which later have feet and leg problems.

VI. BONE COMPOSITION

Bone is not just calcium and phosphorus. It contains about 25% ash or minerals, 20% protein, 10% fat, and 45% water. The fat level in bone will vary with the condition of the horse. The water content decreases with age. The main protein in bone is ossein. The minerals are deposited in the ossein and other bone proteins in a network which gives structure to the bone. The exact mechanism of bone formation and maintenance is still not well understood because of the complex chemistry and nutrient interrelationships involved.

VII. PROTEIN QUALITY

Proper protein nutrition is important in bone formation since bone contains about 20% protein. Studies in Florida (1) showed that inadequate protein intake for the weanling horse limits growth and bone development even when calcium and phosphorus intakes are equal to or above NRC recommendations. A Maryland study showed that the NRC energy and protein recommendations meet the requirements for maximum bone growth and development in well-managed young equines (1a). The weanling horse requires about 0.6–0.7% lysine. Unless a high-lysine source, such as soybean meal, is used as part or all of the protein supplement in the diet to assure an adequate lysine level, lysine supplementation may be needed for best results with the development of the young horse. Amino acid supplementation studies with the horse are in the beginning stages. The possible need for other indispensable amino acids is still not known. Moreover, the effect of indispensable amino acids in proper bone development and maintenance with the horse is still not well understood.

VIII. OTHER MINERALS

Other minerals are also involved. The ash in bone contains about 36% calcium, 17% phosphorus, and 0.8% magnesium. Magnesium is closely associated with calcium and phosphorus but it is not known how it functions in bone metabolism. A deficiency of magnesium in the foal causes skeletal muscle degeneration and a reduction in the bone magnesium level (2). A deficiency of magnesium in the pig causes weak pasterns (3). It is possible that it may cause the same condition in the horse.

A copper deficiency in the pig results in a drawing under of the rear legs and crookedness in the forelegs, a swelling in the region of the hocks, and extreme weakness in the foreleg (4). In another study of copper deficiency in the pig both the hind- and forelegs became severely bent out of shape (5). The chick fed a copper-deficient diet becomes lame in 2–4 weeks and the bones become fragile and easily broken (6). While these studies are with the pig and chick, they indicate the possibility that the same may occur in the horse deficient in copper. A Texas A & M study (7) and another at Ohio State University (8) implicate a copper deficiency as causing bone abnormalities in horses.

Manganese is essential for development of the organic matrix of the bone which is composed largely of mucopolysaccharides (6). With many animals deficient in manganese, bone abnormalities occur. Lameness, enlarged hock, crooked and shortened legs, and "overknuckling" have been obtained. The bone becomes lower in mineral content and density, and breaks occur more easily. Sometimes it takes two generations on a manganese-deficient diet for leg abnormalities to occur (6,9). Since a manganese deficiency in other animal species causes abnormal bone development, it is possible that the same may occur in the horse. Dr. Donoghue, University of Pennsylvania School of Veterinary Medicine, recommends at least twice NRC-recommended levels of manganese for the horse in areas with low-manganese forage (10).

Studies at Oregon State showed that turkey poults fed a 0.5% NaCl diet had significantly greater body weight, bone ash, bone-breaking stength, and tibia weight and lower bone abnormality scores as compared to poults fed lower sodium or chloride levels (11). This study showed that NaCl (or salt) is concerned with bone integrity.

Fluorine is known to be involved in proper bone and teeth development, but its role in the horse is still not known.

Calcium, phosphorus, magnesium, copper, manganese, sodium, chloride, and possibly fluorine are concerned with proper bone formation. It is not known whether any of the other mineral elements are also involved in bone. However, all essential minerals are needed for various body functions and a normal healthy horse. Therefore, proper mineral supplementation is needed to develop a top performing horse.

IX. VITAMINS

Vitamin A is concerned in normal development of bones and teeth. A failure of the spine and some other bones to develop normally results in pressure on the nerves and their resulting degeneration. For example, blindness can occur from a constriction of the optic nerve caused by a narrowing of the bone canal through which it passes. Bone changes may also be responsible for the muscle incoordination and other nervous symptoms shown by vitamin A deficient cattle, sheep, and swine (12). Therefore studies with many animals indicate a relationship between vitamin A and proper bone development. While data are lacking for the horse, it is logical to assume that an adequate vitamin A diet is needed to assure their proper bone development also.

A deficiency of vitamin D results in symptoms similar to those of a lack of calcium or phosphorus or both since all three are concerned with proper bone formation. A deficiency of vitamin D in the horse results in reduced bone calcification, stiff and swollen joints, stiffness of gait, softness of bones, bone deformities, frequent cases of fracture, and a reduction in blood serum calcium and phosphorus.

The role of vitamin D is more complex than previously thought. Recent studies indicate that vitamin D3 is changed to 25-hydroxy-D3 in the liver and into 1,25-dihydroxy-D3 in the kidneys. The 1,25-dihydroxy-D3 is the hormonal form of vitamin D3 which is needed for proper calcium and phosphorus absorption and utilization in the body. This finding raises the question of what happens to calcium and phosphorus utilization if the liver or kidneys are injured by parasites, disease, or other means. It is known, for example, that horses are very susceptible to parasites. Therefore, one needs to monitor closely any situation where horses are fed adequate levels of calcium, phosphorus, and vitamin D3 and still show bone deficiency symptoms. It may be that eventually 1,25-dihyroxy-D3 or similar compounds, instead of vitamin D3, will be used. A number of recent studies indicate that it is preferable to use vitamin D3 instead of vitamin D2 since it is more active in a number of animals including poultry.

One functional role of vitamin C is in the formation and maintenance of intercellular material having collagen or related substances in the bones and soft tissues. A deficiency of this vitamin causes loosening of the teeth and weak bones (12). However, experimental evidence is lacking to indicate that vitamin C supplementation is beneficial for bone problems in the horse. Some horse owners, however, are of the opinion that vitamin C supplementation is beneficial. Conditions under which vitamin C might be beneficial are: (a) during stress; (b) during hot weather (vitamin C supplementation increases egg shell quality with hens during hot weather); (c) during rapid growth or high-level performance; and (d) when something interferes with vitamin C synthesis or utilization.

It is not known which other vitamins may be beneficial for proper bone formation and maintenance in the horse. But, all are needed for a healthy horse and many owners with high-level performance animals supplement their diets with vitamins to ensure adequate vitamin nutrition.

X. STRESS

Many scientists are of the opinion that higher nutrition levels are helpful under high-stress conditions. Some scientists recommend different levels of nutrients for "moderate stress" or for "severe stress" conditions (13,14). University of Missouri scientists believe that pigs from certain strains are more susceptible to stress than others (15). Hysteria, nervousness, and greater density stress in commercial laying hens have been helped by the addition of higher levels of niacin to the diet (16). Almost all nutrients are required to maintain normal immunity and to maximize the body's ability to ward off stress and disease (17). Horses face many forms of stress including anxiety, possible exposure to disease, and stress from the many people and conditions they encounter during training, feeding, trips, racing, performance, crowds, noise, environmental conditions, and others. Therefore, racing and high-level performance horses require individual attention to determine the extra nutrients needed to meet various degrees of stress.

XI. EARLY AGE DEMANDS

A high-level performance horse is comparable to a human athlete and should be treated like one. The horse has the disadvantage of training at a very young age compared to the human. If a human has a life span of approximately 77 years and the horse a life span of 22 years, a 2-year-old horse could be compared to a 7-year-old human. Most performance horses start training at about one year of age, which is comparable to a human athlete starting to train at about $3\frac{1}{2}$ years of age. This early stress of training and performance accounts for some of the breakdown which occurs in the feet and legs of horses. Physiologically, the horse has not yet developed its bone and body systems fully enough to properly withstand all the stress of training and performance at such an early stage of development. Moreover, all horses become one year of age on January 1, regardless of when they were born. The majority of mares foal during March, April, and May. Therefore, horses born before that time have an added advantage in their bone and physiological development, especially if they were born in early January. Foals born later in the year, face more bone and body stress during training, performance, and competition as 2 year olds.

This brief discussion adds support for the recommendation that a well-balanced diet, adequate in all nutrients, is fed to horses to ensure proper bone formation and the minimization of feet and leg problems. Moreover, it stresses the need to have mares foal as early in the year as possible in order to minimize bone problems.

XII. SELECTION NEEDED

Horse owners must pay more attention to selection against breeding stock with a tendency toward poor feet and legs. This has been effective with other classes of livestock where heredity was involved. However, this will not be easy to accomplish since some horse owners find it very difficult to part with certain horses because of pedigree, expense involved, love for the horse, or some other reason. Therefore, feet and leg problems will not be entirely eliminated until more emphasis is placed on not using mares and stallions that have a hereditary weakness for breeding.

XIII. LEVEL OF FEEDING

When young horses are fed for rapid body and skeletal growth, they may develop bone abnormalities and lameness problems (18–20). Overfeeding has been suggested as a possible cause of epiphysitis, osteochondrosis, contracted tendons, and other bone abnormalities, but other conditions may also be involved. Bone growth may be affected by overfeeding, underfeeding, or inadequate nutrient intake due to deficiencies, excesses, or imbalances (10). Moreover, a well-balanced diet may become imbalanced if it is diluted with extra oats, varying quality hay, or other special feeds the feeder may feel the horse should have in addition to the diet well balanced with nutrients.

Unfortunately, there is still a lack of information on the nutritional requirements for different rates of growth in the young horse. It is logical to assume that a young horse growing at a rapid rate needs higher levels of certain nutrients to form bone and body tissue than a horse growing at a slow rate. This puts more emphasis on feeding a total diet properly balanced in energy, protein, minerals, and vitamins. Moreover, it means carefully watching the animals to make sure the level of feed and nutrient intake is adequate to prevent bone abnormalities. If they occur, it may mean that: (1) the feed intake level is too high; (2) the feed does not contain a sufficient level of certain nutrients; or (3) the animal may be genetically predisposed to certain bone abnormalities. In many cases, decreasing the level of feed intake, and thus the rate of growth, is beneficial in decreasing certain bone abnormalities. A recent University of Kentucky study (18) showed that young horses fed for rapid growth may not maximize bone deposition. They also showed that a creep feeding program that supplies NRC-recommended nutrient levels can increase the rate of skeletal growth with little decrease in quality of bone.

XIV. INDIVIDUAL ATTENTION

Horses have individual likes and dislikes for certain feeds, level of feed intake, frequency of feeding, level of hay or concentrate intake preference, and many others. Some horses may dislike certain people who feed, handle, or train them. Essentially, horses are prima donnas and require a certain amount of pampering to get them to consume a properly balanced diet and perform properly. So, high-level performance horses must be carefully observed and studied to learn their peculiarities and individual needs. Therefore, the first requisite is a well-balanced and nutritionally adequate feeding program, but equally important is a good feeder to ensure that the horse eats properly and at a level to meet its nutritional needs for proper development.

The difference in meeting the individual needs of each horse may be the difference between developing a champion or just another horse. Producing high-level performance horses is a very complex and challenging matter.

XV. CONCLUSIONS

Proper bone formation in horses is a very complex problem. It requires not only calcium and phosphorus, but other minerals, vitamins, protein, hormones, and possibly other factors. Bone is not static and is continually being re-formed throughout life in a continuous interchange of calcium and phosphorus between the bone, the blood supply, and other body systems. The early age stress of training for high-level performance accounts for some of the breakdown which occurs in the feet and legs of horses. The pushing of young horses too early and too hard for high-level performance may result in damage to immature, partially ossified bones. More attention also needs to be paid to selection against breeding stock with a tendency toward poor feet and legs. When young horses are fed for rapid body and skeletal growth, they may develop bone abnormality and lameness problems. These problems may be minimized, or eliminated, as more is learned about the optimum level of feeding and the nutritional requirements and management of the horse.

REFERENCES

1. Ott, E. A., and R. L. Asquith. *Fla. Agr, Exp. Stn., Res. Rep.* **AL-1982-2** (1982).
1a. Glade, M. J., N. K. Luba, and H. F. Schryver. *J. Anim. Sci.* **63**, 1432 (1986).
2. Harrington, D. D. *Br. J. Nutr.* **34**, 45 (1975).
3. Miller, E. R., D. E. Ullrey, C. L. Zutant, B. V. Baltzer, D. A. Schmidt, J. A. Hoefer, and R. W. Luecke. *J. Nutr.* **85,**13 (1965).
4. Teague, H. S., and L. E. Carpenter. *J. Nutr.* **43**, 389 (1951).
5. Cartwright, G. E., and M. M. Wintrobe. *Blood* **7**, 1058 (1952).
6. Scott, M. L., M. C. Nesheim, and R. J. Young. "Nutrition of the Chicken." M. L. Scott and Associates, Ithaca, New York, 1982.
7. Bridges, C. H., J. E. Womack, E. D. Harris, and W. L. Scrutchfield. *J. Am. Vet. Med. Assoc.* **180**, 295(1982).
8. Knight, D. A., A. A. Gabel, S. M. Read, R. M. Embertson, W. J. Tyznik, and L. R. Bramlage. *Proc. Am. Assoc. Equine Pract.* **31**, 455(1985).
9. Cunha, T. J. "Swine Feeding and Nutrition." Academic Press, New York, 1977.
10. Donoghue, S. *Proc. 26th Annu. Conv. Am. Assoc. Equine Practi.* p. 65(1980).
11. Egwuatu, C. O., J. A. Harper, D. H. Helfer, and G. H. Arscott. *Poult. Sci.* **62**, 353 (1985).
12. Maynard, L. A., J. K. Loosli, H. F. Hintz, and R. G. Warner. "Animal Nutrition," 7th ed. McGraw-Hill, New York, 1979.
13. Cunha, T. J. *Feedstuffs* **57,**(41), 37 (1985).
14. Cunha, T. J. *Squibb Int. Swine Update Rep.* **3**(1), 1 (1984).

15. Ellersieck, M. R., T. L. Venum, T. L. Durham, W. R. McVickers, S. N. McWilliams, and J. F. Lasley. *J. Anim. Sci.* **48**(3), 453 (1979).
16. Hansen, R. S. *Poult. Sci.* **49,** 1392 (1970).
17. Cunha, T. J. *Horse Rider* **25**(6), 30 (1986).
18. Thompson, K. N., J. P. Baker, and S. G. Jackson. *J. Anim. Sci.* **66,** 1692(1988).
19. Glade, M. J., and T. H. Belling, Jr. *Growth* **48,** 473 (1984).
20. Glade, M. J., and T. H. Belling, Jr. *Equine Vet. Sci.* **6**(3), 151(1986).

22

Exercise Physiology[1]

I. INTRODUCTION

A great deal of emphasis has been placed on feeding, nutrition, breeding, selection, health care, and management in order to obtain optimum results with high-level performance horses. Moreover, high-level performance requirements differ depending on the task the horse is asked to perform. Short-term racing, endurance riding, cutting horse activities, driving, steeple chasing, rodeo and polo activities, show jumping, and a host of other tasks have different energy and other nutrient requirements. The length and intensity of the exercise involved differs, which in turn influences the energy required and the nutrients needed to support the biochemical conversion of energy to muscle action and the maintenance of the total body system to support it. Unfortunately, very little is known about the nutrient requirements of the horse engaged in high-level performance. Likewise, not enough is known about muscle exercise physiology which is the key to how well the horse reacts to differing demands of intensity and level of performance. Fortunately, an increasing number of studies are being conducted on exercise physiology. As this area develops, more knowledge becomes available to help the owner of the high-level performance horse, who faces fierce competition, where fractions of a second may be important under certain performance situations (1). Moreover, the difference between superior or average performance may be very minute. So, developing a champion or just another horse may depend on how well the owner understands exercise physiology and how well the horse is provided with the technology required to optimize his performance capability. This involves proper training, diet, management, and health care. It is clear that the science and technology of properly feeding and training the performance horse is in its infancy.

II. ENERGY

Energy is probably the most important nutrient needed by the high-level performance horse since the exercising muscle needs extra energy. Proper train-

1 This chapter was reviewed by Steven Wickler, Ph.D, DVM, Associate Professor and Associate Director of Equine Research, California State Polytechnic University, Pomona, California.

ing for the specific task required enhances the capacity of the horse for energy utilization and performance. But, storage of the energy required for the different kinds of task the horse is asked to perform is also very important. Energy requirements differ with the distance, duration, intensity, nervousness of the horse, weight carried, condition of the horse, previous training, environmental conditions, skill in handling the horse, and other factors. The daily digestible energy requirement of the high-level performance horse may be considerably greater than the maintenance requirement. The race horse may require as much as three times the maintenance digestible energy requirement (2).

The key behind all types of equine exercise is the conversion of chemically bound energy into the energy required for muscular movement. It appears that in order to sustain a high rate of energy generation and work, ATP (adenosine triphosphate) must be resynthesized at a rate equal to its utilization in muscle activity (3).

A. Energy Metabolism

The entire energy metabolic process is very complex and involves many complicated chemical reactions. An Illinois study (4) strongly suggests that the metabolic data collected using muscle of horses younger than 6 months of age do not accurately reflect adult muscle metabolic enzyme patterns. So the age of the horse needs to be considered in energy physiology studies. Only a brief discussion on energy metabolism will be given here. More detail on the biochemistry involved is given in papers by Miller *et al.* (5) and others (3, 6). Moreover, much more needs to be learned before an accurate picture of energy metabolism in the horse is well understood. ATP is the form of stored energy that is used for muscular contraction. Other energy forms must be converted to ATP before they can be utilized for muscular work (6). Fat, oils, tallow, and the volatile fatty acids (VFA) (primarily acetic, proprionic, and butyric) all need to be broken down into 2-carbon units before they enter the energy pathway which leads toward the formation of ATP. Oxygen is required to transform fat and VFA into ATP. If oxygen is not present, ATP is not produced.

Cellulose, which is a complex carbohydrate found in forages, cannot be digested in the small intestine of the horse since no enzyme is produced there for this purpose. The cellulose passes on to the hindgut area where microorganisms break it down to glucose. Most of the glucose is then broken down into VFA which enter the same pathway as fats for formation of ATP. So oxygen is required to utilize cellulose as a source of ATP for muscular contraction.

Starch and other simpler sugars, found to a large extent in grains, are digested in the small intestine and are absorbed as glucose which can be stored as glycogen. Glucose enters the energy pathway at a different point from fats and can produce ATP in limited quantities without oxygen. This is advantageous to the horse doing high-intensity exercise when oxygen availability is a limiting factor. There is a disadvantage, however. If not enough oxygen is present, all the glucose is not completely broken down to enter the energy pathway and lactic acid (a 3-carbon compound) is formed. Later, when enough oxygen again becomes available, the lactic acid is broken down into a 2-carbon compound which then enters the energy pathway at the same location as fats.

On the basis of this simplified version of energy utilization the following assumptions may be made.

1. Horses performing low-intensity work (aerobic) can utilize fats, VFA, or cellulose as energy sources to produce ATP for muscular contraction since oxygen is not a limiting factor. The aerobic work is that which can be performed within the horse's capacity to supply the tissues with oxygen and for the tissues to utilize the oxygen.

2. Horses performing high-intensity work (anaerobic) need some starch or simpler sugars (such as supplied by the cereal grains) in the diet. How much grain is needed is not known. It might be desirable to use 5–10% or more corn in place of oats as a means of supplying a higher level of starch, other simpler sugars, and energy in the diet. Actually, most tasks which the horse performs involve a combination of aerobic and anaerobic work. Low-intensity work involves the highest percentage of aerobic metabolism whereas maximum racing effort involves the highest level of anaerobic metabolism.

3. It would appear that horses developed for maximum racing performance might be benefited by the use of 5–10% fat in the diet. Whether higher levels of fat might be used and result in better performance is not yet known. In some cases, depending on the diet and level of oats used, oats which are lower in energy and higher in fiber and volume than corn might be decreased 5–10% or more and replaced with a higher energy lower volume grain such as corn. In fact, an Illinois study (7) with Quarter Horses showed that substituting corn for oats decreased diet and intestinal tract volume and improved racing time.

B. Muscle Fiber Types

The horse has three major muscle fiber types. They are: (1) slow-twitch, oxidative; (2) fast-twitch, glycolytic; and (3) fast-twitch, oxidative-glycolytic fibers. The fast-twitch, glycolytic fibers are adapted for rapid contractions and can obtain energy efficiently by anaerobic metabolism. The slow-twitch fibers have slow contractions and obtain energy by aerobic metabolism. The fast-twitch, oxidative-glycolytic fibers have characteristics intermediate between slow and fast twitch and can vary from primarily aerobic to primarily anaerobic (6,8).

The glycogen level and its decrease in horse muscle varies with the muscle fiber type and the kind of exercise. The slow-twitch and fast-twitch oxidative-glycolytic fibers are depleted after slow trotting exercise. The fastest rate of glycogen depletion occurs in the slow-twitch fibers. As glycogen is depleted, the capacity for exercise is decreased. The glycogen level of fast-twitch glycolytic fibers is not depleted at low-speed exercise until the other muscle fibers are depleted. The glycogen level of fast-twitch glycolytic fibers is depleted at maximum or submaximum high-intensity level of exercise after a period of time (6,8).

An Illinois study (9) showed that there are significant variations in the distribution of muscle fiber types within individual muscles, between different muscles, and between the same muscle from adult and fetal origin in the pony. They indicate that caution should presently be used in the interpretation of equine muscle fiber-type data. Another Illinois study (10) with Quarter Horses found no difference in the percentages of fast-twitch glycolytic, fast-twitch oxidative-glycolytic, or slow-twitch oxidative muscle fibers between 6-month-old and mature horses.

A Texas A & M study (11) stated that it is apparent that different types or breeds of horses vary considerably in their muscle fiber distribution. They felt it might be possible to estimate suitability for specific performance by examining a horse's muscle fiber type.

It is important that the muscle groups used for specific tasks be properly trained for the exercise they will perform. So, training programs should entail exercise activities which approximate as closely as possible to those actually performed during competition. It is also important that the training workload be started at a low level and gradually increased in order to prevent injury to the horse. The workloads should eventually reach the competitive level for the specific task in which the horse will be involved.

The challenge ahead is to develop training programs that will further improve the energy systems needed for optimum performance of the activity involved. There are many programs being used but more research information is needed before they will be thoroughly understood. Moreover, the individuality of the horse as well as the trainer may require modification of the training system for optimum results. Regardless of the training system used, it is very important that horses are properly warmed up and cooled down.

C. Effect of Exercise Regime

Training for the kind of work involved increases the efficiency of the horse. When the horse is asked to perform an unaccustomed task, some muscles are used which would not ordinarily be used if prior training for that task had been completed.

A number of studies have shown that exercise and training can influence energy storage. A Texas A & M study (12) showed that the energy stored as glycogen in the muscles of exercising horses can be greatly increased by manipulating the exercise regime and diet.

Conditioning of horses generally results in decreased heart rates and blood lactic acid levels during exercise (17). Both of these changes can improve performance.

Another Texas A & M study (13) showed that cutting horse performance appears to be enhanced in mature, trained cutting horses by training that is specific for that work. They also showed that the work performed by the cutting horses requires the horse to employ anaerobic, glycolytic pathways to produce sufficient energy for cutting horse work.

A West Texas University study (14) suggested that transporting the horse by trailer approximately 1 hour prior to a submaximal exercise event did not affect its physical performance. A Utah study (15) showed that trailering, if done properly, had relatively little effect on racing performance. A Kentucky study (16) showed that transporting horses 60 miles prior to a submaximal exercise event did not affect their physical performance.

A New Jersey study (18) showed that both fat and carbohydrates are used as energy sources in conditioned and unconditioned horses trotting for half an hour but conditioned horses use fatty acids more efficiently than unconditioned horses.

A Kentucky study (19) concluded that exercise training of young growing horses may result in horses with greater athletic potential and durability by the time they enter race training.

A Kansas State University study (20) presents physiological characteristics of an Arabian endurance horse which was the 1985 winner of the Purina 100-mile race of champions. The horse was evaluated on a high-speed treadmill and on a dirt track and compared to two other Arabian horses. The data provide very helpful information on the electrocardiogram, heart rate, blood lactate, plasma volume, hematocrit, muscle fiber composition, and other indices of value for the endurance horse.

An Illinois study (21) suggested that sodium bicarbonate administration may be helpful to some racing Standardbreds. Another study (22) with Thoroughbreds reported no effect of sodium bicarbonate on the time to finish or the horse's metabolic characteristics. More studies are needed to determine the possible need for sodium bicarbonate in horses.

A Michigan study (23) reported that the increased plasma creatine phosphokinase activity (CPK) after 2 days of rest in conditioned horses suggests the importance of gradually increasing work intensity after days of rest to minimize muscle damage. Many horse owners report an increase in the incidence of tying-up as a result of exercise after periods of rest.

An Illinois study (24) involved horses exercised to exhaustion on a motorized treadmill. These workers stated that their results indicate that ammonia accumulation may have a significant effect upon the development of fatigue because of impairment of oxidative metabolism. In their study, glycogen availability did not appear to be a limiting factor, since muscle glycogen levels were decreased by only 25% during the exercise bout.

High ammonia levels seem to accompany high-intensity exercise. The increase in ammonia may be related to the conditioning status. In a study with Standardbred horses which were relatively fit, the blood ammonia and lactate levels were quite elevated when the horses were raced one mile (5). There was a significant correlation between lactate and ammonia levels in this study which suggested that these two metabolites are related in short-term energy-producing mechanisms. They may also be involved in the development of fatigue.

A Texas A & M study (25) suggested that low muscle glycogen concentrations may limit anaerobic work performance. These data also indicate that lactic acid may be a result of, and not a limitation to, the amount of anaerobic work performed. These workers stressed the need for further research on the causes and mechanisms of fatigue and their role in anaerobic work. In a previous study at Texas A & M (12) the data suggested that lactic acid may not be the only factor determining the onset of fatigue, since fatigue was induced at very different levels of lactic acid in the blood.

A Florida study (26) indicated that there is a trend toward increasing mineral deposition with training and exercise. Previously, it had been suggested that some degree of physical activity may be required for normal bone development in young horses (27). A Cornell study (28) reported evidence that training influences calcium retention, bone turnover, and a trend toward increased bone weight.

An Illinois study (7) with Quarter Horses may suggest that 2-year-old horses are not as capable of racing as 3-year-old horses. They also found that replacing oats with corn resulted in a faster racing time. The corn diet occupied less volume per pound and decreased the gastrointestinal mass which may have affected the horse's racing performance.

A Cornell study (29) reported that polo ponies have a combination of high lactic acid production and respiratory alkalosis and so could be used as a model for this condition.

A Connecticut study (30) showed that libido scores of nonexercised 2-year-old Morgan stallions were higher than exercised stallions after 12 weeks of forced daily exercise. Whether the decrease in libido was simply a function of fatigue or a decline in sex drive was not clear.

D. Effect of Fat

Many studies have been conducted on the effect of the addition of fat to horse diets. The horse has been able to utilize efficiently up to 30% fat in the diet (31–

33). Quite a few studies have involved using 5–10% fat with horses under different exercising conditions. It is apparent that adding fat to the diet of exercising horses increases the caloric density and allows a reduction in total feed intake. A lowered feed intake minimizes the possible incidence of founder, colic, tying-up, and digestive disturbances which might otherwise occur with carbohydrate overloading.

The addition of fat to the exercising horse's diet has a sparing effect on muscle glycogen reserves. There is some question as to whether adding fat results in stimulation of muscle glycogen synthesis or a sparing effect on muscle glycogen utilization. Regardless of the exact situation, if a high-level performance horse can be conditioned to utilize fat as an energy source during aerobic work, the resulting increase in glycogen storage can add to the energy supply to the muscles of horses working in a state of oxygen debt. This could delay the onset of fatigue in the exercising horse.

Hard working, heavily stressed horses may require 30–35 Mcal of digestible energy in their total diet (2). This level is much higher than recommended in 1978 by NRC. Horses not fed enough energy to meet high-level performance requirements will utilize body fat stores and lose weight. Since an 1100-lb horse can only comfortably eat 25–30 lb of total feed daily, it is apparent that the diet for the hard working horse should be high in energy density. Therefore, the use of dietary fat is a means of increasing the diet energy density without increasing the volume of feed intake.

Arabian horses ridden for 36 miles required 15% less total feed when fed a diet containing 8% fat (34). Standardbred horses in a study in Sweden (35) fed 18% fat exhibited increased fatty acid metabolism and decreased lactic acid production when performing at 75% of their maximum capacity. A study with Quarter Horses (36) showed that a diet containing 10% fat spared muscle glycogen. Another Texas A & M study (37) showed that adding 10% fat to the diet provided more energy for work and may have increased muscle glycogen storage and its subsequent use for anaerobic work typical of the cutting horse. However, studies to date on adding fat to the diet have shown only minor increases in stamina or delays in the onset of fatigue (34, 38–40). This indicates the need for more studies, especially those which involve determination of the requirement for protein and other nutrients which may be needed at higher levels as the level of fat and calories is increased in the diet. A lack of other nutrients could minimize or decrease the effectiveness of the added fat.

A Texas A & M study (41) showed that 10% fat can be used in high-energy diets for rapidly growing horses to replace some of the carbohydrates, which if fed in excess, have been reported to cause skeletal problems (42). There were no clinical or radiographic indications of skeletal abnormalities in the horses fed 10% fat.

A Virginia study (43) showed no detrimental effects from feeding a 14% fat diet to horses during a period of conditioning. In a Texas A & M study (44) the addition of 5 or 10% fat to the diet of performance horses caused an overall decrease in heart rate, blood lactate, and blood total lipids, and an increase in packed-cell volume. They suggested that the lowered blood lipid concentrations in the horses fed 10% fat might have been due to conditioning for more efficient fatty acid uptake by the working muscle.

A Cornell study (45,45a) reported that the respiratory quotients measured in their experiment indicated that fat is the major substrate for energy generation at slow speeds in the horse. They also showed that more energy was expended by the horse when carrying a rider than when unweighted. The amount of energy expended by the horses was related to the speed involved and appeared to be proportional to the body weight of the animal or the combined weight of the horse plus the rider.

An Illinois study (46) showed that enhancement of muscle glycogen in horses is possible to a small extent by diet and exercise. No apparent muscle stiffness was noted in the horse studies. These workers stated that it has not been established that high glycogen stores in muscles *per se* cause tying-up. Some feel, however, that horses prone to this disorder might be affected more severely when subjected to glycogen loading.

A Kentucky study (47) involved feeding Thoroughbreds 0, 5, 10, or 20% corn oil in the diet. The horses were trained and handled in an effort to duplicate a race track environment. Blood glucose concentrations were higher during and after exercise in the horses fed the high-fat diet which may indicate a glucose-sparing effect. Blood glycerol levels were elevated above resting values throughout exercise and recovery indicating that fat was being metabolized. The corn oil-supplemented diet proved to be very palatable and easily digested.

E. Effect of Protein

A Virginia Polytechnic Institute study (48) found no detrimental effects from feeding an excessive level of protein (20%) to 2-year-old horses which were fitness trained on a treadmill for 5 weeks. They also found no beneficial effects from the 20% protein diet as compared to a 10% protein diet. A Cornell study (49) reported no harmful effects from feeding 24% protein diets to exercising horses, compared to horses fed 12% protein diets.

A Colorado study (50) on endurance-type work showed that increased sweating and higher pulse and respiration rates occurred when excess protein was fed to the horses. These workers also stated that horses may be nutritionally conditioned for endurance-type work and that diets with fairly high levels of fat may be beneficial in this respect. A later Utah study (51) showed that the higher protein levels used in their trial were of no metabolic hindrance to competitive performance. A Maryland study (1) cautioned against using excessive levels of protein in the diet of racing Thoroughbred horses.

A Texas A & M study (52) reported that large amounts of water and possibly nitrogen are lost through sweat in exercising horses. This indicates a need to determine the significance of nitrogen loss in sweat and its long-term effect on exercised muscle metabolism and on protein and/or amino acid requirements. Another Texas A & M study (53) reported that increasing nitrogen intake results in a similar pattern effect on nitrogen retention in conditioned horses to that previously reported in nonconditioned horses. However, nitrogen balances tended to be more positive in conditioned horses as compared to idle horses fed similar levels of nitrogen.

An Illinois study (54) showed that the exercise and training used in their trial caused minimal effects on plasma free amino acids. These scientists felt that horses must be worked at considerably higher intensity or for longer duration, or both, to cause changes in plasma amino acids levels with exercise.

Protein can be used as an energy source, but it is a very inefficient process and is not used unless the horse becomes extremely energy deficient. It is best not to rely on protein as a source of energy. Present evidence indicates no harmful effects from feeding some excess protein, but one should be careful and not exceed NRC requirements to any marked degree. A level of about 12–14% protein in the total diet of high-level performance horses should be adequate. The 2-year-old horse in training can best meet its protein needs with a diet containing a 13–14% protein concentrate plus good-quality hay (55). If the hay used is only average in quality and digestibility, many will use 1 or 2% more protein in the concentrate feed to assure an adequate protein level. It is best to feed a small excess of protein rather than run the risk of a protein deficiency with the high-level performance horse.

F. Effect of Other Nutrients

A Kentucky equine scientist (55) states that the horse in hard training or in racing may have an increased requirement for vitamins A, D, E, thiamin, riboflavin, and B12. A Cornell scientist (56) stated that horses fed poor-quality hay and unsupplemented grain for prolonged periods may respond to vitamins A, E, thiamin, folacin, or biotin supplementation, but further studies are needed before definite guidelines concerning supplements can be made.

A Michigan State study (57) reported a need for supplemental selenium when diets of exercising horses are low in the mineral element. Moreover, exercising horses may need levels of selenium above maintenance requirements.

A Texas A & M study (58) suggested that the NRC-recommended allowance of 1.36 mg of thiamin per lb of diet for maintenance may not be adequate for performing horses. They also reported that large doses of thiamin are largely excreted.

It is apparent that little information is available on the effect of vitamins and minerals on energy utilization in the high-level performance horse. Vitamins and minerals are involved in the complex metabolic and biochemical systems of protein, fat, and carbohydrate utilization in the total biochemical system of providing energy for muscle contraction and performance. Unfortunately, data are lacking on the performances of vitamin and/or mineral-supplemented horses versus those not supplemented under competitive conditions. A detailed discussion on vitamins occurs in Chapter 5 and one on minerals in Chapter 6. The use of vitamin and mineral supplementation is given in the chapters dealing with feeding horses during the various stages of their life cycle.

Exercising horses lose a certain amount of electrolytes. The electrolytes are salts of various minerals which are dissolved in the body fluids. The need for electrolyte supplementation depends on their level in the diet and the amount of sweat produced. A heavily sweating horse may develop a lack of electrolytes and a need for electrolyte supplementation. This is especially the case with endurance horses. If the temperature is high, many of these horses develop excessive fatigue, muscle spasms and cramps, dehydration, and exhaustion. More studies are needed in this area and from them may come information which will help unlock secrets of stamina, speed, and endurance in racing.

One report (59) indicated that the exercising horse may require four times the maintenance requirement of potassium. This is due to a large loss in sweat and other losses from the body.

III. ENERGY REQUIREMENTS

The data shown in Table 22.1 indicate the level of digestible energy required in light, medium, and strenuous race training. It is apparent that the level and density of the energy in the diet increases as the level of exercise or training increases.

The addition of fat to the diet increases its energy level. Moreover, a higher level of grain such as corn which is high in starch and low in fiber may also be added to the diet. The starch and other simpler carbohydrates in grain are required to replace the glycogen used during exercise. An Oklahoma A & M study (60) indicated that high starch levels might cause founder and digestive disturbances. Therefore, there is a need for more studies to determine the optimum level of starch necessary to maintain normal muscle glycogen for the different kinds of exercise.

In trying to increase the energy density of the diet, it should be stressed that the horse should be fed a minimum of 1% of its body weight per day as hay. Hay

TABLE 22.1

Energy Requirements of Horses Used in Various Activities Associated with Race Training (55)

Type of work	Digestible energy required (Mcal)	Digestible energy concentration in total diet (Mcal/lb)
Light[a]	21.89	1.2
Medium[b]	28.69	1.2
Strenuous[c]	34.00	1.3

[a]Includes yearlings being broken, 2 year olds in light training.
[b]Includes horses in medium training.
[c]Includes horses in heavy race training.

has a low-energy level as compared to concentrates. Therefore, the remainder of the diet is concentrates and its level of feeding may vary from 1 to 1.8% of the horse's body weight per day. The level of concentrates fed will vary with its energy density, the hay's energy density and digestibility, the level of work and its intensity, the condition of the horse, and other factors. The diet should be designed to condition the horse and to develop a muscular system for the type of work involved. It should keep the horse in a fit condition. The level of energy fed might allow a minimal weight gain but not an accumulation of fat. It should be stressed that horses vary greatly in their energy needs. One set of energy recommendations may cause a horse to become overweight and another to lose weight. So, one must continually observe the horse's condition and feed it accordingly.

A survey of feeding practices at two Standardbred race tracks (61) by Cornell scientists showed that the 1978 NRC estimates of energy requirements were not adequate for racing Standardbred horses. They also stated that the effect of environmental conditions at the track and temperament of the horses need to be considered in energy requirements.

A Cornell study (61,62) showed that the 1978 NRC report significantly underestimated the energy needs of racehorses. If those NRC recommendations are followed in detail, energy requirements were underestimated by 49%. They calculated that the digestible energy intake of Thoroughbred racehorses at two tracks averaged 35.7 Mcal of digestible energy per day which is considerably above 1978 NRC recommendations and in line with 1989 NRC recommendations.

IV. CONCLUSIONS

The science of properly feeding and training for optimum results with the high-level performance horse is in its infancy. Proper training for the specific task required enhances the capacity of the horse for optimum energy utilization and performance. But, equally important is the storage and availability of the energy required for the different kinds of tasks the horse is asked to perform. Exercise, training, and diet can influence energy storage. The key to top performance is optimum conversion of the chemically bound energy into the energy required for muscular contraction.

Oxygen is needed to transform fat, cellulose, and VFA into ATP. Starch and other simpler sugars which are stored as glycogen can produce ATP in limited quantities without oxygen. ATP is the form of energy which is used for muscle contraction. Low-intensity work (aerobic) can utilize fats, FVA, and cellulose as energy sources. High-intensity work (anaerobic) can utilize starch and simpler sugars as energy sources. Most tasks the horse performs require a combination of aerobic and anaerobic work: low-intensity work involves the highest percentage of aerobic metabolism whereas maximum racing effort involves the highest level of anaerobic metabolism.

The horse has three major muscle fiber types. It is important that the muscle groups used for specific tasks be properly trained for the exercise they will perform.

Training and exercise are beneficial to proper bone development.

The addition of fat to the exercising horse's diet has a sparing effect on muscle glycogen reserves. It may be due to a sparing effect on muscle glycogen or possibly to stimulation of muscle glycogen synthesis. The added fat increases the caloric density of the diet and allows a reduction in total feed intake which may minimize digestive disturbances in the high-level performing horse. Many scientists recommend the addition of 5–10% of a high-quality fat to the diet of high-level performance horses.

Horses performing high-intensity work (anaerobic) may need a higher level of grain such as corn to supply more starch and simpler sugars. How much grain will satisfy their need is not known. Excess grain is to be avoided, however, since it may cause digestive disturbances because of carbohydrate overloading. One study with Quarter Horses showed that substituting corn for oats decreased diet and intestinal tract volume and improved racing time.

A number of studies have shown no harmful effects from feeding a reasonable amount of protein above the NRC requirement levels.

Vitamins and minerals are needed for muscle metabolism and energy utilization. Little information is available on the need for their supplementation and the levels to use for the high-level performance horse.

REFERENCES

1. Glade, M. J. *Equine Vet. J.* **15**, 31 (1983).
2. Householder, D. D., and G. D. Potter. *Tex. A & M Mimeo Rep.* (1988).
3. Pagan, J. D., B. Essen-Gastavsson, A. Lindholm, and J. Thornton. *Proc. Equine Nutr. Physiol. Symp., 10th,* p. 425 (1987).

4. Kline, K. H., and P. J. Bechtel. *Proc. Equine Nutr. Physiol. Symp., 10th,* p. 403 (1987).
5. Miller, P., L. Lawrence, K. Kline, R. Kane, E. Kwicz, J. Smith, M. Fisher, A. Siegal, and K. Bump. *Proc. Equine Nutr. Physiol. Symp., 10th,* p. 397 (1987).
6. Topliff, D. R. *Okla. A & M Mimeo Rep.* (1987).
7. Burke, D. J., W. W. Albert, and P. C. Harrison. *Proc. Equine Nutr. Physiol. Symp., 7th,* p. 197 (1981).
8. Wood, C. R. *Univ. Ky., Equine Data Line,* March, p. EL-2 (1986).
9. Raub, R. H., P. J. Bechtel, K. H. Kline, and L. M. Lawrence. *Proc. Equine Nutr. Physiol. Symp., 8th,* p. 211 (1983).
10. Kline, K. H., and P. J. Bechtel. *Proc. Equine Nutr. Physiol. Symp., 8th,* p. 217 (1983).
11. Sigler, D. H., G. D. Potter, and T. R. Dutson. *Proc. Equine Nutr. Physiol. Symp., 7th,* p. 179 (1981).
12. Topliff, D. R., G. D. Potter, T. R. Dutson, J. L. Kreider, and G. T. Jessap. *Proc. Equine Nutr. Physiol. Symp., 8th,* p. 119 (1983).
13. Webb, S. P., G. D. Potter, J. W. Evans, C. A. Schwab, and B. D. Scott. *Proc. Equine Nutr. Physiol. Symp., 10th,* p. 351 (1987).
14. Beaunoyer, D. E., and J. D. Chapman. *Proc. Equine Nutr. Physiol. Symp., 10th,* p. 379 (1987).
15. Slade, L. *Proc. Equine Nutr. Physiol. Symp., 10th,* p. 511 (1987).
16. Beaunoyer, D. *Univ. Ky., Equine Data Line,* March p. EL3 (1987).
17. Erickson, H. H., W. L. Sexton, R. M. DeBowes, J. R. Coffman, and D. H. Sigler. *Proc. Equine Nutr. Physiol. Symp., 9th,* p. 194 (1985).
18. Goodman, H. M., G. W. VanderNoot, J. R. Trout, and R. L. Squibb. *J. Anim. Sci.* **37,** 1 (1973).
19. Raub, R. H., S. G. Jackson, and J. P. Baker. *Proc. Equine Nutr. Physiol. Symp., 10th,* p. 409 (1987).
20. Erickson, H. H., B. K. Erickson, G. L. Landgren, M. K. Hopper, H. C. Butler, and J. R. Gillespie. *Proc. Equine Nutr. Physiol. Symp., 10th,* p. 493 (1987).
21. Lawrence, L., K. Kline, P. Miller, J. Smith, A. Siegel, K. Kurcz, R. Kane, M. Fisher, and K. Bump. *Proc. Equine Nutr. Physiol. Symp., 10th,* p. 499 (1987).
22. Kelso, T., D. Hodgson, E. Witt, W. Bayly, B. Grant, and P. Gollnick. *Proc. Int. Conf. Equine Exercise Physiol. 2nd* p. 64 (1986).
23. Shelle, J. E., W. D. VanHuss, J. S. Rook, and D. E. Ullrey. *Proc. Equine Nutr. Physiol. Symp., 9th,* p. 206 (1985).
24. Miller, P. A., L. M. Lawrence, and A. M. Hank. *Proc. Equine Nutr. Physiol. Symp., 9th,* p. 218 (1985).
25. Topliff, D. R., G. D. Potter, J. L. Kreider, T. R. Dutson, and G. T. Jessup. *Proc. Equine Nutr. Physiol. Symp., 9th,* p. 224 (1985).
26. Mills, D. L., S. Lieb, and E. A. Ott. *Proc. Equine Nutr. Physiol. Symp., 8th,* p. 50 (1983).
27. Owen, J. M. *Equine Vet. J.* **7,** 40 (1975).
28. Schryver, H. F., H. F. Hintz, and J. E. Lowe. *Am. J. Vet. Res.* **39,** 245 (1978).
29. Craig, L., H. F. Hintz, L. V. Soderholm, K. L. Shaw, and H. F. Schryver. *Cornell Vet.* **75,** 297 (1985).
30. Dinger, J. E., and E. E. Noiles. *J. Anim. Sci.* **62,** 1220 (1986).
31. Hambleton, P. L., L. M. Slade, D. W. Hamar, E. W. Keinholz, and L. D. Lewis. *J. Anim. Sci.* **51,** 1330 (1980).
32. Kane, E., J. P. Baker, and L. S. Bull. *J. Anim. Sci.* **48,** 1379 (1979).
33. Rich, G. A., J. P. Fontenot, and T. N. Meacham. *Proc. Equine Nutr. Physiol. Symp., 7th,* p. 30 (1981).
34. Hintz, H. F., M. W. Ross, F. R. Lesser, P. F. Leids, K. K. White, J. E. Lowe, C. E. Short, and H. F. Schryver. *J. Equine Med. Surg.* **2,** 483 (1978).

35. Pagan, J. D., B. Essen-Gustafson, A. Lindholm, and J. Thornton. *Proc. Equine Nutr. Physiol. Symp., 10th,* p. 425 (1987).
36. Meyers, M. C., G. D. Potter, L. W. Greene, S. F. Crouse, and J. W. Evans. *Proc. Equine Nutr. Physiol. Symp., 10th,* p. 107 (1987).
37. Webb, S. P., G. D. Potter, and J. W. Evans. *Proc. Equine Nutr. Physiol. Symp., 10th,* p. 115 (1987).
38. Glade, M. J. *Large Anim. Vet.* **43,** 30 (1988).
39. Hambleton, P. L., L. M. Slade, D. W. Hamar, E. W. Kienholz, and L. D. Lewis. *J. Anim. Sci.* **51,** 1330 (1980).
40. White, K. K., F. R. Lesser, H. F. Hintz, and J. E. Lowe. *J. Equine Med. Surg.* **2,** 525 (1978).
41. Scott, B. D., G. D. Potter, J. W. Evans, J. C. Reagor, G. W. Webb, and S. P. Webb. *Proc. Equine Nutr. Physiol. Symp., 10th,* p. 101 (1987).
42. Glade, M. J., and T. H. Belling, Jr. *Growth* **48,** 473 (1984).
43. Worth, M. J., J. P. Fontenot, and T. N. Meacham. *Proc. Equine Nutr. Physiol. Symp., 10th,* p. 145 (1987).
44. Meyers, M. C., G. D. Potter, J. W. Evans, L. W. Greene, and S. F. Crouse. *Proc. Tex. A & M Agric. Ext. Serv., Horse Short Course* p. 20 (1987).
45. Pagan, J. D. and H. F. Hintz. *Proc. Equine Nutr. Physiol. Symp., 9th,* p. 182 (1985).
45a. Pagan, J. D. and H. F. Hintz. *J. Anim. Sci.* **63,** 822 (1986).
46. Kline, K. H., and W. W. Albert. *Proc. Equine Nutr. Physiol. Symp., 7th,* p. 186 (1981).
47. Duren, S., C. Wood, and S. Jackson. *Univ. Ky., Equine Line,* Sept., p. EL-1 (1987).
48. Frank, N. B., T. N. Meacham, and J. P. Fontenot. *Proc. Equine Nutr. Physiol. Symp., 10th,* p. 579 (1987).
49. Hintz, H. F., K. H. White, C. E. Short, J. E. Lowe, and M. Ross. *J. Anim. Sci.* **51,** 202 (1980).
50. Slade, L. M., L. D. Lewis, C. R. Quinn, and M. L. Chandler. *Proc. Equine Nutr. Physiol. Symp., 4th,* p. 114 (1975).
51. Slade, L. M. *Proc. Equine Nutr. Physiol. Symp., 10th,* p. 585 (1987).
52. Freeman, D. W., G. D. Potter, G. T. Schelling, and J. L. Kreider. *Proc. Equine Nutr. Physiol. Symp., 9th,* p. 230 (1985).
53. Freeman, D. W., G. D. Potter, G. T. Schelling, and J. L. Kreider. *Proc. Equine Nutr. Physiol. Symp., 9th,* p. 236 (1985).
54. Russell, M. A., A. V. Rodiek, and L. M. Lawrence. *Proc. Equine Nutr. Physiol. Symp., 8th,* p. 125 (1983).
55. Jackson, S. G. *Proc. Univ. of Ky. Conf. Racehorse* p. 40 (1987).
56. Hintz, H. F. *Anim. Health Nutr.* **42**(4), April (1987).
57. Shelle, J. E., W. D. Van Huss, J. S. Rook, and D. E. Ullrey. *Proc. Equine Nutr. Physiol. Symp., 9th,* p. 104 (1985).
58. Topliff, D. R., G. D. Potter, J. L. Kreider, and C. R. Creagor. *Proc. Equine Nutr. Physiol. Symp., 7th,* p. 167 (1981).
59. Meyer, H., C. Gurer, and A. Linder. *Proc. Equine Nutr. Physiol. Symp., 9th,* p. 130 (1985).
60. Topliff, D. R., S. F. Lee, and D. W. Freeman. *Proc. Equine Nutr. Physiol. Symp., 10th,* p. 421 (1987).
61. Ignatoff, J., and H. F. Hintz. *Feedstuffs* **52**(41), 24 (1980).
62. Winter, L. D., and H. F. Hintz. *Proc. Equine Nutr. Physiol. Symp., 7th,* p. 136 (1981).

23

Miscellaneous Topics

There are a number of areas in which limited research information is available on the horse. These are important areas and will be discussed in this chapter.

I. ANTIBIOTICS

There is little research information available on the value of antibiotics for horses. Many horse owners and veterinarians are using them, however, and are satisfied with the results obtained.

It has been shown with other farm animals that antibiotics are the most helpful when the subclinical disease level on the farm is high. The same should be true for horses. The antibiotics may be especially helpful for young foals that suffer setbacks because of infections, digestive troubles, lack of milk, poor weather, or other stress factors. Horses being moved from one race track or horse show to another in different localities and under varying weather conditions may also be benefited by antibiotics. The experience obtained by good horse managers indicates that antibiotics are beneficial under many conditions encountered on the farm.

Presently, there is enough research information available to enable the U.S. Food and Drug Administration (FDA) to allow the use of 85 mg of chlortetracycline per head daily for horses up to 1 year of age to stimulate growth and feeding efficiency. This is based on research at Washington State University (1) and in England (2). W. E. Maderious, a practicing horse veterinarian in Madera, California, reported that the use of antibiotics in his practice was beneficial for horses (3). Many others also use antibiotics for horses, although more studies are needed to determine more adequately the role played by antibiotics in horses (1a,1b).

II. EFFECT OF COLD WEATHER ON HORSES

Very little information is known on the effect of cold weather on horses, but an interesting study on this subject was reported by the University of Alberta in

TABLE 23.1

Measurements at 20° and 70°F

Horse type	Temperature (°F)	Hair coat depth (mm)[a]	Skin thickness (mm)[a]	Skin temperature (°F)	Rectal temperature (°F)
Brood mare	70	13.5	1.7	85.3	98.6
	20	18.1	1.8	57.2	98.4
Yearling	70	18.7	1.7	81.3	99.5
	20	22.8	1.8	52.6	99.3

[a]A millimeter is 0.03937 of an inch. An inch has 25.4 mm.

Edmonton, Canada (4). Measurements were made on horses exposed to cold temperatures of 20°F and higher temperatures of 70°F. The information in Table 23.1 gives data on a brood mare weighing 1218 lb and a yearling weighing 630 lb. This information, and other data obtained at the University of Alberta, indicate the following.

1. Horses exposed to cold temperatures have an increase in hair depth. Some of the increase is due to contraction of the muscle at the base of each hair, which causes it to stand more upright in the skin.

2. Yearlings had skin that was 1.6 mm thick whereas the skin of older horses was generally 2 mm or greater in thickness. Therefore, as horses get older their skin becomes thicker. Cold weather also causes a slightly thicker skin. The temperature of the skin of the lower part of the limbs decreases much more during cold exposure than the skin of the upper part. At 0°F the skin temperature of the lower part of the limbs was usually between 35 and 40°F, while that on the upper part of the limbs ranged from 55 to 60°F. The higher the skin temperature the less heat loss occurs through the skin. This makes it possible to maintain normal temperature more easily around the internal body organs. The temperature of the skin surface is controlled by regulation of blood flow through the skin.

3. There was very little difference in the rectal temperature at external temperatures of between 20 and 70°F.

Table 23.2 shows the critical temperature of a brood mare (not pregnant) and a yearling horse. The critical temperature represents the temperature below which a horse must produce extra heat to maintain its body temperature. An increase in heat production during cold weather means an increased requirement for feed energy for the maintenance of body temperature and body energy reserves.

From the information shown in Table 23.2, and from other data obtained at the University of Alberta, the following can be concluded.

TABLE 23.2

Critical Temperature of Horses

Time of year	Body weight (lb)	Chamber temperature (°F)	Critical temperature (°F)
Early winter			
Brood mare	1108	21	32
	1108	7	31
	1108	0	28
Late winter			
Brood mare	1218	−4	15
Yearling	630	−4	30

1. Mature horses, in reasonably fat condition, have a critical temperature of about 30°F during the first part of the winter. However, after developing a winter hair coat (about 18 mm or almost 1 in long) and gaining about 100 lb the critical temperature is reduced to 15°F.

2. It is probable that (a) young horses, (b) horses in thin condition, and (c) horses that have been kept in heated barns (and have not developed any winter hardiness) would have critical temperatures of 30–40°F.

3. The University of Alberta scientists state that it can be calculated from their studies that horses require 15–20% more feed for each 10°F the air temperature falls below their critical temperature. However, thin horses or horses with short hair may need even greater increases in their dietary intake.

This information is very interesting and shows that more feed is required by the horse at lower temperatures. Extra feed is needed during the winter as compared to the summer. In swine, for example, the sow is usually fed 1 lb more feed (about 20–25% more) during the winter as compared to the summer.

Another Canadian study (5) indicated that young horses housed outdoors in cold weather may need to be fed readily digested diets to compensate for their reduced feed intake. In this study, weanling horses fed readily digested diets *ad libitum* gained weight at or above expected values even at severely cold ambient temperatures. They observed that feed intakes by horses 6–12 months of age kept in a heated barn were negatively related to environmental temperature, whereas intakes by horses aged between 18 and 24 months were unaffected by cold temperatures. Weight and age-scaled digestible energy intakes were reduced by 6.1% at temperatures below −10°C compared with temperatures above −10°C.

III. LEARNING ABILITY

Horse producers are interested in the learning ability of their horses. The young foal needs to learn quickly in order to develop a strong parent–offspring bond. It also needs to learn to discriminate between the aspects of its environment that are beneficial and those that are harmful. It also needs to recognize individuals with whom it has conflicts and those it does not. Learning ability in horses is also important since it will affect their trainability. An excellent study on this subject was conducted and recently reported by the University of Kentucky (6). In these experiments, 37 yearling geldings of Quarter Horse breeding were used. A maze (a series of tall, solid panels through which a horse takes a series of turns in order to go through them) was developed to test the learning ability of the horses. The horses were kept in individual stalls during the night. They were released, one at a time each morning, to pass through the maze to reach the outside pens. This allowed an observer, positioned above the maze, and out of sight of the horses, to determine how quickly they learned to make the correct turns to get through the maze. The reward was the horse's escape to the outside pens where water and other horses were available. Thirst was probably the driving force to get outside. In their studies they found the following.

1. Previous thinking that horses prefer to have things on the right side was not confirmed by the study. Some horses prefer to go to the right and some to the left. This should be considered during training procedures.

2. The belief that horses never return to the place where they have been frightened was not supported in this study. Of the 29 horses that received adverse stimulus, 14 returned at least once to the place where they had been frightened. One horse returned twice.

3. Taller horses of comparable weight tended to go to the left side. The horses that preferred to make left choices made more errors when a right choice was necessary to escape through the maze.

4. Heavier horses tended to make right-side choices as their first choice when coming in the maze. Heavier horses also took longer to escape through the maze.

5. In the pig and other animals, a deficiency of protein affects learning ability. Some scientists think it also affects learning ability in the human. Therefore, the Kentucky scientists studied the affect of the level of protein in the diet on learning ability. The levels of protein used were not in a range expected to cause malnutrition. However, they wanted to determine if differences would occur with protein levels of 10, 13, 16, and 19% protein in the diet. There was some evidence that the horses fed the 19% protein level learned faster in certain phases of the study but not in others. Therefore, they concluded that judgment on the effect of protein on learning ability in the horse should be reserved until more studies are conducted.

This study by the Kentucky scientists is important and more research is needed on this subject. It is known that the horse has intelligence. This is evidenced by their ability to open latches, undo knots, and find their way back to their stables or home after traveling long distances over previously untraveled areas. In the wild state, horses remember where the water holes and best grazing areas are, and return to them each year. Moreover, they can sense danger and are able to communicate it to other horses in the band. In fact, stallions will warn their group of mares and help round them up and lead them out of danger. Horses also respond to kindness and gentle handling and react adversely to persons who mistreat or mishandle them. Therefore, it is important to understand the learning ability of the horse in order to improve further their training and response during racing and performance. Some horses never overcome their highly strung nature, which is counterproductive to becoming a winner or a champion. It may be that these horses became nervous individuals during the training process because of some peculiar aspect of their learning process. Research may find new methods to help the master–trainer do an even better job in developing horses.

There is enough evidence obtained with other animals to indicate that nutritional deficiencies can affect learning ability. Therefore, horses should be fed a well-balanced diet with adequate levels of protein, minerals, vitamins, and energy to prevent a lack of nutrients interfering with learning ability. Moreover, proper nutrition is needed for optimum growth, reproduction, and performance.

Fig. 23.1 Horses are used to separate cattle. It requires a well-trained horse and rider to do this job properly. (Courtesy of T. J. Cunha, Cal Poly University, Pomona.)

The diet that the mare receives during gestation will influence the development of the foal *in utero* and later may also affect its ability to grow and learn properly. Therefore, there is no time when one can relax from feeding well-balanced diets in a production program for valuable horses being developed for high-level performance and racing.

A Texas A & M study (7), using the Hebb-Williams closed field maze, showed that horses are more prone to error in this maze, but eliminate errors and negotiate the problems more quickly than beef cattle. Also, horses made more efficient use of visual cues. In another Texas A & M study (8) the data suggested that individual differences in performance in a learning situation are detectable and these differences, at least for colts and geldings, can be correlated to individual differences noted in trainability.

A Cornell study (9) showed that the ponys' rank in the social hierarchy was not correlated with ranks obtained in the maze and shock-avoidance tests. Their results also suggested that long-term learning and learning permanence are related and that better learners in an avoidance-conditioning situation employing an aversive stimulus also tend to learn more quickly in a maze-learning situation with positive reinforcement.

A University of California study (10) found no correlation between social dominance and learning scores. Their study showed that Quarter Horses learned significantly faster than Thoroughbreds, and learned progressively more rapidly for both breeds in a second discrimination task. They also found that Quarter Horses tended to be less reactive than Thoroughbreds, but individual reactivity ratings and learning scores were not correlated.

A Cornell study (11) showed that a lack of normal maternal care does not appear to adversely affect the maze-learning ability of foals. The orphan foals initially moved through the maze more slowly, but made no more errors than mothered foals. Adult mares were more variable and made significantly more errors—that is, took more trials to learn a simple reversal—than their foals. Young horses appeared to learn more rapidly than older horses. They also reported that rank in a social dominance order was not related to maze-learning ability, but heavier mares tended to be socially dominant.

IV. FEEDING BEHAVIOR

During their evolution, horses were selectively adapted to be grazing animals. Their sources of feed were grasses, legumes, and shrubs. These were available throughout the year but there were seasonal variations in both the quality and quantity of the grazing forages available. The horse was developed in an environment that caused their natural selection for the consumption of small and frequent meals of different forages. Even today, under free-ranging conditions, the horse eats frequently and when grazing takes a few bites at one locality and moves forward chewing while walking to eat more forage at a nearby location.

A Kentucky study (12) involved obtaining information on the feeding behavior of horses. Table 23.3 shows the percentage of time spent by horses with the various feed treatments indicated.

The horses fed the hay diet spent significantly more time eating feed and less time chewing wood, in coprophagy (eating feces), in searching for feed, and in standing than did the horses fed the concentrate diet. The horses fed the concentrate diet spent a significantly greater percentage of their time chewing wood. Increased wood chewing has been reported in other studies where a completely pelleted diet was used (13,14). The addition of sodium carbonate (Na_2CO_3 to the concentrate diet decreased the amount of time spent wood chewing, but the difference was not significant. It did significantly decrease coprophagy and increased the time spent standing. Table 23.3 provides some interesting information concerning the behavior of the horse and the percentage of time it spends on various activities.

A Cornell study (15) involved the feeding and drinking behavior of 11 pony mares and 15 foals living on pasture with free access to water. Lactating mares on pastures spent about 70% of the day feeding. Foals began feeding on their first day of life. The foals grazed primarily during the early morning and evening. Almost all grazing by foals was done while their mothers were grazing. Drinking by mares increased as the temperature rose. Frequency was greatest at 86–95°F at which temperature the mares drank once every 1.8 hours. Drinking by foals was very rare. The youngest age at which a foal was observed to drink was 3

TABLE 23.3

Percentage of Time Spent in Various Activities[a]

| Activity | Time spent by horses fed only: | | |
	Hay (%)	Concentrates (%)	Concentrates + Na_2CO_3 (%)
Eating feed	39.5	3.37	3.73
Drinking water	1.0	1.20	1.13
Chewing wood	2.13	10.67	8.10
Coprophagy	0.77	2.73	1.13
Licking salt	0.53	1.76	2.43
Searching	7.50	13.13	12.43
Lying down	3.50	5.07	3.83
Standing	44.67	61.67	66.90

[a]Data from Willard et al. (12).

weeks, and 8 of 15 foals were never observed to drink before weaning. It may be that frequency of drinking by foals can be used as an indicator of the adequacy of the dam's milk supply to meet the fluid as well as the energy requirements of the foal.

A Cornell study (16) showed that when fresh hay was supplied in the mornings, the ponies spent similar amounts of time eating whether visual contact was allowed or not, but in the afternoon significantly more time was spent feeding when visual contact with other horses was allowed (73 ± 4%) than when it was not (60 ± 7%). They also found that these stalled pony mares spent the majority of their time eating but preferred to eat from the floor rather than from a trough or manger. They frequently interrupted feeding to lift their heads. They avoided contaminating hay with urine or feces, but not on all occasions.

Another Cornell study (17) found that the only difference in behavior between ponies fed one meal per day and six meals daily was that the ponies eating six meals daily spent less time eating and more time standing, walking, drinking, and eating bedding than the ponies fed one meal per day. They suggested that horse owners might consider decreasing the size and increasing the frequency of their horse's meals in order to mimic patterns and maintain homeostasis.

University of Pennsylvania studies (18,19) indicated that, in ponies, intravenous glucose loads can prolong the duration of satiety experienced after a meal.

An Oregon State University study (20) involving the taste reactions of young horses to sweet, sour, salty, and bitter solutions indicated that the immature horse reacts quite similarly to sheep in taste behavior.

Dr. Ralston recently published an excellent review on the patterns in the feeding of horses and ponies (21). His review indicates that free-ranging horses and ponies eat for 10–12 hours daily in grazing periods which may last from 2 to 3 hours each. The frequency and duration of the grazing periods are influenced by the quality and quantity of the forages available as well as climatic factors. Horses will graze longer on patches of preferred forages and less in areas of less palatable forages. The grazing periods are separated by resting periods and by locomotor or social activity with other horses.

If horses are given free access to feed in confined stalls they eat in patterns that are similar to those under free-ranging conditions. Ponies fed a completed pelleted feed free choice spent about 38% of a 24-hour period in eating activity and consumed about 10 meals per day. A study with Thoroughbred horses fed grain and free-choice loose hay ate about 11 meals in a 24-hour period and devoted about 53% of the available time to eating activities.

Horses and ponies do not fast voluntarily for more than 3–5 hours. Their most prolonged fasts occur between 1 A.M. and 6 A.M. Toward the end of a meal, horses will decrease their eating rate and are more easily distracted by movements and noises in their surroundings. During this period they may sniff the ground, bedding, or other materials. They may often engage in social activity if other horses are present. During resting, horses and ponies assume a position in which they stand with their head down, ears back, eyes half closed and one hind leg flexed. They may lie down before eating again, especially at night.

Scientists studying horse behavior define meals as a period of eating lasting more than 10 minutes in which more than 0.33 lb of pellets are consumed. Termination of an eating period is defined as a minimum of 15 minutes when no feed is consumed. Studies to date indicate there is a strong similarity between the eating behavior of horses and ponies.

Coprophagy (eating of feces) and wood chewing has been reported when horses and ponies are fed completely pelleted diets. Some work indicates, however, that this does not occur in ponies permitted free access to a pelleted feed and bedded on wood shavings. They nibbled at the wood shavings and consumed an unmeasured quantity. In another study, it was reported that coprophagy is eliminated and wood chewing reduced when crude protein content in the experimental diet is raised from 6.2 to 10%. It appears, therefore, that coprophagy and wood chewing is a complex matter which may be influenced by a number of factors including boredom and even aggressive interactions between neighboring horses. There are also indications that wood chewing may decrease as horses become used to consuming a nutritionally complete pelleted diet.

The appearance, smell, taste, and texture of the feed will influence feed intake. A preference for certain feed mixtures by the horse may also occur. Horses also show preferences for certain forages or forage mixtures. When a pasture has too many different forages, the horses may selectively eat what is present in the order of their preference and may avoid certain forages or weeds altogether. If more than one forage is used on a horse farm to meet seasonal and moisture needs, it is best to plant each kind of forage in a different field. Therefore, the horse placed in a field has no choice but to consume the forage which is there. Cattle are similar to horses in this regard. If three different forages are in one field, they will eat the preferred forage close to the ground before going to the second and then to the third preferred forage. But, if each forage is planted in a separate field, the cattle will eat all of them well and perform equally well on each forage. When horses eat one preferred forage and leave the other alone, the second forage continues growing, becomes more mature, increases in fiber level, becomes lower in feeding value and usually less palatable. Therefore, when the horse runs out of the preferred forage and starts consuming the second preferred forage, it is lower in feeding value.

Horses avoid feed and pasture areas contaminated with horse feces, possibly on the basis of odor alone but it may be more complicated than this. Horses and ponies separate out and refuse to eat clumps of dirt attached to the roots of forages. Horses and ponies lack the ability to consume voluntarily a specific forage based on its energy content. The preferred forage may not be uniform in both its fiber and soluble carbohydrate level.

Ponies fed a calcium-deficient diet did not show a preference for calcium supplements in a Cornell study. However, they appear to have a specific appetite for salt if the diet is deficient in it. This would indicate that meeting mineral needs of horses under grazing conditions would best be met with a complete mineral mixture which includes salt at a level to help ensure an adequate intake of minerals.

From these studies it is apparent that the horse eats small meals frequently during a 24-hour period. The first priority for the feed consumed is its use by the body to meet its daily maintenance requirements. The more a horse eats above maintenance needs, the more is left over for growth, development, reproduction, lactation, or performance. Having a high-quality, well-balanced diet nutritionally is only the beginning of a good feeding program. The next phase is to get the horses to consume the proper amount of feed needed to meet the work function involved. It is very important for the owner of racing or high-level performance horses to study their feeding behavior to ensure an adequate feed intake. More-over, many horses are prima donnas and require extra individual study of their feeding behavior to meet their feeding needs adequately. Therefore, a good horse feeding program may depend as much on the feeder as it does on the diet being used. As more is learned about the feeding behavior of the horse, the easier it will be to develop individual horses to their full potential.

V. SLEEPING BEHAVIOR

Present thinking is that foals sleep lying flat and stretched out on their side with head to ground contact (this is called lateral recumbency). Foals spend considerable amounts of time sleeping, but as they grow older, they lie down less and sleep less. The mature horse rests and sleeps standing up. Most do not lie down regularly, but some do sleep and rest lying down. When the mature horse sleeps, the head droops and the eyes are closed. It is thought that horses may sleep for 6–8 hours during a 24-hour period. Most of the sleep will occur during the warmest part of the day. Not all of the sleep will occur at one time, however. It will be irregular, short, and influenced by hunger, environmental conditions, noise, its sense of security in the area, whether it is alone or part of a group of horses, and other factors. There may also be differences between horses under relatively confined conditions as compared to horses kept in large pastures or horses under range or mountainous conditions in a wild state. In a wild band of horses, not all the horses will sleep at the same time whether it is night or day. This allows at least one horse or more to serve as a sentry. This is a means of providing security for the remainder of the horses. It appears that the sentry will not lie down or go to sleep until another horse starts sentry duty. In the wild state this is very important since predators can be a serious problem. Unfortunately there is very little information available on the sleeping behavior of horses during the various stages of their life cycle. Sentry behavior as well as sleeping patterns are important to understand under the various conditions in which horses are kept. This is especially so in city environments where many factors may be threatening to horses and their ability to sleep and rest properly. Most horses are now located in cities or the outskirts of large cities. An understanding of sleeping and sentry behavior may allow the manipulation of these behaviors by different management practices. This would be especially helpful with young horses for whom sleep is necessary and important. However, older horses being worked or exercised heavily also require a specified amount of rest and sleep for optimum performance.

A study by Dr. Glade (22) involved the sleeping and resting behavior of six young Thoroughbred foals. The foals were observed from 9 A.M. to 5 P.M. during group confinement in large (75 × 75 feet) and small (40 × 40 feet) corrals and during individual confinement in rectangular pens (75 × 15 feet) and square box stalls (12 × 12 feet). The observers never entered the enclosures or otherwise interfered with the animals, although outside interactions (such as occasional pedestrian petting attempts, erratic traffic noise, etc.) were not aggressively prevented. The foals were observed for 5 days which were sunny with an average temperature ranging between 85 and 95°F. The animals were fully familiar with each enclosure from previous experience. They spent an entire observation period in only one type of enclosure and had been there at least one hour before observation was initiated.

In the study, he found that lying down was most frequent when an individual horse was kept in a 75 × 15 foot rectangular pen. The foals each averaged 6.9 lying down periods (which averaged 16 minutes in duration) during the 8 hours of observation. In the smaller 12 × 12 foot stalls with only one horse, each animal averaged about the same lying down periods (5.7) but their duration was about twice as long (30 minutes). Dr. Glade thought that this probably reflected the contrast between the relatively quiet and protected smaller box stall arrangement and the greater exposure to increased distractions in the larger rectangular pens.

Lying down was less frequent when the foals were kept in groups in the large 75 × 75 foot enclosure (4.2 times during an 8 hour period) and even less (2.2 times) in the smaller pen (40 × 40 feet). The average time during recumbency (lying down) was longer (24 minutes) in the large pen than in the smaller pen (8 minutes).

Dr. Glade assumed that grouping horses in a pen encouraged longer periods of relaxation than did semi-isolation. Perhaps the presence of others to provide a warning if necessary resulted in an increased sense of security. One other study has suggested that herd behavior encourages lying down. It also suggests that deep sleep and perhaps dreaming occurs more readily when a horse is lying on its side with its head resting on the ground. This may indicate that the amount of time spent in this position might be a useful indicator of the effectiveness of resting behavior in fulfilling the need for sleep.

The amount of time spent in lateral recumbency (lying on their side with head to ground contact) totaled 3–4% of the 8 hours in all pens except the small corral where it was almost nonexistent. The two pens with the longest single dimension had the greatest percentage of lying down periods spent with some head to ground contact. These pens provided the greatest separation from the rest of the group. Dr. Glade stated that this may suggest that when space is limited a rectangular pen may be preferable to a square pen, even when the total square footage is equal.

Sentry behavior by horses kept in groups was indicated by the fact that at least one foal remained standing, although occasionally the five others were all lying down. The role of the sentry never changed smoothly. No sentry ever lay down until one other animal had been standing for over 5 minutes. During this short study no relationship was found between the frequency or duration of sentry behavior and the known dominance hierarchy of the group. This contrasts with one report that dominant adults are the first among a group to lie down.

VI. SOCIAL DOMINANCE

An excellent review paper has been written by Dr. W. D. McCort on the behavior of feral horses and ponies. A perusal of this paper provides information which may be of value in managing horses under intensified conditions (23).

An excellent review paper measuring social behavior and social dominance with many species of animals by Dr. Craig (24) indicates the complexity of dominance in animals. Sexual hormones and sensitivity to them can influence the onset of aggression and status attained. After dominance orders are established, they tend to be stable in female groups but are less so in male groups. Psychological influence can affect dominance relationships when strangers meet and social alliances within groups may affect relative status of individuals. Social dominance develops more slowly when young animals are kept in intact peer groups where they need not compete for resources. Learned generalizations may cause smaller and weaker animals to accept subordinate status readily when confronted with strangers that would be formidable opponents.

One study (25) showed that pony foals of high-ranking dams were especially likely to attain high rank. Dominant mares may come to the aid of their progeny. Another study (26) with horses indicated that higher social rank gives priority to resource access when horses are competing for scarce resources. There appear to be some exceptions to this under pasture conditions.

VII. WEIGHT EQUIVALENTS

Table 23.4 can be used for information on weight equivalents. Table 23.5 gives information to use when converting from one unit to another.

TABLE 23.4

Weight Equivalents

1 lb = 453.6 g = 0.4536 kg = 16 oz
1 oz = 28.35 g
1 kg = 1000 g = 2.2046 lb
1g = 1000 mg
1 mg = 1000 μg = 0.001 g
1 μ = 0.001 mg = 0.000001 g
1 μg per g or 1 mg per kg is the same as ppm

VIII. CONCLUSIONS

Many horse owners are using antibiotics for horses but additional studies are needed to determine their role more adequately.

More feed is required by the horse at lower temperatures. A Canadian study indicated that horses require 15–20% more feed for each 10°F the air temperature falls below their critical temperature.

Young horses appear to learn more rapidly than older horses. It appears that rank in a social dominance order is not related to learning ability. It is important to understand the learning ability of horses as a means of improving their training and response during performance.

Free-ranging horses eat for 10–12 hours daily in grazing periods which may last from 2 to 3 hours each. Horses given free access to feed in confined stalls eat

TABLE 23.5

Weight-Unit Conversion Factors

Units given	Units wanted	For conversion multiply by	Units given	Units wanted	For conversion multiply by
lb	g	453.6	μg/kg	μg/lb	0.4536
lb	kg	0.4536	Mcal	kcal	1,000.
oz	g	28.35	kcal/kg	kcal/lb	0.4536
kg	lb	2.2046	kcal/lb	kcal/kg	2.2046
kg	mg	1,000,000.	ppm	μg/g	1.0
kg	g	1,000.	ppm	mg/kg	1.0
g	mg	1,000.	ppm	mg/lb	0.4536
g	μg	1,000,000.	mg/kg	%	0.0001
mg	μg	1,000.	ppm	%	0.0001
mg/g	mg/lb	453.6	mg/g	%	0.1
mg/kg	mg/lb	0.4536	g/kg	%	0.1

in patterns similar to those under free-ranging conditions. A study with Thoroughbred horses fed grain and free-choice loose hay ate about 11 meals in a 24-hour period and devoted about 53% of the available time to eating activities. It is important to study the feeding behavior of high-level performance horses to ensure an adequate feed intake.

Foals spend considerable amounts of time sleeping. As they grow older they lie down less and sleep less. The mature horse sleeps standing up but some sleep and rest lying down. Sleeping behavior and sentry duty, to watch for predators or intruders, is important to understand in order to improve horse production practices.

After dominance orders are established, they tend to be stable in female groups but are less so in male groups. Sexual hormones and sensitivity to them can influence the onset of aggression and the status attained.

REFERENCES

1. Schneider, B. H., W. E. Ham, J. T. Rose, J. S. Lucas, and M. E. Ensminger. *Wash., Agric. Exp. Stn., Circ.* **263** (1955).
1a. Miller, E. D., E. R. Miller, and D. E. Ullrey. *Mich. Ext. Bull.* **537** (1975).
1b. Hintz, H. F., J. P. Baker, R. M. Jordon, E. A. Ott, G. D. Potter, and L. M. Slade. "Nutrient Requirements of Horses." *N.A.S.–N.R.C.,* Washington, D.C. (1978).
2. Taylor, J. H., W. S. Gordon, and P. Burrell. *Vet. Rec.* **66,** 744 (1954).
3. Maderious, W. E. *West. Livest. J.* September, pp. 87–88 (1960).
4. Young, B. A., and J. Coote. *Univ. Alberta, Feeders Day Rep.* June, pp. 21–23 (1973).
5. Cymbaluk, N. F., and G. I. Christison. *J. Anim. Sci.* **67,** 48 (1989).
6. Kratzer, D. D., W. M. Netherland, R. E. Pulse, and J. P. Baker. *J. Anim. Sci.* **46,** 896 (1977).
7. Ingram, R. S., C. McCall, G. D. Potter, and T. H. Friend. *Proc. Annu. Meet. Am. Soc. Anim. Sci.* p. 147 (1979).
8. Fiske, J. C., and G. D. Potter. *J. Anim. Sci.* **49,** 583 (1979).
9. Haag, E. L., R. Rudman, and K. A. Houpt. *J. Anim. Sci.* **50,** 329 (1980).
10. Mader, D. R., and E. O. Price. *J. Anim. Sci.* **50,** 962 (1980).
11. Houpt, K. A., M. S. Parsons, and H. F. Hintz. *J. Anim. Sci.* **55,** 1027 (1982).
12. Willard, J. G., J. C. Willard, S. A. Wolfram, and J. P. Baker. *J. Anim. Sci.* **45,** 87 (1977).
13. Haenlein, G. F. W., R. D. Holdren, and Y. M. Yoon. *J. Anim. Sci.* **25,** 740 (1966).
14. Willard, J., J. C. Willard, and J. P. Baker. *J. Anim. Sci.* **37,** 277 (1973).
15. Davis, S. L. C., K. A. Houpt, and J. Carnevale. *J. Anim. Sci.* **60,** 883 (1985).
16. Sweeting, M. P., C. E. Houpt, and K. A. Houpt. *J. Anim. Sci.* **60,** 369 (1985).
17. Youket, R. J., J. M. Carnevale, K. A. Houpt, and T. R. Houpt. *J. Anim. Sci.* **61,** 1103 (1985).
18. Ralston, S. L., and C. A. Baile. *J. Anim. Sci.* **54,** 1132 (1982).
19. Ralston, S. L., and C. A. Baile. *J. Anim. Sci.* **55,** 243 (1982).
20. Randall, R. P., W. A. Schurg, and D. C. Church. *J. Anim. Sci.* **47,** 51 (1978).
21. Ralston, S. L. *J. Anim. Sci.* **59,** 1354 (1984).
22. Glade, M. J. *Equine Pract.* **6**(5), 10 (1984).
23. McCort, W. D. *J. Anim. Sci.* **58,** 493 (1984).
24. Craig, J. V. *J. Anim. Sci.* **62,** 1120 (1986).
25. Tyler, S. *J. Anim. Behav. Monogr.* **5,** 85 (1972).
26. Sereni, J. E., and M. F. Bouissou. *Biol. Behav.* **3,** 87 (1978).

Appendix

PLEASURE HORSE MANAGEMENT RECOMMENDATIONS[1]

A. Feed and Water

1. Place top of manger or feed box at about two-thirds of the height of withers—normally 30 in for foals and 40 in for mature horses.

2. Approximate feed box dimensions are: foals—18 in × 12 in × 8 in deep; yearlings and mature horses—24 in × 18 in × 8 in deep.

3. Manger or roughage rack space per animal:

> Foals: 24 linear inches
> Yearlings: 30 linear in
> Mature horses: 36 linear in

4. Provide plenty of fresh, clean water. Allow for 1 linear ft of open water space or one automatic watering bowl for every ten horses.

B. Housing and Lots

1. Geography and climate determine housing requirement. A shed open on one side is sufficient for all but the most severe weather.

2. Horses being fitted for show or sale should be housed individually.

3. Provide the following square ft of housing space:

Age:	Stalls:
Foals	90–110
Yearlings	110–125
Mature Horses	140–150

4. Daily exercise should be given.

5. Provide a minimum of 500 sq ft per horse for an outside lot. This varies with geography, available space, and whether or not additional exercise space is available.

1 These were developed by the Nutrition Council of the American Feed Industry (AFIA) in cooperation with many scientists in Universities, USDA, Industry, and other agencies and are reproduced from the AFIA Management guide. They were reproduced through the courtesy of Lee H. Boyd, Vice President AFIA.

6. To avoid possible injury, use paved lots only where mud and sanitation make paving necessary.

7. When soil and lot conditions dictate, hardpacked gravel or earth should extend for approximately 10 ft around waterers.

8. Provide artificial shade where necessary.

9. Minimum shade requirements are:

Foals: 20–30 sq ft

Yearlings: 40–50 sq ft

Mature horses: 50–60 sq ft

10. Board, plank, or pole fences are preferable to wire fencing. When wire fences are used, add a top rail to make fence readily visible and to prevent horses from mashing and breaking down the fence. Height of fence should be 5–6 ft for dry lots and 4–5 ft for pastures.

C. Special Equipment

1. The need for special facilities and equipment for handling horses is dependent on local practices and the number of horses involved. Equipment to be considered will include a squeeze, breeding chute, loading chute, hobbles, nose twitch, teasing walls, cross ties, casting or throwing harnesses, grooming equipment, and isolation quarters. The chutes should be cleated, of sound construction and adequate height, and free of sharp projections.

D. General Management

1. Hand-feed grain and other concentrates.

2. Unless otherwise provided, feed salt and other needed minerals free choice.

3. Exercise horses kept in confinement a minimum of $\frac{1}{2}$ hr daily and groom daily.

4. Shoe horses that are worked or exercised on hard surfaces.

5. Castrate males to be used for gelded pleasure horses by 12–18 mos.

6. Foals should be trained to lead by weaning time and trained to ride by 2 yrs of age. Daily workouts are recommended during training.

7. Stallions may be used for breeding two or three times weekly by 3 years of age. Mature stallions may be used more frequently.

8. Treat for internal parasites at least twice yearly, or as directed by a veterinarian.

9. Vaccinate for such diseases as tetanus and encephalitis according to veterinarian's recommendation.

10. Avoid oversupplementation due to use of multiple sources of nutrient supplements.

11. Light management—day length and light intensity affect hair coat in the winter. Use 16 hours of light at 6–10 candles of intensity (200 watt, 10 ft over stall).

Index